Phyto

Principles and resources for site remediation and landscape design

Phyto presents the concepts of phytoremediation and phytotechnology in one comprehensive guide, illustrating when plants can be considered for the uptake, removal or mitigation of on-site pollutants. Current scientific case studies are covered, highlighting the advantages and limitations of plant-based cleanup. Typical contaminant groups found in the built environment are explained, and plant lists for mitigation of specific contaminants are included where applicable.

This is the first book to address the benefits of phytotechnologies from a design point of view, taking complex scientific terms and translating the research into an easy-to-understand reference book for those involved in creating planting solutions. Typically, phytotechnology planting techniques are currently employed post-site contamination to help clean up already contaminated soil by taking advantage of the positive effects that plants can have upon harmful toxins and chemicals. This book presents a new concept to create projective planting designs with preventative phytotechnology abilities, 'phytobuffering' where future pollution may be expected for particular site programs.

Filled with tables, photographs and detailed drawings, Kennen and Kirkwood guide the reader through the process of selecting plants for their aesthetic and environmental qualities, combined with their contaminant-removal benefits.

KATE KENNEN is a landscape architect, and the founder and president of Offshoots, Inc., a Boston, Massachusetts landscape architecture practice focused on productive planting techniques and phytotechnology consulting. Offshoots has won numerous awards for projects integrating plantings to clean up polluted sites. Having spent her childhood at her family's garden centre in central Massachusetts, Kate is well versed in the plants of the Northeast. She completed her undergraduate studies in Landscape Architecture at Cornell University, and received her master's degree in Landscape

Architecture with distinction from the Harvard Graduate School of Design. Kate also currently teaches in the Landscape Architecture department at the Harvard University Graduate School of Design. Previous to opening Offshoots, Kennen worked as an Associate at Design Workshop in Aspen, Colorado.

NIALL KIRKWOOD is a landscape architect, technologist and Professor at the Harvard University Graduate School of Design, where he has taught since 1992. He teaches, carries out research and publishes on a range of topics related to landscape architectural design, the built environment and the sustainable reuse of land including urban regeneration, landfill reuse, environmental site technologies and international site development. His publications include *Manufactured Sites: Rethinking the Post-Industrial Landscape* (Routledge), *Principles of Brownfield Regeneration* (Island Press) also published in Korean and Chinese translations, *Weathering and Durability in Landscape Architecture* (Wiley) and *The Art of Landscape Detail* (Wiley). Professor Kirkwood is a Fellow of the American Society of Landscape Architects and in addition is the Gerard O'Hare Visiting Professor, University of Ulster, Belfast, Northern Ireland, Distinguished Visiting Professor, Korea University, Seoul and Visiting Professor, Tsinghua University, Beijing, China.

In two words: "Beyond comprehensive". *Phyto* is by far the most comprehensive compilation of phytotechnologies out there. It truly goes beyond by tying together this broad set of plant technologies for cleaning the environment with the necessary form and functionality of landscape design. As an advocate and trainer in phytotechnologies, I especially appreciate the illustrative graphics and easy-to-understand descriptions that clearly convey the science, engineering, design, and planning to the technical and artisan alike.

– David Tsao, Ph.D, BP Corporation North America, Inc.

Phyto is a fantastic resource, not just to landscape architects but also to engineers and scientists as well. As phytoremediation developed, advancement efforts focused on the biochemical science of the processes, and while the field was cognizant of 'ancillary benefits' valuation was not considered, mostly due to lacking knowledge and resources. *Phyto* brings the social and physical science into a common meeting place, and provides much needed discussion, fantastic visualizations and cross cultural presentation of plant-based technologies that can be incorporated into our urban spaces to serve both public health and the quality of life itself.

– Joel G. Burken, Missouri University of Science and Technology

This book closes a very important gap between phytotechnologies and practice. Through creative design, the authors succeed in translating a comprehensive subject matter into accessible information. A special merit is that the book predicts vegetation strategies becoming an anticipatory tool in the hands of the landscape architect in advance of potential future contamination preventing human exposure to soil, water and air contamination.

– Jaco Vangronsveld, Centre for Environmental Sciences of Hasselt University, Belgium

Phyto

Principles and resources for site remediation and landscape design

Kate Kennen
and Niall Kirkwood

Routledge
Taylor & Francis Group

LONDON AND NEW YORK

First published 2015
by Routledge
2 Park Square, Milton Park, Abingdon, Oxon OX14 4RN

and by Routledge
711 Third Avenue, New York, NY 10017

Routledge is an imprint of the Taylor & Francis Group, an informa business

British Library Cataloguing-in-Publication Data
A catalogue record for this book is available from the British Library

Library of Congress Cataloging in Publication Data
Phyto : principles and resources for site remediation and landscape design / Kate Kennen and Niall Kirkwood. – First
edition.
  pages cm
Principles and resources for site remediation and landscape design
Includes bibliographical references and index.
1. Phytoremediation. 2. Landscape design. I. Kirkwood, Niall. II. Title. III. Title: Principles and resources for site
remediation
and landscape design.
TD878.48.K46 2015
628.4–dc23
2014019290

ISBN: 978-0-415-81415-7 (pbk)
ISBN: 978-1-315-74666-1 (ebk)

Typeset in Garamond and Franklin Gothic by
Servis Filmsetting Ltd, Stockport, Cheshire

Contents

vii

Acknowledgements

Professor John Stilgoe of the Graduate School of Design and the Faculty of Arts and Sciences at Harvard University initially suggested that this book should be written as a much-needed addition to the literature of landscape architecture, landscape technology and planting design. The authors acknowledge Professor Stilgoe's clarion call and at the same time recognize that it has been some years in the making, since his original request. During these years, the phytotechnology field has advanced and reshaped itself to take account of new scientific research, emerging design ideas in landscape architecture regarding ecological processes and more open and time-related projects with a general sympathy towards site approaches that engage natural processes in a more direct and visceral way.

Therefore this book project would not have been possible without the advice, knowledge and general encouragement provided by a wide range of colleagues, friends and experts in the design, planning and engineering fields and scientific community. This includes active collaboration over a number of years between the authors and a wide group of phytotechnology scientists, landscape architects, ecologists, environmental engineers and federal regulators, and current and former graduate students in landscape architecture and urban design. In particular we would like to acknowledge the significant role played in the evolution of this book by Dr. Jason White of the Connecticut Agricultural Experiment Station, President of the International Phytotechnology Society and Managing Editor of the *International Journal of Phytoremediation*, Steven Rock of the US Environmental Protection Agency (EPA), Dr. Lee Newman, SUNY College of Environmental Science and Forestry, Immediate Past President of the International Phytotechnology Society and Co-Editor in Chief of the *International Journal of Phytoremediation*, Dr. Joel G. Burken, Missouri University of Science and Technology, Environmental Research Center, Dr. David Tsao of BP Corporation North America, Inc., and Dr. Jaco Vangronsveld, University of Hasselt Center for Environmental Sciences, Belgium. In addition, many other scientists and consultants contributed significantly to the text, case studies, research, review and editing, and the names and addresses of these individuals are included in the list of contributors.

Special thanks are due to colleagues Michael Lindquist and Andrew Hartness, who worked tirelessly on the illustrations for the book, Eammon Coughlin, Jennifer Kaplan, Michael Easler, and Jennifer Haskell, who contributed significantly to the plant list research, and Jenny Hill, Renee Stoops and Stevie Falmulari for their review and editing suggestions.

The authors have also received support from a number of local academic colleagues: Harvard Graduate School of Design faculty, including Charles Waldheim, Chair of the Department of Landscape Architecture, Peter Del Tredici, Gary Hilderbrand, Laura Solano and Alistair McIntosh. Other faculty colleagues who have provided support at Harvard University include Jack Spengler and Eric Chivian of the Center for Health and the Global Environment.

We would like to recognize our graduate students from GSD 9108 and 6335 Phyto Research Seminars who addressed the subject of phytotechnologies as well as carried out case studies and practicum exercises in the spring of 2011, 2012 and fall 2013 at the Harvard Graduate School of Design. The origins of this book lie partly in the discussions and case studies that arose from these classes. Participants included: Julia Africa, Rebecca Bartlett, Alexis Delvecchio, Kenya Endo, Christina Harris, Nancy Kim, Inju Lee, Amy Linne, Pilsoo Maing, Lauren McClure, Kathryn Michael, Alpa Nawre, Alissa Priebe, Soomin Shin, Patchara Wongboonsin (2011 Class), Christine Abbott, Naz Beykan, John Duffryn Burns, Amna Chaundhry, Michael Easler, Melissa How, Michael Luegering, Eva Ying, Hatzav Yoffe, Shanji Li (2012 Class), Kunkook Bae, Edwin Baimpwi, Vivian Chong, Karina Contreras, Jennifer Corlett, Omar Davis, Stephanie Hsia, Takuya Iwamura, Jungsoo Kim, Ronald Lim, Leo Miller, Gabriella Rodriguez, Miree Song, Patrick Sunbury, Kyle Trulen (2013 Class) and Megan Jones (Teaching Assistant, 2013).

In addition, we would like to recognize a group of research and illustration assistants who contributed during the summer of 2013 and spring of 2014. These include Harvard Design School faculty member Zaneta Hong and interns, who include Shuai Hao, Megan Jones, Geunhwan Jeong, Kara Lam, Ronald Lin, Cali Pfaff, Michele Richmond, Thomas Rogalski, Kyle Trulen and Arta Yazdanseta.

Both authors would like to acknowledge the guidance and support throughout the entire development and production of this book from our editor Louise Fox, Associate Editor – Landscape of Routledge/Taylor and Francis, Sadé Lee, Editorial Assistant, Ed Gibbons, Production Editor and contributing Routledge staff.

Finally, the authors would like to thank their families for their continued support during the writing process and development of the manuscript. Niall would like to acknowledge the constant support of wife, Louise, and daughter, Chloe, each in their own way. Kate would especially like to thank her husband, Chris Mian, for years of patient listening about phytotechnologies, and for his endless help throughout the writing of this book.

Cambridge, Massachusetts

Spring 2015

x

Contributors

Scientific content, case studies, and review

Alan Baker, Ph.D.
The University of Melbourne Australia
9 Victoria Road
Felixstowe, SFK IP11 7P, United Kingdom

Michael Blaylock, Ph.D.
Edenspace Systems Corporation
210 N 21st Street, Suite B
Purcellville, VA 20132 USA
www.edenspace.com

Sally Brown, Ph.D.
University of Washington
School of Environmental and Forest
Sciences
Box 352100
Seattle, WA 98105, USA

Joel Burken, Ph.D.
Missouri University of Science and
Technology
1401 N Pine Street
224 Butler Carlton Hall
Rolla, MO 65409-0030, USA

Rufus Chaney, Ph.D.
US Department of Agriculture
Environmental Management and Byproducts
Utilization Laboratory
10300 Baltimore Blvd., Bldg. 007
Beltsville, MD 20705, USA

Andy Cundy, Ph.D.
University of Brighton
School of Environment and Technology
Lewes Road
Brighton BN2 4GJ, United Kingdom

Alan Darlington, Ph.D.
Nedlaw Living Walls
250 Woolwich St. S Breslau,
ON N0B 1M0, Canada
www.naturaire.com

Mark Dawson, M.S.
Sand Creek Consultants, Inc.
151 Mill St.
Amherst, WI 54406, USA
www.sand-creek.com

Bill Doucette, Ph.D.
Utah State University
Utah Water Research Laboratory
8200 Old Main Hill
Logan, UT 84322-8200, USA

Stephen Ebbs, Ph.D.
Southern Illinois University
Department of Plant Biology, Center
for Ecology
420 Life Science II, Mailcode 6509
1125 Lincoln Drive
Carbondale, IL 62901, USA

Walter Eifert
ELM Site Solutions, Inc.
209 Hunters Woods Lane
Martinsburg, WV 25404, USA

Stephanie Eisner
City of Salem, Willow Lake Water Pollution
Control Facility
5915 Windsor Is. Rd. N.
Salem, OR 97303, USA
www.cityofsalem.net

Stevie Famulari
North Dakota State University
Landscape Architecture Department
620 10th Avenue North
Fargo, ND 58102, USA

John Freeman, Ph.D.
Phytoremediation and Phytomining
Consultants United
1101 Mariposa St.
Gilroy, CA 95020, USA
www.phytoconsultants.com

Wolfgang Friesl-Hanl
AIT Austrian Institute of Technology
Health & Environment Department
Environmental Resources & Technologies
Konrad-Lorenz-Straße 24 3430 Tulln, Austria

Edward G. Gatliff, Ph.D.
Applied Natural Sciences, Inc.
7355 Dixon Dr
Hamilton, Ohio 45011, USA
www.treemediation.com

Stanislaw Gawronski, Ph.D.
Warsaw University of Life Science
Laboratory of Basic Research in
Horticulture
Faculty of Horticulture, Biotechnology and
Landscape Architecture
Nowoursynowska 159
Warsaw 02-787, Poland

Ganga M. Hettiarachchi, Ph.D.
Kansas State University
Department of Agronomy
2107 Throckmorton Plant
Sciences Center
Manhattan, KS 66506, USA

Jenny Hill
University of Toronto
35 St George Street, Room 415A
Toronto, ON M5S 1A4, Canada

Jim Jordahl
CH2M HILL
709 SE 9th St.
Ankeny, IA 50021, USA
www.CH2M.com

Mary-Cathrine Leewis, M.S.
Ph.D Candidate
University of Alaska – Fairbanks
211 Irving I, PO Box 756100
Fairbanks, AK 99775, USA

Mary Beth Leigh, Ph.D.
University of Alaska Fairbanks
Institute of Arctic Biology
Department of Biology and Wildlife
902 N. Koyukuk Dr.
Fairbanks, AK 99775, USA

xii

Lou Licht, Ph.D.
Ecolotree Inc.
3017 Valley view Ln NE
North Liberty IA 52317, USA
www.ecolotree.com

Matt Limmer, Ph.D.
Missouri University of Science and
Technology
1401 N. Pine
Rolla, MO 65409, USA

Amanda Ludlow
Roux Associates, Inc.
209 Shafter Street
Islandia NY 11749, USA
www.rouxinc.com

Michel Mench, Ph.D.
INRA (UMR BIOGECO)
University Bordeaux 1, ave. des Facultés
Talence 33170, France

Jaconette Mirck, Ph.D.
Brandenburg University of Technology
Soil Protection and Recultivation
Konrad-Wachsmann-Allee 6
D-03046 Cottbus, Germany

Donald Moses
US Army Corps of Engineers,
Omaha District
1616 Capitol Avenue
Omaha, NE 68102-4901, USA

Lee Newman, Ph.D.
State University of New York
College of Environmental Science
and Forestry
248 Illick Hall, 1 Forestry Drive
Syracuse NY 13210, USA

Elizabeth Guthrie Nichols, Ph.D.
North Carolina State University
College of Natural Resources
2721 Sullivan Drive
Raleigh, NC 27695, USA

David J. Nowak, Ph.D.
USDA Forest Service
Northern Research Station
5 Moon Library, SUNY-ESF
Syracuse, NY 13215, USA

Genna Olson, P.G.
CARDNO ATC
2725 East Millbrook Road, Suite 121
Raleigh, NC 27604, USA
www.cardnoatc.com – www.cardno.com

Charles M. Reynolds, Ph.D.
US Army Soil Science-Soil Microbiology
ERDC-Cold Regions Research and Engineering
Laboratory
72 Lyme Road
Hanover, NH 03755, USA

Steven Rock
US Environmental Protection Agency (US EPA)
5995 Center Hill Ave.
Cincinnati, OH 45224, USA

Christopher J. Rog, P.G. CPG
Sand Creek Consultants, Inc.
108 E. Davenport St.
Rhinelander, WI 54501, USA
www.sand-creek.com

Liz Rylott, Ph.D.
University of York
CNAP, Department of Biology
Wentworth Way
York YO10 5DD, United Kingdom

xiii

Jerald L. Schnoor, Ph.D.
The University of Iowa
Department of Civil and Environmental
Engineering
Iowa City, Iowa 52242, USA

Julian Singer, Ph.D.
Formerly of University of Georgia
Savannah River Ecology Laboratory
Currently with CH2MHill
540 – 12 Avenue SW
Calgary, AB T2R 0H4, Canada
www.CH2M.com

Jason Smesrud, PE
CH2M HILL
Water Business Group
2020 SW 4th Ave., Suite 300
Portland, OR 97201, USA
www.CH2M.com

Renee Stoops
Plant Allies
1117 NE 155th Ave.
Portland, OR 97230, USA

David Tsao, Ph.D.
BP Corporation North America, Inc.
150 W Warrenville Rd.
Naperville, IL 60563, USA

Antony Van der Ent, Ph.D.
Centre for Mined Land
Rehabilitation
Sustainable Minerals Institute
The University of Queensland
Brisbane, QLD, 4072, Australia

Jaco Vangronsveld, Ph.D.
Hasselt University
Centre for Environmental Sciences
Agoralaan, buidling D
Diepenbeek BE-VLI B-3590, Belgium

xiv

Timothy Volk
State University of New York
College of Environmental Science and Forestry
346 Illick Hall
Syracuse, NY 13210, USA

Jason C. White, Ph.D.
Connecticut Agricultural Experiment Station
123 Huntington Street
New Haven, CT 06504, USA

Ronald S. Zalesny Jr., Research Plant Geneticist
US Forest Service – Phytotechnologies, Genetics
and Energy Crop Production Unit
Northern Research Station, 5985 Highway K
Rhinelander, WI 54501, USA

Barbara Zeeb, Ph.D.
Royal Military College of Canada
Department of Chemistry & Chemical Engineering
13 General Crerar, Sawyer Building, Room 5517
Kingston, ON, Canada K7K 7B4

Illustrations

Special thanks to the following practitioners,
who significantly contributed to the publication's
illustrations.

Andrew Hartness
HartnessVision LLC
Cambridge, MA, USA
hartnessvision.com

Michael G. Lindquist
38 Sewall St. Apt. 2
Somerville, MA 02145, USA

Preface

Toxic trees, virus-bearing vines, plants from outer space that eat poisons for lunch, then snack on young adults. B-movies from the 1950s vividly portrayed freakish vegetation to scare their audiences. In the last reel, mankind destroys the mutant greenery through the ingenuity of the scientist hero.... But recent eye-grabbing headlines – "Lead-eating mustard plants," "Pint sized plants pack a punch in fight against heavy metals," and "Pollution-purging poplars" – seem to have brought B-movies to life and unsettled our comfortable view of vegetation as benign and green. However, these seemingly freakish plants are in fact our good friends.

(Kirkwood, 2002)

The excitement and expectations of the B-movie "toxic trees" and "virus-bearing vines" are vividly brought to light in the developing field of phytotechnology, or 'phyto' for short. The more sensational aspects of this vegetation are tempered by the scientific basis and practical uses of plants in confronting pollutants in the contemporary environment. However, we can still continue to be entranced by both the processes taking place inside the plants, roots and surrounding soils and the good work plants can carry out on our behalf.

Phytotechnology applications have the capacity to play a significant role in transforming contaminated urban land, providing a more sustainable choice for remediation when combined with short- and long-term land planning. In some cases, plants can take up, break down or hold pollutants in place. However, the science that lies behind phytotechnology has been found by the authors to be challenging to comprehend for the non-specialist reader, and therefore difficult to implement. The ambition of this book is to bridge the critical science and engineering associated with phytotechnology site applications and its creative design use in the field.

I Background

This publication is the first targeted towards the spatial design, form, structure and aesthetics of this technology, rather than simply the science behind it. The authors' intent is to translate current research and field studies carried out by scientists into a format useful for the design practitioner in addressing site pollutants. Chapters 1–3 of the book delve into the science and regulatory issues around phytotechnology, including the nature of particular site contaminants and field case studies. Chapters 4–6 focus on projecting the potential environmental, spatial, cultural and aesthetic qualities of these productive vegetation types matched to site programs and specific contaminants.

The content of the book utilizes diagrams to illustrate a basic understanding of how the science related to phytotechnology functions, when it may work in site applications and when it does not. The great majority of the background information has been collected from individuals, institutions or agencies who have undertaken research or site installations. Detailed information on the relevant plant species for potential contaminants commonly found on sites is included. Additional diagrams and charts illustrate typical contaminants present on various types of programmatic sites (for example gas stations, road corridors, railroad corridors). Innovative plant combinations involving treatment solutions for these site programs provide practical design ideas where aesthetics and social functions have also been considered. Preventative planting palettes for certain site programs, such as railway corridors, dry cleaners, parks and urban homes are also created, thus allowing landscape design to propose vegetation strategies in advance of future potential contamination. In this way phytotechnologies become projective, anticipatory and a creative tool for the landscape architect and site owner to create landscape amenities for the citizens and communities in which these sites occur.

The origins of this book lie in two areas. First, noting the increase in urban landscape reclamation projects and environmental engineering practices, particularly on brownfield and contaminated land, there is an increasing need for clear guidance for landscape site design applications of plant-based remediation, in anticipation of further non-remediation-based planting installations. Second is the desire on the part of the authors to build on the earlier scientific research work of pioneers in the phytotechnology field and make their work and other current research work comprehensible and therefore usable to a larger range of stakeholders and participants. The book is the outcome of continuous study over the last 15 years by the authors, around the challenges, opportunities and techniques of phytotechnology and the activities of plant selection, landscape design and monitoring to reclaim post-industrial land and landscapes.

II Book structure

The structure of the book is as follows.

Chapter 1 provides an overview of phytotechnologies and their current application in the environment and their anticipated future use. This includes coverage of the definition and evolution of the subject, discussion of the legal framework in which they are used, review of their efficacy and finally an outline of potential innovative applications.

Chapter 2 reviews the fundamentals of the scientific processes involved in phytotechnology remediation and provides a summary of these processes. The activities of soil enhancement and plant cultivation are also addressed.

Chapter 3 offers the reader a survey of the groupings of contaminants commonly addressed by phytotechnology approaches and plant selection. These are related to the following chapter on planting typologies and application to polluted sites.

Chapter 4 outlines the interrelationship of specific contaminants with specific planting types, illustrating 18 different phytotechnology planting typologies.

Chapter 5 applies the phytotechnology planting types developed in Chapter 4 to a range of 16 commonly found land-use programs such as gas stations, road corridors, military sites and agricultural uses.

Chapter 6 is a listing of additional resources for those interested in following up on specific areas of the phytotechnology field.

Through an understanding of new research on phytotechnology vegetation and the potential opportunities and, conversely, limitations of their application, the reader will comprehend the range of topics that engage with this emerging technology. These topics taken together constitute the core knowledge of phytotechnology. The authors strongly support the notion that they will form a new core subject within the study and practice of landscape architecture and land regeneration in the future.

III Why the book is needed

This book is unique among publications on phytotechnology, as it links the scientific basis of the topic with its application in the planning and design fields. This involves addressing the reality of contemporary sites and their former and current programs, and the range of contaminants that are likely to be found. The need to remediate particular chemicals in soils and groundwater using plants is connected with neighborhood and community health and sustainability. In the authors' estimation, there exists a large gap between the publications and journals covering the scientific research, including laboratory experiments and field testing, and its general understanding and subsequent application on current brownfield sites.

One scientific journal, *The International Journal of Phytoremediation*, and several research-based books listed in the bibliography which are edited collections of research papers have been published on the subject of phytotechnologies. However, all of these publications are presentations of research and field studies, and the information presented is scientific, text-based and difficult to decipher by the landscape design professional. *Phyto* is different from these publications. This book is first and foremost a design-based guide utilizing simple charts and diagrams to clearly explain the science. Lists of potentially applicable plant species are provided. Plants hardy and suited for the northeast US climate are emphasized but the introduced design typologies and overall site strategies can be applied globally.

In addition five further issues for why this book is needed can be identified.

A Issue one: the contaminated environment

Major urban centers of population continue to lose industrial and manufacturing companies, and the land that supported them is left vacant with the contaminants and former infrastructure. This has created numerous abandoned, derelict and contaminated sites of varying sizes within communities, many of them located next to community amenities such as playgrounds, schools, recreational fields, daycares and senior centers. One of the central priorities of city planning initiatives is the urban reclamation and regeneration of inner city land such as degraded river edges, railroad yards, ports, harbors and piers that are central to the revitalization of city districts and neighborhoods. There is a growing need for innovative, sustainable, low-cost methods to address the remaining contamination of soils, groundwater, sediments and surface water over a wide range of site locations.

B Issue two: the productive use of contaminated land

With concerns for sustainable planning and quality of life entering public discourse and the diminishing amount of available urban sites for development, there is a need to regenerate the middle-scale and smaller sites found within the city fabric. Driving the cleanup of polluted sites is the desire for higher and best use of the land, for economic development opportunities and community facilities. Of interest to the landscape architect is how phytotechnology can inform more progressive and creative planning and design work to achieve these productive uses, and conversely, to what extent new programs and uses can direct the regeneration of these sites through phytotechnologies. In addition, the potential for phytotechnologies to produce an economic product such as biomass for energy, paper or wood products while also cleaning sites is of primary interest.

C Issue three: available information on phytotechnologies

Current information available on phytotechnology is very widely dispersed and varied, appearing in a range of media including books, magazines, websites and technical manuals. Research presented may be contradictory, and it is difficult to determine which research is the most current. In addition, test projects and field applications have generally occurred on inaccessible places in remote locations out of the public eye, such as Department of Defense sites or large-scale industrial or resource extraction lands (Stoops, 2014). This easy-to-read handbook targeted at the design and landscape professions summarizes the wealth of research and field case studies produced in recent years. It not only includes case studies that have worked, but describes many misconceptions in the field where phytotechnologies likely are not very effective.

D Issue four: design, performance and landscape architecture

There has been a changing mood in the development of landscape sites where functional ecologies, community, public health, urban design and sustainable development concerns are a driving force in addition to aesthetic factors. More recent projective design approaches have included the consideration of what the landscape can 'do' in addition to the human programs layered on sites. Phytotechnologies can enhance the landscape functionality of green corridors, vegetation patches, new woodlands, hedgerows, urban agriculture and wetlands.

E Issue five: future projections of domestic site contamination

Homeowners have an increased awareness of site contamination issues in gardens, yards and surrounding house properties through local reporting on the concerns of existing site urban fill, past pesticide use, leaking oil tanks and lead poisoning of children through playing in the polluted soil of home gardens and yards. Enough research now exists to apply phytotechnology concepts at a residential scale. This strategy is based on the need to make communities more livable for residents and to provide safer landscapes for all citizens – but particularly for children and seniors – and a healthier approach to community resources such as playgrounds, schools, pocket parks and homes in the neighborhood.

xix

IV Audience for the book

Phyto is conceived as a practical and easy-to-use handbook for college-level instruction and in continuing education programs, and as a reference for design professionals and those in the horticultural and construction industries. It also is envisioned as contributing to the advancement of discussion in the design and planning fields about the way designers conceive and construct phyto landscape design work in a variety of site conditions. The following are examples of who might read it and how the authors expect it to be used.

A Landscape architecture and other design students

As a textbook or guide for planting design and site remediation for landscape architecture, landscape planning, urban design and site design students in a plants or technology class or in a planning and design studio.

B Landscape architects, other design and engineering practitioners

In the professional offices of landscape architecture, site engineering, environmental engineering and ecological engineering consultants as a reference for planting design and remediation for urban sites.

C Urban designers and planners

In the private, professional and municipal offices of planners, urban designers and municipal employees to assist in the initial planning and research on sustainable cleanup alternatives for brownfield site projects with polluted soils and groundwater.

D Horticultural industry

For plant growers, nurserymen and members of the horticultural community as a handbook of some plant varieties able to be grown and available to the phytoremediation industry.

E Landscape construction companies

As a reference guide for landscape and engineering construction companies involved in the implementation and maintenance of the landscape, from project, site supervision and procurement managers to plant installation job captains and field workers.

F Organic land care associations

As a reference guide in order to instruct industry professionals and community planners, groups and educators in a range of productive planting techniques for urban and ex-urban sites.

The subject of phytotechnology brings together the disciplines, professional worlds and knowledge of landscape design, science, engineering, horticulture, site planning and cultural and social programs, whether through initiatives such as land regeneration, urban gardens, energy creation, greening of local neighborhoods, new recreational venues or local stormwater management action, and often including intensive local community interest and involvement. Even as the "seemingly freakish plants" identified in the opening quote appear to be independently and single-handedly tackling the polluted lands that dot the current environment of communities, they are actually there as a result of a larger multidisciplinary effort carried out by concerned teams of scientists, engineers, government officials, academic researchers and independent research groups, design and planning professionals in private practice and community volunteers. Community organizers, local planning offices, members of government agencies and non-profit environmental groups are all critical players in the implementation of these technologies.

It is the authors' intention firstly that this book should guide all of these players to an understanding of the potential of phytoremediation-based design and of the promise of these "freakish plants" not only to remediate sites but to act within a pre-emptive approach to address the future evolution of these sites. Secondly, that residents and community members will be able to access and enjoy a new range of landscape spaces and outdoor planted places that are indeed healthier, less polluted and full of lessons about the power of natural processes of remediation. Through the art and science of phytotechnologies and on an increasing range of design sites, landscape architecture will possess a swift and sure means of touching the greater world and creating a more ecologically sustainable, resilient and responsive means to shape the future constructed environment.

Foreword

Steve Rock

People have deliberately grown plants to alter their environment for at least millennia. The Roman roads were lined with poplar trees both to provide shade and to keep the roads' foundations dry by consuming water along the edges, thus making the roads last longer.

The broadest definition of phytotechnologies includes any plantings that enhance the environmental goals for the planet. The field has grown from narrow beginnings to widespread applications and has moved from hopeful but ultimately unfounded expectations to a mature set of techniques and technologies that are commonly accepted as a part of the environmental cleanup toolbox.

Even before the field was named, people have been using plants to enhance their work. In the 1930s bioprospecting was used as a way to predict the presence of minerals subsurface. Prospectors, particularly in newly opened lands of Siberia, discovered that they could search for plants that grew only in areas rich in certain minerals. It was noted that some plants were reliable indicators of minerals and that leaves and twigs could contain quantities of metals much higher than those in others of the same type in other locations.

In the 1970s several research groups began systematically studying and classifying the relationship between metals and plants, and found some plants growing in metal-rich soils to have extraordinary properties. Three researchers in particular, Drs. R. R. Brooks, R. D. Reeves, and A. J. M. Baker and their teams, crisscrossed the globe finding and cataloging plants that grew on metal-rich soils and took up unusual quantities of those metals. Some plants were found to take up more metals than normal plants and eventually were named as accumulators and hyperaccumulators.

Increasing general environmental awareness at the time spurred traditional agricultural research to study the effects of environmental contaminants on food-production crops, particularly the uptake of potentially harmful heavy metals. The practice of using biosolids from sewage sludge as fertilizer brought rural crop plants into contact with all the industrial pollutants that were flushed down urban drains. It was found that some contaminants did move into some crop plants. One USDA researcher, Dr. Rufus Chaney, suggested that while planting metals excluders might help to protect the food supply, it might also be possible to clean soil by raising crops that extract and accumulate metal which could be harvested not for food but for remediation.

Also in the 1970s the new field of bioremediation was exploring how to use microbes to attempt environmental cleanup of degradable contaminants. Research began into whether and how much plants enhanced the microbial degradation of pesticides and petroleum products. It was soon clear that planted systems did remediate certain contaminants sooner, deeper, and in some cases more completely than microbial systems alone.

Such fundamental research continued into the 1980s, attracting attention from university research teams, government agencies, and private industry. The new-found national and international environmental awareness of the time and the creation and passage of foundational environmental legislation such as the Clean Water Act and the CERCLA (Superfund) led to increased funding into many possible remediation strategies. Municipalities and corporations were under pressure to reduce the discharge of toxins into the air and water, and onto the land. Cleaning of historical contamination became a new and large industry. Consulting and contracting companies sprang up everywhere, industrial and commercial enterprises started in-house remediation divisions, and government agencies were started or became larger. It is no wonder that by the late 1980s some people were turning their thoughts to commercializing this new process of using plants for remediation.

The earliest definitions of phytoremediation in the 1990s in publications and presentations refer to environmental protection via metal uptake by plants. Terms and definitions quickly proliferated as firms tried to distinguish and differentiate themselves and their processes. Phyto hyphen anything became a way to classify increasingly specific uses of planted systems. Phyto-degradation, -extraction, -enhanced bioremediation, etc. were used to describe and differentiate aspects of the field. Other terms like rhizofiltration and hydraulic control were invented and used in specific circumstances. Phytotechnologies to this day is an umbrella term that aims to encompass all uses of plants for environmental goals.

The 1990s were a time of proliferation of patents as well as companies and invented words. Some patents were for inventions, some for techniques, and some for practices that had been widely used but never patented.

One successful phyto-based patent was introduced when Edd Gatliff patented a TreeWell system that in part uses a downhole sleeve and air tube to induce and enable tree roots to penetrate deeper than they would naturally. The system allows trees to be targeted at a particular depth below ground surface and to tap into contaminated groundwater while bypassing clean water-bearing layers. This innovation combines several known and novel techniques and devices, and allows remediation to depths and in places otherwise unobtainable.

Other patents were not so specific and had a chilling effect on the deployment of the some remediation practices. The numbers of on-site applications and field experiments fell at the end of the decade, in part due to legal concerns over patents and in part because expectations caught up with reality.

It was widely hoped that phytoremediation would solve the problem of widespread low-level contamination of heavy metals in soil. Many heavy metals of concern, and particularly lead, persist in soils for many decades from spills, dumping, or atmospheric deposition. Large areas of land have soil that poses a risk to residents and workers but there are few tools that are economical, non-invasive, and effective. Phytoremediation for metals (phytoextraction) was hoped to be all those things and quite literally a green technology into the bargain.

There are some plants that will naturally accumulate some metals under some circumstances. These natural accumulators are often small, grow slowly, and are difficult to cultivate outside of their native and often narrow range. It was hoped and claimed that some plants that grew faster, larger, and using standard agricultural equipment and practices could be induced to take up enough metal to clean soil. Unfortunately, induced phytoextraction of metals has several flaws that have to date proved insurmountable. These include the fact that the most widely used technique relies on chemically altering the contaminant to become much more soluble than in its natural state. This more soluble metal is then more likely to be taken up by the planted phytoextraction crop; the soluble metal is also more likely to be washed away into surface water and groundwater, where it poses an even greater risk than when it was bound into the soil, which is both morally and regulatorily unacceptable.

There were a number of highly publicized demonstration projects with optimistic reports and enticing pictures. Phytoremediation entered the public lexicon via popular articles, usually featuring a picture of a field of sunflowers. After a few careful experiments it was determined that indeed the plants could be induced to take up quantities of metal that could lead to a respectable cleanup in a matter of years, but that the need to prevent the escape of the mobilized metals would prevent the process from ever becoming economically feasible.

First the industrial boosters repurposed their staff and resources. Then the contractors and consultants changed their focus. Phytoextraction remains a popular academic topic of study in the search for the plant that might naturally extract and accumulate enough contaminant to be an effective tool, or to find a safe way to induce uptake. There have been some attempts at genetic modification. Currently, phytoextraction of metals has not lived up to its early promise, and despite continued academic and public interest is not a mainstream tool in the remediation toolkit.

However, at the same time that phytoextraction of metals was enjoying a lot of press, discussion, and also some failures, other phytotechnologies to mitigate contaminated groundwater plumes and treat organic pollutants such as petroleum and solvents were quietly maturing and taking their place in the toolkit. One of the processes that plants do naturally and quite well is move water. This has been used widely and effectively in landfill covers to prevent precipitation penetration, and in subsurface applications to control contaminated groundwater plumes, and in phytoforensics, where plants are used to track subsurface contaminants.

It was soon shown that planted cover systems for landfills are generally as effective as conventional covers in many parts of the US. Like all plant-based systems, the actual effectiveness will be a function of location. A nationwide field study from 1999 to 2011 showed how to determine equivalency for landfill cover systems. Now that there are hundreds of plant-based covers in place and enough more on engineering firms' drawing boards, such covers are no longer considered experimental or innovative and regulatory approval is regularly given.

Planting trees not only to control water but also to enhance bioremediation of organics and light solvents is also common enough to be included in many cleanup plans. Although most metals do not move easily into plants, several other organic contaminants of interest are soluble enough to move or translocate into plants, where they are often degraded, without the need for harvesting the plants.

This ability of plants in general and trees in particular to take soluble contaminants from groundwater has allowed an interesting and potentially very useful technique called phytoforensics. Since 2000 Drs. Don Vroblesky, James Landmeyer, and Joel Burken have pioneered and refined the techniques needed to remove tree cores and analyze the chemical content of the sap. Side-by-side studies have shown that phytoforensics can reveal the origin and direction of groundwater contamination with as great accuracy as and considerably less expense and disruption than conventional testing and monitoring-well drilling.

No discussion of phytotechnologies is complete without including wetlands. In use for cleaning wastewater since at least the 1880s, wetland technology continues to be developed and improved. Many large environmental firms have some capacity to size, specify, and install constructed wetlands to treat industrial or municipal outflow. It is one of the most robust and frequently applied uses of planted systems to achieve such diverse environmental goals as organics degradation, metals sequestration, and wildlife habitat creation – often at the same time.

Since the first meetings to discuss these topics, like the "Beneficial Effects of Vegetation in Contaminated Soil" meeting hosted by Kansas State University in 1992, to the now annual conferences of the International Phytotechnology Society, researchers, consultants, regulators, and contractors meet and talk about the what works and what does not. The field has undergone a tremendous shift from fringe idea, to highly touted silver bullet, to the current state of reasonable expectations for successful application on a local site-by-site basis.

Phytotechnologists, landscape architects, and site designers share an overlapping toolbox with plants, soils, and water as the pieces to build the constructs we are called upon to create. Often a site will employ both sets of professionals – one to clean the canvas and one to provide the finishing touches once the site's structures are complete. This book provides a means to bridge those task areas so the means to remediate a site may be part of the final landscape site design. Each profession has a specific and distinct vocabulary, as is appropriate for fields that come from widely different origins and have individual project goals and deadlines. This book will help to overcome that language gap, for the landscape architecture community and for any scientists and engineers who want to understand this design discipline.

Planting any given vegetation is neither difficult nor complex, but planting for a particular outcome that sometimes won't be realized until years or decades later requires experience, and patience. Practitioners in both fields recognize the need for time on a plant scale, although the site owners and regulators sometimes don't share that view.

In conclusion, the future of phytotechnology and its application to a wide number of sites and over a range of timescales is still evolving. Designers and scientists working in collaboration can help create the correct environments to advance the range and type of plants to be used, as well as create phased projects that can begin to demonstrate the value of phytotechnologies over time.

Ultimately, phytotechnology is about using specifically selected plants, installation techniques, and creative design approaches to rethink the landscapes of the post-industrial age. It is less about simply the beauty of plants, less about gratuitous site planning and design and the creation of individual design ideas; rather, it is to focus through design on plant characteristics to sequester, take up or break down contaminants in soils and groundwater. The purpose is to understand and include the margins of scientific research and invention to employ broader boundaries, where plant-based remediation can be used for improvement and renewal, and to plan beyond the short term for a longer vision for the contemporary environments of cities, towns and communities.

Icons

Contaminants

Organic contaminants

- Petroleum
- Chlorinated solvents
- Explosives
- Pesticides
- Persistent Organic Pollutants (POPs)

Inorganic contaminants

- Nutrients
- Metals
- Salts
- Radionuclides

Organic and inorganic mechanisms

Phytodegradation
Rhizodegradation
Phytovolatilization
Phytometabolism

Phytoextraction

Phytohydraulics

Phytostabilization/Phytosequestration

Rhizofiltration

Abbreviations

Abbreviation	Icon for contaminant	Name	Description
Al	■	Aluminum	Inorganic metal(loid) associated with metals mining, production and smelting.
As	■	Arsenic	Inorganic metal(loid) commonly found in pesticides and pressure-treated lumber and naturally occurring in high concentrations in some soils and groundwater.
B	■	Boron	Inorganic metal(loid) commonly associated with glass manufacturing, pesticide use and leather tanning.
BOD		Biochemical Oxygen Demand	Biochemical oxygen demand or BOD is the amount of dissolved oxygen needed in a body of water to break down organic material present. It is used to gauge the organic quality of the water.
BTEX	●	Benzene, Toluene, Ethyl benzene, and Xylene	Volatile organic compounds (VOCs) found in petroleum products.
Cd	■	Cadmium	Inorganic metal commonly contaminating agricultural fields and derived from soil amendments and mining and smelting activities.
Ce	■	Cesium	Inorganic radionuclide associated with nuclear energy production and military activities.
CERCLA		Comprehensive Environmental Response, Compensation and Liability Act	Commonly known as 'Superfund', this was a Federal US Law enacted in 1980 that established a trust fund used by the government to clean up contaminated sites on the National Priorities List (NPL).
Co	■	Cobalt	Inorganic metal commonly used as a colorant in glass and ceramic production, as well as alloy and aircraft manufacturing.
CO		Carbon monoxide	Toxic gas created by automobiles and the incomplete combustion of hydrocarbon fuels; component of air pollution.
CO_2		Carbon dioxide	Component of air pollution, greenhouse gas.
Cr	■	Chromium	Inorganic metal commonly associated with the electroplating, automotive and tannery industries as well as the production of pressure-treated lumber.

Abbreviation	Icon for contaminant	Name	Description
Cu	■	Copper	Inorganic metal commonly used in metals, pipe and wire production, pesticides and fungicides.
DDT, DDE	●	Dichlorodiphenyltrichloroethane, Dichlorodiphenyldichloroethylene	Highly toxic persistent organic compound used as a pesticide and banned in the US since 1972. DDE is a common toxic breakdown product of DDT.
EDTA		Ethylene Diamine Tetra-acetic Acid	Chemical added (chelant) to make pollutants more bioavailable to plants for uptake.
EG		Ethylene Glycol	Organic compound commonly used in de-icing fluids.
EPA		Environmental Protection Agency	Federal government regulatory agency in the United States responsible for enforcing laws pertaining to the natural environment and regulating the cleanup of contaminated sites.
Carbon tet Halon 104 Freon 1	●	Carbon tetrachloride	Organic chlorinated solvent compound denser than water. Used as a refrigerant, fire suppressant, industrial degreaser and in the cleaning industry.
DNAPL		Dense Non-Aqueous Phase Liquid	Oily type of pollution that lies beneath water.
DRO	●	Diesel Range Organics	Organic compounds typically found in diesel fuel.
F	■	Fluorine	Inorganic metal associated with phosphate fertilizer production as well as smelting, coal-fired power plants and mining.
Fe	■	Iron	Inorganic metal widely found and usually not considered a contaminant except when in high concentrations in water.
GRO	●	Gasoline Range Organics	Organic compounds typically found in gasoline.
Hg	■	Mercury	Inorganic metal associated with coal-burning power plants, metals and paint manufacturing.
HMX	●	1,3,5,7-tetranitro-1,3,5,7-tetrazocane	Explosive organic compound, commonly associated with military uses.
LNAPL		Light Non-Aqueous Phase Liquid	Oily type of pollution that floats on water.
Log K_{ow}		Octanol-Water Partition Coefficient	A dimensionless constant which provides a measure of how an organic compound will partition between an organic phase and water.
LSP		Licensed Site Professional	An engineer, environmental scientist, or geoscientist licensed by the State, who is qualified to assess contamination and conduct cleanups.
LUST(s)		Leaking Underground Storage Tank(s)	Tanks found below ground that are typically leaking fuel. Common on former and current industrial sites and old gas stations.
Mn	■	Manganese	Inorganic metal, widely found and usually not considered a contaminant except when in high concentrations in water.
Mo	■	Molybdenum	Inorganic metal most often associated with mining operations.
MTBE	●	Methyl Tertiary Butyl Ether	Organic compound that is an additive to gasoline. Can exist in both liquid and gas phases.
N	■	Nitrogen	Essential inorganic nutrient needed for plant growth that can become an environmental pollutant from agricultural activities and wastewater.
NASA		The National Aeronautics and Space Administration	The agency of the United States government that is responsible for the nation's civilian space program and for aeronautics and aerospace research.

Abbreviation	Icon for contaminant	Name	Description
Ni	▪	Nickel	Inorganic metal commonly generated from mining and battery-production operations.
NO$_X$/NO$_2$		Nitrogen oxides	Component of air pollution (smog and acid rain) created by fossil fuel combustion and automobile engines.
NPL		National Priorities List	List of national priorities among the known releases or threatened releases of hazardous substances, pollutants or contaminants throughout the United States. The NPL is intended primarily to guide the EPA in determining which sites warrant further investigation (US EPA, 2014).
O$_3$		Ozone	Component of air pollution created by reactions between VOCs and nitrogen oxides as they are exposed to sunlight.
P		Phosphorus	Essential inorganic nutrient needed for plant growth associated with agricultural activities and roadways.
PAHs	●	Polycyclic Aromatic Hydrocarbons	Class of petroleum organic hydrocarbons that contain difficult-to-break-down benzene ring structures. Associated with fuel spills, coal processing or petroleum manufacturing.
Pb	▪	Lead	Persistent inorganic metal causing widespread contamination in urban areas. Formerly added to paint and gasoline until the 1970s.
PCE/Perc	○	Perchloroethylene, Tetrachloroethene	Organic chlorinated solvent compound denser than water. Commonly associated with dry-cleaning or metal-working facilities.
PCBs	●	Polychlorinated Biphenyls	A class of persistent organic pollutants banned in the US since 1979 that do not break down easily. Associated with many types of manufacturing or industrial processes.
PG		Propylene glycol	Organic compound commonly used in de-icing fluids.
Phyto		Phytotechnologies	Abbreviation for phytotechnologies.
PM (2.5)		Particulate matter (small)	Small liquid and solid particles found in the air. Very harmful to human respiratory systems.
PM (10)		Particulate matter (large)	Larger liquid and solid particles found in the air.
POPs	●	Persistent Organic Pollutant(s)	A group of 24 toxic organic contaminants that do not break down in the environment and exist for a very long time.
RAO		Response Action Outcome	A classification given to a site to designate the level to which significant risks or substantial hazards have been mitigated at the conclusion of remedial action.
RDX	●	Cyclo-Trimethylene-Trinitramine, 1,3,5-Trinitroperhydro-1,3,5-Triazine	Explosive organic compound commonly associated with military uses.
Se	▪	Selenium	Inorganic metal(loid) naturally occurring in high concentrations in some soils and groundwater.
SO$_X$/SO$_2$		Sulfur oxides	Component of air pollution (smog and acid rain) created by fossil fuel combustion and automobile engines.
Sr	▪	Strontium	Inorganic radionuclide associated with nuclear energy production and military activities.
T/^{3}H	▪	Tritium	Inorganic radioactive isotope of hydrogen associated with military activities.

Abbreviation	Icon for contaminant	Name	Description
TCE		Trichloroethylene	Organic chlorinated solvent compound, denser than water. Commonly associated with dry-cleaning or metal-working facilities.
TNT		Trinitrotoluene	Explosive organic compound commonly associated with military bases and some mining activities.
TPH		Total Petroleum Hydrocarbons	A combined measure of all the organic hydrocarbon compounds (can be hundreds) found in a petroleum sample at a given site.
TSS		Total Suspended Solids	A measurement of the amount of particles suspended in water. As TSS increases, a water body begins to lose its ability to support a diversity of aquatic life.
U		Uranium	Inorganic radionuclide associated with nuclear energy production and nuclear energy.
VC		Vinyl chloride, chloroethene	Organic chlorinated solvent compound used to produce PVC (polymer polyvinyl chloride), a type of popular plastic.
VOC		Volatile Organic Compounds	Synthetic organic chemical capable of becoming vapor at relatively low temperatures.
Z		Zinc	Inorganic metal commonly associated with mining, smelting and industrial operations.

xxxii

1: Phytotechnology and the contemporary environment: an overview

In this chapter, the key background issues and design topics surrounding plant-based remediation are introduced and an overview of the potential use of phytotechnologies in site design work is given. Chapter 1 also includes the current definition of the term phytotechnologies, and outlines the opportunities and constraints of vegetation-based remediation. Previous and current research on the subject is described and an outline of the legal framework in which phytotechnologies are used is provided. Finally, potential innovative applications are summarized as well as projected areas of future scientific and field research.

I   What is phytotechnology?

The authors have proposed a broader definition of phytotechnology than is currently available, so as to engage and integrate this work with contemporary site design practices:

> Phytotechnology is the use of vegetation to remediate, contain or prevent contaminants in soils, sediments and groundwater, and/or add nutrients, porosity and organic matter. It is also a set of planning, engineering and design tools and cultural practices that can assist landscape architects, site designers, engineers and environmental planners in working on current and future individual sites, the urban fabric and regional landscapes.
>
> Definition by Kirkwood and Kennen as an expansion of previous definitions (Rock, 2000; ITRC, 2009)

The major focus of phytotechnology in this book is on the plant-based remediation of soils and groundwater. Planted systems for stormwater and wastewater treatment are already commonly integrated in landscape design practice, therefore the book will only briefly address these topics. This work will also touch on air pollution as it relates to the natural ability of plants to bioaccumulate or degrade airborne pollutants or render them less harmful. Phytotechnology implements on-site scientific and engineering solutions to

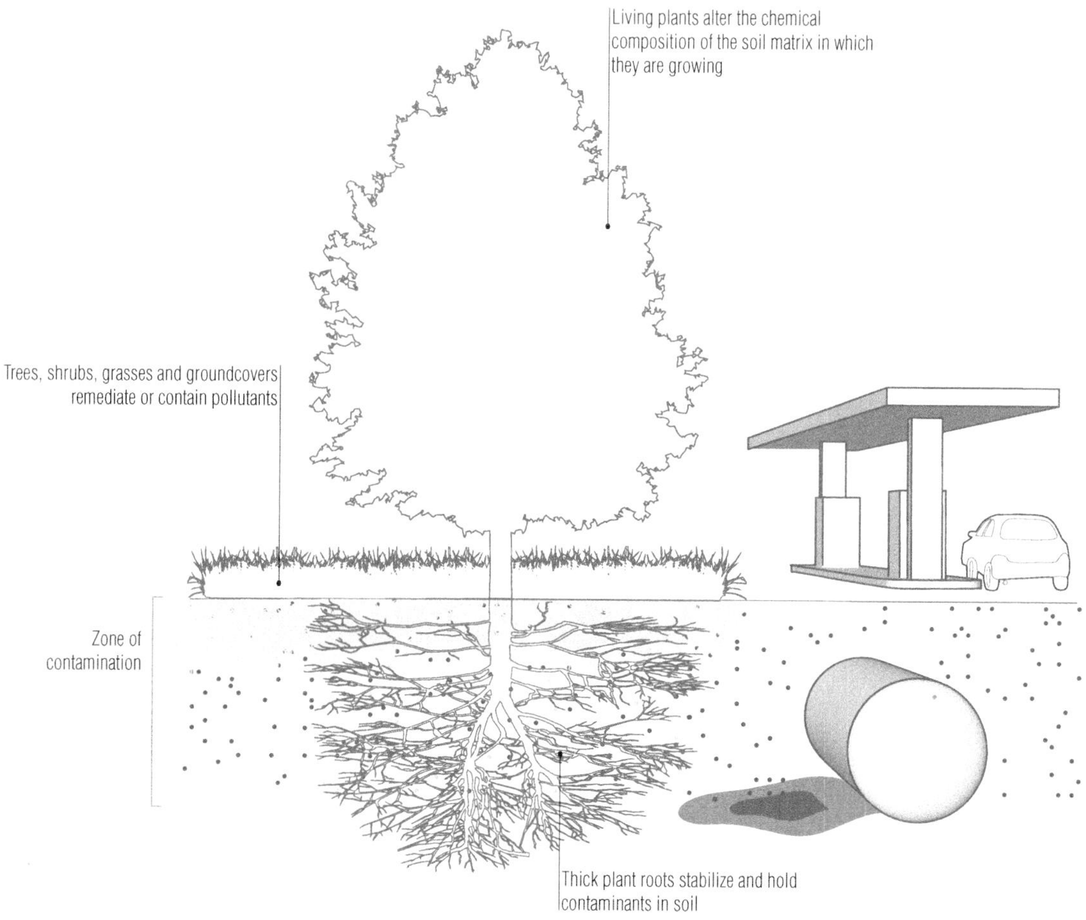

Living plants alter the chemical
composition of the soil matrix in which
they are growing

Trees, shrubs, grasses and groundcovers
remediate or contain pollutants

Zone of
contamination

Thick plant roots stabilize and hold
contaminants in soil

Figure 1.1 Phytotechnologies

contaminants found predominantly in soils and groundwater, via introduced vegetation that is targeted to be self-sustaining and integrated in the site design.

Following the above definition, the background to the term phytotechnology, and in particular its evolution over recent years, is worth describing. Confusion can result from a slight difference between the terms 'phytoremediation,' which is more traditionally used in scientific papers, and 'phytotechnology,' which is seen more in recent literature. Instances are not uncommon where the two terms have been used interchangeably, leading to further confusion about the subject.

II   The difference between phytotechnologies and phytoremediation

The term phytoremediation, or remediation by plants, simply describes the degradation and/or removal of a particular contaminant on a polluted site by a specific plant or group of plants. However, in addition to the degradation and/or removal of contaminants, phytotechnology also includes techniques such as the stabilization of pollutants within the surrounding soil or root structure of a plant and the pre-emptive installation of plant-based approaches so as to treat a pollutant or mitigate an ecological problem before it actually occurs. Stabilization utilizing plants does not actually remediate or break down the pollutants but renders them immobile in the soil, thus allowing no further contact to take place between the occupants of the site and subsurface contamination. In addition, the term phytotechnology may also include prophylactic advance plantings on a site that can help prevent contamination that could arise in the future from site activities. Where phytoremediation is typically known to focus on upland plantings for soil and groundwater cleanup, phytotechnology includes all plant-based pollution-remediation and prevention systems, including constructed wetlands, bioswales, green roofs, green walls and planted landfill caps. Taking an even broader view, parks, community gardens and greenways often have phytotechnology components designed into these landscapes, such as protective riparian buffers and vegetated filter strips, where introduced vegetation addresses a range of environmental constraints and pollution control.

Phytotechnologies are based on ecological principles and consider the natural systems as an integral component of human and societal interventions. It is this that makes the use of phytotechnologies integral with evolving landscape architectural design practices. For the remainder of the book the term phytotechnology will be used to describe the comprehensive application of plants on contaminated land and its relationship to the field of landscape architecture and site design.

III   Why do we need phytotechnologies?   5

Recently 'greenfield development' or building on formerly undeveloped or agricultural land has tended to overshadow the reclamation, regeneration and reuse of polluted brownfields. In particular, sites that by virtue of past industrial uses are today contaminated, environmentally disturbed, ecologically threadbare and perceived as economically and socially dysfunctional need remediation to become habitable again.

A The brownfield problem and the need for cost-effective solutions

The class of site known as 'brownfield' is universal and gaining more attention. The term is not only found across every part of the country, but in almost every nation and across each continent. The sites are often the most contentious type, politically, ecologically, culturally, economically and aesthetically. They include those with leaking or obsolete underground oil storage tanks, such as gas stations, former industrial sites and former manufactured-gas plants. They also include landfills in varying stages of use from active to closure, railroad corridors, burial grounds and Department of Defense (DOD) military lands. Twenty percent of all real estate transfers in the United States are brownfield sites (Sattler et al., 2010), with the current value of these lands in 2010 in the range of US$2 trillion. The US Environmental Protection Agency (US EPA) estimates there are approximately 450,000–600,000 identified sites located across the country, although this number has been considered unrealistically low (US Accounting Office, 1992). More than 16% of global land areas, equivalent to about 52 million hectares, are impacted by soil pollution worldwide (Anjum, 2013). All these sites, whether large or small, nationally or internationally, need a wider range of cost-effective solutions to clean up or mitigate the risks from soils, groundwater, sediments and existing infrastructure of canals, pools, lagoons and buildings found there.

Remediation technologies are, however, very costly, preventing cleanup of contaminated brownfield sites. The majority of traditional remediation approaches are expensive and energy intensive in their approach to quickly correcting an environmental problem that was decades in the making. Among the remediation methods that are under review by regulatory agencies are phytotechnologies, used either singly or in combination with other industry methods such as removing or capping of polluted soils, and mechanical pumping and treatment of groundwater plumes. The cost-effectiveness of phytotechnologies versus traditional remediation-industry approaches is often a significant advantage and the long-term energy required is often less, since phytotechnologies typically do not require mechanical pumping systems, utility power or much supporting infrastructure and equipment. Plant-based cleanup methods can be as little as 3% of the cost of traditional cleanup costs. Examples provided by author David Glass in his reports (Glass, 1999) demonstrate that phytotechnologies are significantly cheaper than the remediation-industry standard methods. For example, pump-and-treat for groundwater, or incineration of polluted soils, are cheaper by a factor of up to one to thirty; and more specialized methods such as thermal desorption, by a factor of one to ten; soil washing, by a factor of one to four; and bioremediation, by a factor of one to two. These figures, however, do not take into account differences based on individual existing site conditions, location and externalities caused by pollutant types and intensity, climate and human factors, such as the needs for ongoing monitoring, maintenance and site security.

The above-mentioned extent of contaminated sites is based on those already discovered, inventoried and being addressed in some fashion in either the short or long term. There still remains the larger number of landscapes and sites that are in private ownership and currently occupied by industry and manufacturing and that can still produce site pollutants or will be occupied in the future with all the potential for further site contamination. The scale of industrial activities and the number of individual sites may be never ending, with attendant levels of pollution. The potential for plant-based remediation to contribute to the larger cleanup work will increasingly include the design professional, by providing a new set of tools for site regeneration and a source of continual new project work.

B The limitations of conventional remediation practice

Conventional industry remediation practices, such as the 'pump-and-treat' (cleaning polluted groundwater through extraction, filtration and recharge methods) and 'dig-and-haul' (where polluted soils, as the name suggests, are dug up and shipped off site), are not only expensive but are single-outcome technologies and have limited site-design potential beyond treatment. Additionally, these traditional remediation approaches are often extremely invasive and disruptive and, by destroying the microenvironment, even leave the soil infertile and unsuitable for agricultural and horticultural uses.

C The build-up of everyday pollutants

The pollution potential of ubiquitous everyday landscapes and installations such as roadways, septic systems and lawn-care applications has recently come to the forefront as a concern. The build-up of contaminants not only affects the surrounding natural resources of an area but also puts a major strain on the local and, in some cases, regional ecosystems. This is due either to prior ignorance of the persistent effects of contamination in the environment, by lack of long-term vision, or to carelessness on the part of local and municipal governments in first legislating for and then administering the overview of environmental regulations. There is a critical need to prevent daily releases of small amounts of pollutants from these widespread land uses.

IV   Opportunities and constraints

In phytotechnology the natural properties and mechanisms of living plants are used to accomplish defined environmental outcomes, especially the reduction of chemicals in soils and groundwater. The diversity of available plants also gives versatility to the application of phytotechnology across a range of landscape locations and types. However, many site conditions and pollution situations render phytotechnologies 'not viable' for implementation. A review of the opportunities and constraints for using these technologies is provided below.

A Opportunities

The advantages of the application of phytotechnologies are as follows.

7

- Plant-based systems are natural, passive, solar energy-driven methods of addressing the cleanup and regeneration of several types of pollution-impacted landscapes.
- The process leaves the soil intact, even improved, unlike other, more invasive methods of site remediation used by the industry, such as removal and disposal, soil washing and thermal desorption.
- Phytotechnologies have the potential to treat a wide range of organic contaminants in the soil and groundwater in low to moderately contaminated sites. However, there are also many cases where phytotechnologies are not applicable. Very specific plant and soil interactions must be considered and monitored for their effectiveness, as detailed in Chapters 2 and 3.
- The use of phytotechnology in a variety of landscape-restoration and environmental installations is attractive to scientists, engineers and designers. These include hybrid technologies combining chemical, physical and other biological processes with plant-based methods.
- Vegetation-based remediation, when applicable, has been found to be less expensive in comparison with other, more conventional, industry-based technologies and approaches.
- Public acceptance is considered high, particularly if the site is located close to or within residential neighborhoods, as phytotechnology is a natural, low-energy, visually and aesthetically pleasing remediation technology.
- The application of phytotechnologies can be integrated into other vegetation and landform design strategies and programs proposed for the site. The expansion of the range of natural cleanup technologies at the disposal of the landscape architect during the planning and design of post-industrial sites can be a starting point for design.

- Pollution prevention: Plantings can prevent the spread of future pollution releases and the further environmental degradation of urban land and waterways.
- Indicator species: Vegetal indicators of ecosystem health can be integrated into monitoring and assessment strategies for sites.

1 Ancillary potential benefits

Ancillary potential benefits from phytotechnology applications include such factors as community use, educational use and habitat creation.

- Community use: The involvement of stakeholders and adjoining neighborhoods can offer opportunities to engage local communities with phytotechnology installations and to provide an additional amenity in areas where green space is limited. This can lead to informal policing and security for the site as well as greater community involvement in the planning and design stages.
- Educational use: Closely related to community use is the potential role for phytotechnology installations in providing an outdoor classroom experience for local students at all levels. In addition, they can be a living-experiment setting for non-students and students alike that can educate residents about the dangers of post-industrial land, polluted soils and groundwater and the natural techniques that are available to provide remedies for the contamination.
- Habitat creation: The introduction of vegetation as a natural remediation technique increases the amount and variety of habitat on a formerly polluted and abandoned site. If this is carefully considered during the design process, phytotechnology applications can increase the canopy cover, nesting sites and potential food available to wildlife without exposing animals to toxicity.
- Biomass production: Remediation plantings may also be harvested and utilized for the production of biomass and energy sources, creating an economic product that may have the potential to offset remediation costs.
- Climate change: Long-term phyto sites can assist in creating microclimates, mitigating climate change and controlling environmental disease.
- Benefit to agricultural systems: New installations of plants on marginalized land can support nutrient cycles, crop pollination and long-term improvement of soils.

B Constraints

The disadvantages of the application of phytotechnologies include the following.

- Many contaminants cannot be remediated with phytotechnologies, or soil/climate conditions are not favorable for their application. In addition, some soils may be too toxic or infertile for any plants to be grown. Contaminant and plant interaction details are provided in Chapter 3.
- The process is limited to relatively shallow contamination sites and is dependent on the adaptability and climate zone of the plants that can be used.
- In some cases, plants may need to be harvested and disposed of as a waste to remove a pollutant; this can be costly and energy intensive.

8

- Contaminants stored by plants could potentially be released through either transpiration or uncontrolled incineration of harvested plant materials. In addition, when pollutants are taken into plant stems and leaves, a risk of exposure may be created. Humans, animals or insects could consume or be exposed to plants that have stored pollutants, although this is typically not a significant risk factor.

- Once plantings are installed, ongoing maintenance operations may be costly and must take account of adequate drainage, watering, testing and protection for the introduced planting.

- Monitoring may be required and soil- and groundwater-testing practices may be costly or inaccurate.

- The elongated timescale of the phytotechnology installation may preclude its use in short-term site-regeneration projects. Many phytotechnologies take at least 5 years or more to reach maturity and some could be designed as legacy projects, with lifespans of 50 years or more. They require a long-term commitment to management and maintenance by the site owner, as the process is dependent on a plant's ability to grow and thrive in an environment that is not always ideal for normal plant growth.

- Natural systems are variable, and weather, browsing by animals, disease and insect infestations can be devastating or produce unanticipated results.

- Suitable plant stock may not be available from local growers or may require installation during specific seasons of the year.

9

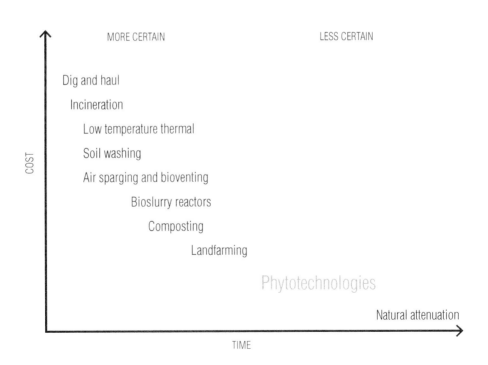

Compared to other remediation options, phytotechnologies can provide a significant cost savings when feasible, but requires longer treatment time frames. In addition, there is some degree of performance uncertainty in all plant-based systems.
Source: Graphic redrawn from original provided by (Reynolds, 2011)

Figure 1.2 Phytotechnologies Cost Benefits vs Treatment Time

- A lack of understanding of the science, and implementation without the participation of experienced researchers, engineers and scientists, can lead to many improperly designed and implemented projects.
- In more temperate climates the systems may become inactive or much less active in the winter months, and they may not be usable at all in more extreme environments such as are found in Alaska or the Arctic region.
- Current legal, regulatory and economic conditions surrounding remediation using plants may be difficult to navigate. Regulators may not be aware of potential phytotechnology opportunities, and extensive precedent research and explanation may be required.

V The current state of phytotechnologies

A Market share

In the remediation industry, phytotechnologies still account for less than 1% of the market share of remediation techniques carried out (Pilon-Smits, 2005). The lagging use of phytotechnologies is a result of several factors, including

- an inadequate amount of core and applied research and field testing, mostly due to a lack of research funding sources, which have been dwindling in the US since 2001
- consequently diminishing support from federal, state and local regulators
- lack of a proven track record, due to variability in field experiments and lack of metrics of success for novel technologies that do not map well to the traditional metrics of assessment and monitoring for older technologies
- general uncertainty inherent in natural and biological systems
- insufficient available information on systems that do work, preventing their integration into engineering and design practices
- typically, within environmental engineering firms responsible for the design of remediation systems, a lack of staff versed in agronomics or plants, and greater reliability placed in traditional engineering practices, since plant-based systems are not their area of expertise and indeed may be more uncertain
- typically, greater monitoring costs and a longer time frame for remediation.

B Perceptions

In addition, there is some dissent both in the sciences and in the field of landscape architecture that either criticizes phytotechnologies as ineffective and not worthy of the community's attention and resources or, alternately, hypes phytotechnologies as the 'silver bullet' for pollutants. The truth lies somewhere in the middle, as some good opportunities certainly exist where phytotechnologies can be applied and there are also many cases where they are not feasible and should be avoided. Typically, misinterpretations arise when old or outdated studies are referenced and utilized, which commonly happens when web-based research is re-referenced over and over again and is not updated for years on end. In addition, many practitioners are not aware of the highly specific plant and soil-based interactions required in phytotechnologies, and applicability and success are extremely site and contaminant specific. Field studies where the environmental conditions, plant and toxins were mismatched and the goal was not reached can harm the reputation of the entire field (US EPA, 2002). Additionally, short-term projects and lab and greenhouse studies may produce successful results that do not effectively translate into field applications, where weather patterns, pests, competition and poor project management are common failure modes that are absent in greenhouse or laboratory studies. Longer time frames than are the norm in conventional methods are often needed for plant treatments to reach performance metrics and remediation goals, but these are not available, due to academic schedules or grant-funding rules. For all these reasons, misinformation surrounding plant applications and the capability of site remedies can quickly spread and generate unjustified preconceptions about phyto.

C History

As described in the Foreword to this book, the field of phytotechnologies in the US was generally named and formally established in the 1980s. In the 1990s a large number of phytotechnology greenhouse and lab experiments were published, leading to a belief in the applicability of plant-based cleanup approaches to a broad range of pollutants in groundwater and soils and site contexts. In addition, a number of plants found to 'hyperaccumulate' metals were discovered during the same time period, and speculation arose that these plants could potentially be used for the remediation of metals-contaminated sites. Unfortunately, this generated a lot of excitement and, in a 'boom and bust' scenario, advocates of the science overestimated and oversold the technology (White and Newman, 2011). Its performance was mixed, however, and failures in the field outnumbered successes because implementation in the field occurred before the science had been substantiated in the laboratory (White and Newman, 2011). Uptake and remediation of lead by sunflowers, for example, was hailed as an exemplar without the biology and mechanisms being fully understood. When actually applied on real-world sites, it was found that this previously 'hyped' technology was unsuccessful at field scale. Interestingly, sunflowers for lead remediation continued to appear in landscape architecture renderings as a phytoremediation application well past 2010, even though they essentially failed in field trials in the late 1990s.

As has been documented by Dr. Alan Baker and Dr. Charles Reynolds, among others, this led in the late 1990s to what was termed an "unmasking phase," where the unjustified claims led to a crash of the phytotechnology field both in credibility and in funding (Reynolds, 2013). This is commonly termed the "Trough of Disillusionment" that follows the "Peak of Expectations" for new technologies (Burken, 2014). Since then, a very slow but steady flow of scientific research work has been carried out, with growing acceptance by a more cautious and prudent remediation industry and regulatory authorities. Although the concept of phytotechnology is simple, research is slow, complicated and does not map well to the need for quick site remediation. Nonetheless, in many cases phytoremediation still may be the most suitable alternative, compared to the very abrupt, disruptive and expensive process of soil removal and physical extraction of contaminants.

11

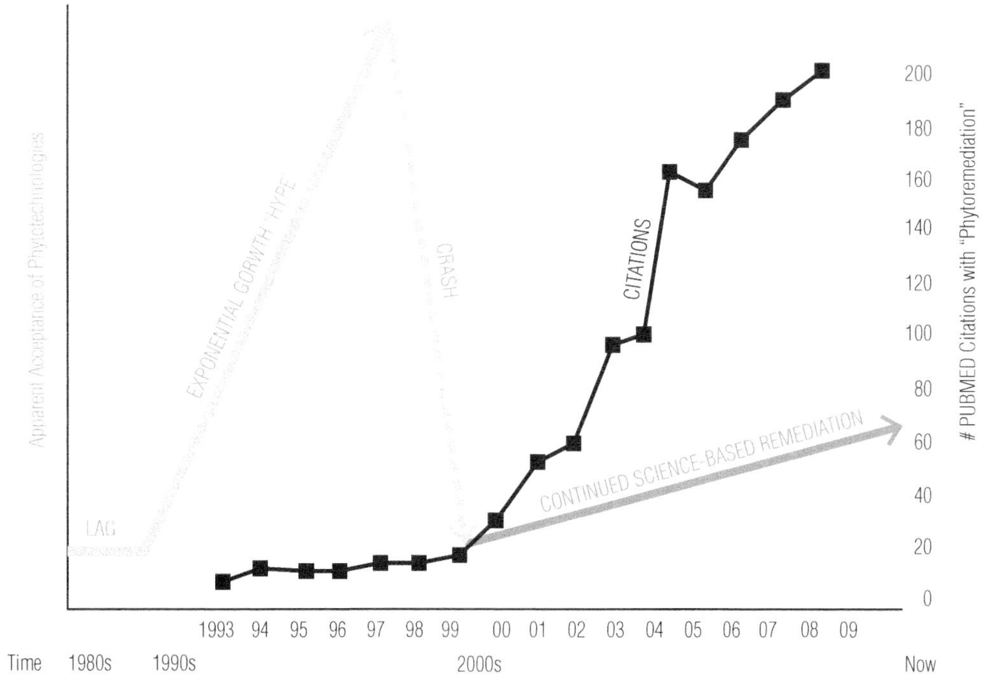

In the early 1990s, the phytotechnology field was just beginning and there was a huge surge in interest that led to overspeculation and 'hype'. Failures in on-site projects outnumbered successes since implementation in the field occurred before the science was substantiated in the laboratory (White and Newman, 2011). The field of phytotechnologies is slowly rebuilding its reputation based on projects rooted in science rather than speculation. Source of Graph: Redrawn from (Reynolds, 2012) (Baker, 2013)

Figure 1.3 Perceived History of Phytotechnologies

D Research funding

For the field of phytotechnology to advance, research funding must be available so that academic institutions can apply for grants to conduct experiments and field trials. In the 1990s, environmental research funding was prevalent. The US EPA was actively funding and participating in phytoremediation projects, and large government departments such as the Department of Energy, Department of Defense, US Army and the US Department of Agriculture (USDA) were sponsoring projects to further phytoremediation research. However, the failed field trials conducted in the late 1990s tarnished the reputation of the phytoremediation field and, with the onset of the Bush administration in 2001, remediation of government-contaminated land was no longer a political priority; government-sponsored environmental research funding in the US started dwindling. In 2005, the Bush administration started cutting the EPA's budget by about 4% per year (Schnoor, 2007). Two important research-funding programs administered by the EPA (the Ecosystem Services Research program and Science To Achieve Results grant funding) have been reduced by millions of dollars in recent years (Schnoor, 2007). The US EPA Hazardous Substances Research Center was eliminated entirely (Burken, 2014). Unfortunately, even under more recent administrations, the EPA's budget has continued to be cut, due to the economic recession and congressional calls for spending cuts (Davenport, 2013). Since 2004, the funding for the EPA's Office of Research and Development has declined by 28.5%, and the

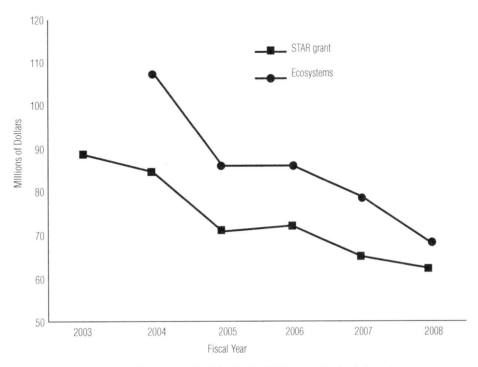

Two important research funding programs administered by the EPA, the research budget for Ecosystems and Science To Achieve Results (STAR) grant funding have been reduced by millions of dollars since 2004 (Schnoor, 2007). This has contributed to a difficult financial environment for phytotechnology scientists looking to fund research work.

Figure 1.4 Environmental Protection Agency's Research Budget

Ecosystem Services Research program has declined by 58% (AIBS, 2013). The current research funding climate for phytotechnologies at the publication date of this book is poor. Funding exists for plant-based biofuels and nano-particles research, so many of the former institutions focused on phyto research have now shifted to those areas or other funded areas of interest. Many formerly active researchers from the1990s are no longer actively conducting research in the phyto field. With the engagement of landscape architects and other professions, hopefully, interest in the field will reemerge and government agencies again will reprioritize appropriate funding for this important area of science.

E Current research

The current state of phytotechnologies research and application includes laboratory, field testing and small-scale site experiments carried out by several US and international institutions and industries. In addition, some US government departments such as the US Geological Survey (USGS), US Military and National Aeronautics and Space Administration (NASA) continue to conduct field trials. A list of sample projects follows in Chapter 3.

The development of phytotechnologies has been aided by growing interest in the subject from a variety of sources, including those planners and engineers focused on sustainable development, city greening and community urban agriculture. There still remain, however, a number of outstanding issues to be overcome.

With the reuse of polluted sites on the rise, and redevelopment a critical component in sustainably moving forward, the potential of utilizing plants to cleanse soil on site is more economically and environmentally attractive than removing contaminated soils off site. Developers, municipalities and site planners are actively looking for alternatives to traditional remediation; however, they do not always have the resources to understand when phytotechnologies might be an option for them. To this end, the International Phytotechnology Society (IPS), a non-profit worldwide professional society with members in Europe, Africa, India, Asia and North America, is leading the development of the core scientific research with guidance from the US EPA. In this way the earlier mistakes of the late 1990s and the 'unmasking period' can be avoided.

VI   Legal and regulatory framework

A Remediation projects

Contaminants that may be found on sites are well documented and their properties understood and exposures regulated. The legal and cleanup process in the US is rigorous and the strict legal framework for remediation work must be adhered to in the redevelopment of brownfield sites. Within a regulatory framework there initially arise a number of key questions regarding the use of phytotechnologies as a remediation method.

14

- What regulatory program is the site designer working with?
- Who are the stakeholders and professionals involved?
- What is the implementation process?
- Where does any waste or biomass produced go?
- What are the overall risk and long-term project goals?

Irrespective of the federal, state or local cleanup program that covers the project site, the extent of effort, the requirements of management and bureaucracy and the final result of remediation should all be the same. The program will depend on whether the project is involved with permitting a discharge or cleaning up a site of pollutants and whether there is a single contaminant or a 'cocktail' of contaminants in a range of media such as soils, sediments and groundwater. A number of the cleanup programs arise from the following:

- Federal Comprehensive Environmental Response, Compensation and Liability Act (CERCLA) 1980, commonly known as 'Superfund'
- Federal Resource Conservation and Recovery Act (RCRA) 1976
- Federal Toxic Substances Control Act 1976
- State Resource Conservation and Recovery Act (RCRA)
- State Site Cleanup
- State Solid Waste

Landscape and redevelopment sites may involve a single entity or a number of these programs in an overlapping fashion. The major difference between the programs is that the federal ones are generally larger, more complex and therefore slower, according to the site conditions. The state programs are smaller, simpler and faster, involving semi-privatized Licensed Site Professionals programs to move project cleanups along in a timely manner, while the federal programs contain specialized programs for leaking underground storage tanks (LUSTs), brownfields, landfills and RCRA Corrective Action, which can produce another layer of complexity and management.

Cleanup of site contamination using phytotechnologies is generally 'risk based', that is, site specific and chemical specific using risk-management techniques based on the future end-uses for the site. The goal is for the site to pose 'no significant risk of harm' to health, safety, public welfare and the environment. Such cleanup can be accomplished by meeting generic cleanup standards and employing a site-specific risk-assessment method tied to end-use, usually referred to as Risk-Based Corrective Action. For example, whether the site will become a park, as opposed to housing, or will remain a light industrial complex will determine the levels of cleanup required.

The stakeholders involved in this work include the following:

- regulators
- other government entities such as planning boards
- the principal responsible parties for the pollution
- development agencies or developers for the site
- landscape architects and site designers and engineers
- future owners and occupants of the site, if known
- adjacent site owners, stakeholders and occupants

The cleanup process works using phytotechnology in five phases.

- Phase 1: Preliminary assessment of Site (Phase 1 Site Assessment)
- Phase 2: Comprehensive site assessment (Phase 2 Site Assessment)
- Phase 3: Assessment and remedial alternatives*
- Phase 4: Implementation of phytotechnology remediation
- Phase 5: Operation and maintenance of phytotechnology installation

* *Note*: In this phase an assessment of alternative remediation methods and techniques is carried out. The assumption is that phytotechnology will be selected as one of the appropriate remedies on the site, and may often be used in a treatment train with other remediation methods. For example, hot spots may first be excavated and removed or chemicals may be injected to encourage contaminant breakdown before the phytotechnology planting is installed. Often, during this phase phytotechnologies may be found not to be applicable on a site. The selection of the remedy is based on the following:

- overall protection of human health and the environment
- long-term reliability and effectiveness
- the ability to be carried out on site

15

- cost
- attainment of risk-based goals, such as reduction of toxicity, mobility or volume of contamination
- compliance with all federal and state applicable or relevant and appropriate requirements.

Stated more simply, will phytotechnology meet the remedial goals, is it technically reasonable, is the cost reasonable, does it fit with reuse plans and timetables and will it do the job and provide certainty? When the remediation project is completed, a 'Response Action Outcome' or a 'letter of no further action' is issued, or a site delisting from the National Priorities List (NPL) can take place for a CERCLA site, documenting that cleanup requirements have been met.

Impediments to the use of phytotechnologies on a range of sites include: regulatory uncertainty and delays; liability risk in relation to the general public; stigma; and perception of risk by local stakeholders and adjacent landowners attached to sites with phytotechnology remediation approaches and proposals. Among the issues raised by their use are the use of species within the same genera and monocultures, genetically engineered cultivars, ownership, storage, distribution and risk of commercial exploitation. In particular, the following issues have arisen.

- Bioaccumulation: When certain contaminants are present on a site, the opportunity exists for the accumulation of contaminants in the biomass of plants and the potential for toxicity and food-chain transfer. The bioaccumulation of pollutants can create a perception of risk to adjacent communities and stakeholders, particularly in residential neighborhoods. This perception may not be based on scientific data that may be currently available; rather, it may be based on values related to deeply held convictions within the community.
- Contaminated biomass collection: Stakeholders are often significantly concerned about the possible accumulation of contaminants in the biomass of the plants and the future disposal of contaminated biomass off site through landfilling or incineration. Close control of site operations may be required, including the collection of biomass from off the ground, such as leaf litter, fallen bark and branches, and the complete harvesting of plants on a defined schedule.
- Waste disposal: What becomes of the waste generated during remediation, such as vegetation harvested and disposed of as part of the removal process? There are three types of waste: solid, remediation and hazardous; two of these, remediation and hazardous, are relevant to phytotechnologies. In many cases, plants may completely degrade contaminants and they need not be harvested from sites. However, in some cases plant material that has absorbed inorganic contaminants such as metals into roots, shoots and leaves may be classified as hazardous waste for the purposes of disposal. Remediation waste covers those materials from a phytotechnology installation that are not hazardous but still exceed reportable concentrations and come from a listed site.

B Pollution-prevention projects

Phytotechnologies have always been considered a tool for site remediation; however, it has been seen that phytotechnologies can be also utilized as a preventative measure rather than solely as a remediation or post-remediation tool. An example of this is the early use of phyto-buffering, where plants are grown to contain anticipated spills or plumes that might occur below ground at some time in the future. This moves the application of phytotechnologies away from the laws and regulatory frameworks surrounding remediation activities and the control of waste, and towards a horticultural and planning-related set of concerns.

Preventative phytotechnology planting strategies fall under the normal set of planning and local engineering regulations regarding site works, as would any other type of landscape installation. This is similar to the introduction and use of bioswales on a site as a landscape element to divert, capture and cleanse stormwater runoff in a parking lot. The design and installation of phytotechnology buffers can be based on anticipated pollution events.

VII   Designer checklist for phytotechnology

A detailed phytotechnology project checklist and decision tree has been developed by Dr. David Tsao and the Interstate Technology and Regulatory Council in the free document, PHYTO 3 (available for download at: www.itrcweb.org). This document details a logical, step-by-step process and series of questions to ask so as to determine if phytotechnologies may be a viable approach on a particular property. The authors refer you to this reference for a detailed project guidance document (ITRC, 2009). However, in order to give the landscape architect every chance of success in executing a phytotechnology installation, a listing of the main items that a practitioner should know before starting follows below.

17

A Preplanning phase

1   Defining a 'phytotechnology project vision'
Establish a project vision for demonstration, experimental or full-installation sites, to include short-term issues relating to the cleanup or longer-term sustainable goals for the land and its surrounding environments.

2   Site selection
While sites are selected by virtue of their previous or future uses, local knowledge about sites from local municipalities and the community can help to determine priority issues in terms of suitable access points, circulation and adjacency concerns.

3   Data collection to be carried out prior to phytotechnology design

- Soil sampling and environmental testing
- Subgrade and groundwater conditions
- Existing vegetation and wildlife
- Microclimate and weather
- Existing utilities and water supply
- Site boundaries for access and security
- Storage of equipment for maintenance and testing

4   Economic values

Identify loan and grant programs (if available) to assist in both testing and remediation activities.

5   Partnerships with local stakeholders

Consider organization and outreach for educational enrichment, maintenance and public acceptance.

6   Phytoremediation education

Expose clients and stakeholders to the considerable applications of technology, capital and labor at work in the transformation of polluted sites through phytotechnologies. This could involve on-site documentation of remediation and redevelopment processes; interviews and tours with engineers involved in a cleanup; development of individual and group media and arts projects; creation of an interactive technology curriculum that can be used by other communities facing severe environmental threats; and establishment of a project website that gathers the resources and research developed through the project and makes them available to other communities, educators, municipal leaders and environmental professionals nationally and internationally.

B Phytoremediation design and protocol phase

1   On-site remediation

Develop protocols regarding access, security protection, environmental engineering-site practices, supply of plant stock, irrigation/water supply, installation techniques, monitoring and tabulation of results, documentation of site installation and ongoing maintenance. This should be done in conjunction with an experienced phytotechnology expert and may involve several different types of engineers and scientific consultants, including agronomists and hydrologists, biologists, chemists, micro-horticulturalists, foresters, ecologists and civil and environmental engineers.

2   Environmental opportunities

• Consider habitat creation
• Animal shelters/microclimate
• Other productive landscape or biomass opportunities

C Implementation phase

1   Establishment of installation protocols

Once the landscape architect has agreed on a solution for the phytotechnology application an installation protocol needs to be developed. Close work with soils specialists, agronomists, forestry practitioners and maintenance operators will be required.

D Post-implementation phase

1 *Maintenance and ongoing operations*

- Security of installation area
- Graphics and warning signage for installation area
- Regular weeding, watering and management of area
- Ongoing monitoring and repeating as required

2 *Publication of results: best practice workbook or manual*

3 *Disposal of vegetation material (if required)*

4 *Reoccupation of phytotechnology site area*

VIII   Innovative applications

A Integration into daily design practice

As has been mentioned above, phytotechnology planting techniques are currently employed *post* site contamination, to help clean up and remove already contaminated soil or groundwater for some form of reuse or reclamation. Where future pollution may be expected from particular site programs such as gas stations or industrial manufacturing sites, there is potential for planting approaches such as vegetation buffers with preventative phytotechnology abilities. Current phytotechnology plantings are typically of one plant species (monocultures) installed in a field application. Plant combinations can be considered to both treat toxins and create aesthetic and functional compositions. The overall goal is to expand awareness of phytotechnology so that it is not only employed to remediate contaminants that already exist, but proactively used in everyday landscapes. The creation of productive landscapes is the ultimate objective, with plantings that not only have aesthetic functions but also enhance environmental and human health conditions.

B Biomass production

Cultivation of short-rotation willow and poplar coppice was introduced after the oil crisis of the 1970s, with the intention of replacing fossil fuels with new energy sources. Extensive research to identify fast-growing species that could be grown intensively for use in energy production suggested that willows grown in coppice systems were the most suitable. Nutrient utilization and stand management were seen to be more cost efficient for willow than for other woody species, and short-rotation willow coppice proved to be a sustainable way of producing fuels that were carbon dioxide neutral, since burning of the biomass would release into the atmosphere the carbon dioxide that the plants had taken from the air. Willows in short-rotation coppice systems are currently grown, consisting mainly of different clones and hybrids that have been specifically bred for this purpose. In the initial phase, approximately 15,000 cuttings per hectare are planted in double rows, to facilitate future weeding,

19

Field of four-year-old willow biomass crops just prior to being harvested in the late fall (Volk, SUNY ESF, 2014).

Willow (*Salix*) biomass crops resprouting in the spring after being harvested the previous winter. This willow is about a month old above ground on a four-year-old root system (Volk, SUNY ESF, 2014).

Four-year-old willow biomass crops being harvested with a New Holland forage harvester and coppice header. This system cuts and chips the willow biomass in a single pass (Volk, SUNY ESF, 2014).

Figure 1.5 Willow Biomass/Coppice Planting and Harvesting

20

fertilization and harvesting. The willows are harvested every three to five years, during winter when the soil is frozen, using specially designed machines. After harvest, the plants coppice vigorously, and replanting is therefore not necessary. The estimated economic lifespan of a short-rotation willow coppice stand is 20 to 25 years.

In recent years other fast-growing species have been used for biomass ethanol production, including grasses such as *Miscanthus* and *Panicum* as well as corn. Poplars may also be cultivated for energy production, including for production of wood pellets. Additional uses for poplars include cardboard and hardwood production.

Recently, many phytotechnology sites have been proposed where willows, poplars or other biomass-production species are simultaneously cleaning soils or groundwater while producing these economic products. This is often an attractive proposition, since utilizing marginal land frees up quality agricultural land that might otherwise have been used for this purpose.

C Economic development

Phytotechnology installations provide opportunities for workforce development in a specialized market. More recently, workers in the general construction industry have retrained in the remediation and reclamation work that is connected to post-industrial sites. Funding for this was provided by the US EPA Brownfield Program, and more opportunities and higher compensation are available to workers in the large number of brownfield lands.

D Carbon sequestration

Poplars are among the fastest-growing tree species in North America and are a central tree used in phytotechnology. They are capable of accumulating enormous amounts of wood and biomass in a relatively short period of time. With selection of appropriate varieties and proper care in site projects, poplars sequester enormous amounts of carbon dioxide in a short period of time. Research on this topic includes screening and selecting poplar clones for growth, and understanding processes of below-ground carbon movement and storage. A future hope is to establish a basis on which poplar trees and other plant species used in phytotechnology projects can be credited for the amount of carbon their plantations are accumulating and to document the chain of custody for a variety of products made from poplars.

E Plant sentinels

In some cases, phytotechnology plants may have potential to serve as detectors for environmental contaminants via aerial and satellite photography with remote sensing. When plants take up certain contaminants, changes in their chemical structure can occur that may have the potential to be read through sensing equipment.

21

F Phytoforensics

As plants collect water and nutrients from the subsurface, they also collect pollutant molecules and atoms. Developed from the work of Dr. Don Vroblesky of USGS and Dr. Joel Burken of Missouri S&T, phytoforensics uses existing on-site plants – most commonly trees – to identify and delineate subsurface contaminants. This primary sampling method inserts a thin probe (increment borer) horizontally into the trunk to collect cores of trunk tissues (i.e. xylem) that can be analyzed in the laboratory for the presence of pollutants in the tree. The amount of contaminants in the tree correlates to the amount of contaminants to which the roots are exposed. Tree cores that show elevated contamination can point to hot spots below the surface of the ground and provide clues as to the original source of contamination and current spatial extent of pollution. Tapping into available trees in an area that is suspected to be contaminated can help engineers to better and more rapidly delineate contaminants in the subsurface. Traditional groundwater sampling requires the use of heavy equipment to drill into the ground and the creation of sampling wells to draw water from the site. Individual wells can take days to install and months to sample. Phytoforensics is fast and inexpensive (compared to traditional well-testing techniques) and can quickly provide field information about the underlying conditions (Burken, 2013).

The process may involve either coring trunks of trees to gather small samples or inserting sampling devices into the trees in the field. A thin filament called a solid-phase microextraction fiber, or SPME, can detect traces of chemicals at minute levels, right at the tree, down to parts per trillion for many

Core samples of trees are taken with a standard arborist's tree coring device and brought back to the lab in vials for analysis. The samples are analyzed to determine if contaminants are present in the tree (Burken, 2014).

Figure 1.6 Phytoforensics Sampling Tools

compounds (Sheehan et al., 2012). Field portable analysis is beneficial, providing results for the level of contamination in the tree in real time. The use of field gas chromatography mass-spectrometers (GC-MS) has proven to be effective (Limmer et al., 2014), but the instrumentation is expensive to operate. Novel methods to provide similar analysis at less cost are under development.

In Sedalia, MO, for example, well drilling and testing for the solvents trichloroethylene (TCE) and perchloroethylene (PCE) near an abandoned section of railroad took 12 years and placed 40 traditional engineering sampling wells. Working with the environmental consulting firm Foth Engineering, Burken and a team of students spent one day at the site and took 114 tree samples. Their work more accurately determined the extent and locations of contamination, for a fraction of the cost (Burken et al., 2009). In Rolla, MO, Burken's students have used phytoforensics to determine the extent of contamination from the Busy Bee Laundry, which is adjacent to Schuman Park and just two blocks from Missouri S&T and the building where the phytoforensic research was initiated. By testing trees in Schuman Park, the Missouri S&T team determined that solvents from the dry-cleaning operation had seeped into the groundwater of the park, but not at levels hazardous to human health. The USGS is working with the city of Rolla and with funding from the National Science Foundation to remediate the area by planting additional trees to extract the contaminants from the groundwater. The project is expected to remove the contaminants at increased rates and decrease any potential release into Frisco Lake. By deploying a new technique whereby a probe is left in the plant and a small measuring device is brought to the tree, concentrations of pollutants were able to be easily assessed in the trees over a period of years. This project validates the concept of monitoring phytotechnologies through novel plant sampling, versus traditional, expensive groundwater sampling.

Recently, funding from the US Army's Leonard Wood Institute helped to develop new approaches to current phytoforensics methods in order to analyze more water-soluble and nonvolatile compounds. The current methods detect molecules as gases, but explosives constitute a different contaminant type that requires detection as liquid. A method for collecting aqueous samples was developed to aid the military in detecting areas where explosives may have leaked or been spilled on military bases. The method is also being used for perchlorate and several other nonvolatile compounds not previously tested in phytoforensic methods. Through these new efforts and findings, researchers can now detect the presence of trace amounts of explosives, in addition to chlorinated solvents, petroleum, and other organic volatiles.

Recent breakthroughs in the field have included directional analysis of the plume from the coring of a single tree (Limmer, 2013), advances in solid-phase sampler development (Shetty, 2014), real-time/in-field GC-MS analysis and long-term monitoring advances (Limmer, 2014). In addition to detecting the presence of individual contaminants from the soils and groundwater, trees can also be used to detect the timing of contaminant releases into the ground through analysis of the differential concentration of some elements in the growth rings of the tree, termed dendrochemistry (Balouet, 2012). Phytoforensics has been accepted by the USGS as a reliable testing and monitoring technology, and a technology transfer document for more information is available and listed in Chapter 6.

In the future, it may become commonplace to use the existing trees on a site to determine the location of contaminant plumes below the surface and then to monitor the efficacy of phytotechnologies. Monitoring of phytotechnology impacts may have great benefit in gaining acceptance from regulators and site owners who need proof of efficacy in order to validate claims of remediation and pollutant removal rates. Advanced methods in phytoforensics and phytomonitoring mean that contaminants can be detected more rapidly and treated at far lower costs, using a method that offers multiple ecological benefits along with pollutant attenuation.

IX Conclusions

23

In conclusion, the major barrier to utilizing phytotechnologies in the field remains the lack of complete knowledge about: the process; accepted metrics of monitoring and success; molecular genetics; and biochemical mechanisms of adaptive tolerance in plants to organic and inorganic contaminants. This requires both laboratory and field tests to continue to address contaminant types matched with plant species and to follow remedial treatments over prolonged periods. Research in phytotechnologies has enhanced our understanding in the fields of plant and soil sciences; however, more effective and commercially feasible techniques are still required.

To advance the field, the following needs to happen:

- clearly distinguish those sites, phytotechnology processes and techniques that have been studied the most, and identify those with the best opportunities for potential success
- improve communication and cooperation with the private commercial sector responsible for implementing, maintaining and monitoring these technologies
- exploit new economic opportunities such as the production of bioenergy and bio-fortified crops; long-term remediation activities also provide economic gain
- secure more funding for phytotechnology research for the coming years and exert pressure on federal and private sources to ensure that the primary and applied research activities will continue to be developed, building on the work of previous years.

Large remediation operations usually come in association with significant commercial or urban development projects. Developers and regulators are to be encouraged, along with the design and planning professions, in certain situations to consider phytotechnologies as a viable alternative in site remediation, regeneration and reuse. Commercial soil remediation occurs in relation to: the growth of urban areas or significant infrastructure, where low levels of contamination must be reached; change of soil use; or toxic spills, where urgent solutions are needed to maintain socio-political acceptance. In these cases, conventional remediation options are often the best, due to their rapidity, despite their high initial cost and often detrimental ecological and property impacts. The desire for rapidity makes it difficult for phytotechnologies to compete. Therefore, phytotechnologies are often relegated to projects with low economic value and the following profile: (a) a long-term period is possible; (b) current use of the soil does not imply risks to humans or the environment. These kinds of projects are usually restricted to marginal areas without short-term economic value, such as former mining areas, landfills, DOD lands or post-industrial sites.

For the landscape architect the main lesson here is to understand clearly what the use of these installations can achieve and what it cannot, and to realize that all phytotechnology projects have to be pollutant specific and plant specific. Science-based efforts will avoid repetition of the 'overselling' period where planted installations failed to carry out the specific task of remediation. Projects must also be monitored appropriately to show impacts and results accurately. In addition, the landscape architect has to become familiar with the current research and literature on the subject. In fact, each project site can be considered a unique condition with regard to soils, groundwater and location (climate and context), yet employ standard approaches to design and vegetation that are based on site typologies and an understanding of the history of the site. In this respect they appear to be similar to conventional landscape project sites, yet the true conditions underlying the site – pollutants and their type, location, intensity and state – make these contemporary landscapes both unique and important to confront and address.

Following this introduction to phytotechnology and the contemporary environment, let us now move on to review in Chapter 2 the fundamentals of how plants carry out the remediation of soils and groundwater, and the precise nature of their interaction with organic and inorganic contaminants.

24

2: Fundamentals

I Introduction

For phytotechnology systems to be successful, it is imperative for the reader to gain an understanding of the fundamental science supporting these installations. There are many contaminated sites where phytotechnologies will not be applicable, and other opportunities that hold more promise. This chapter will cover the basic science behind plant-based remediation systems to help provide an understanding of where plants may be useful in remediation. A clear understanding of the fundamentals, including mechanisms, basic contaminant types and general planting considerations, is important in the conception, design and implementation of these productive landscapes.

II A short overview of plant functions

This section will review basic plant functions that contribute to remediation mechanisms. For a plant to grow, it needs several basic resources including energy (sunlight), nutrients and water. In the processing of these resources, contaminants can get taken up, transformed or broken down. Three transformations in plant growth are essential to consider.

A Energy transfer

Leaves transform energy from the sun via photosynthesis to generate plant biomass. About 20–40% of total photosynthetic products produced by the plant are sugars that are transported down to the root zone and are leached out into the soil through the roots (Campbell and Greaves, 1990). The sugars, oxygen and other root exudates released around the root zone (organic acids, amino acids and enzymes, to name a few) both can help to transform contaminants and also attract many microorganisms to live there (Lugtenberg and Dekkers, 1999). The presence of these root exudates causes the soil to be populated with 100 to 1000 times more living microorganisms around plant root zones than in the soil alone (Reynolds et al., 1999). The microbes create a protective barrier around the root zone and play a large part in breaking down potentially harmful substances such as pathogens. This high microbe-populated area with a symbiotic relationship to the plant is called the "rhizosphere" (Lugtenberg and Kamilova, 2009). The rhizosphere creates a rich environment for contaminants to potentially be modified, by either the plant itself or the microbes.

B Nutrient transfer

In exchange for the free provision of sugars and phytochemicals from the plant, the microbes help the plant to obtain nutrients. There are 13 essential nutrients required by plants. Many of these are in forms not available to the plants, and microbes can unlock them into forms for uptake. This is similar to digestion in humans, where microbes process food and nutrients internally in our stomachs; however, in plants much of this occurs externally in the root zone. Rather than internal digestion, plants have something akin to external stomachs, processing many nutrients they need outside the plant

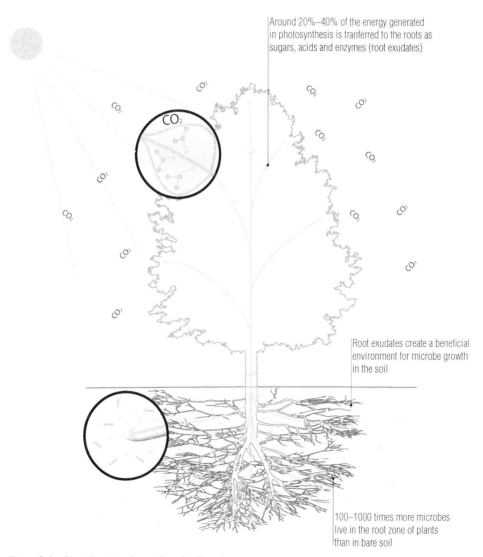

Around 20%–40% of the energy generated in photosynthesis is tranferred to the roots as sugars, acids and enzymes (root exudates)

Root exudates create a beneficial environment for microbe growth in the soil

100–1000 times more microbes live in the root zone of plants than in bare soil

Figure 2.1a Plant Function: Energy Transfer (Down)

28

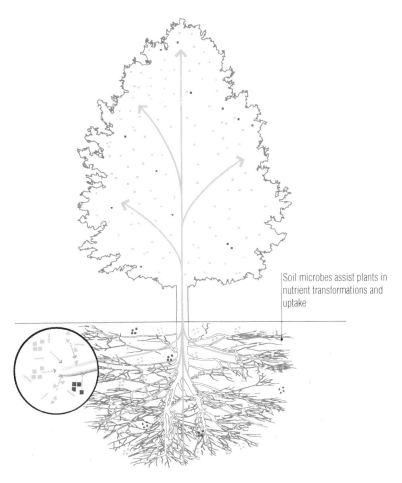

Soil microbes assist plants in nutrient transformations and uptake

29

Figure 2.1b Plant Function: Nutrient Transfer (Up)

(Rog, 2013). Once the nutrients are made available by soil biology, enzymes, acids and other exudates released by the plant, transport pathways exist for the plant to take up the needed nutrients. Pollutants in soils can sometimes have a similar chemical structure to the resources a plant requires, and can be inadvertently taken up by plants in the process (ITRC, 2009).

C Water transfer

Plants act as pumps, extracting water from the soil, moving it through the stems and leaves, using it in photosynthesis and transpiring extra water to the air through the leaves. It is estimated that only 10% of the water taken up by plants is used by the plant and the rest is evapotranspired into the air. Plants move an incredible amount of water; each year, plants in North America move more water than all the rivers in North America combined (Burken, 2011). In fact, 75% of water vapor over land worldwide is a result of plant transpiration (Von Caemmerer and Baker, 2007). As the plants take up water, they create a hydraulic pull towards the plant. In phytotechnologies, plants may take up polluted water, potentially degrading or extracting and storing away pollutants during the process. This pull of water can also potentially slow the migration of contaminants in water below the surface.

Water not used in plant growth is
evapotranspired into the atmosphere

O_2

30

Plants act as solar-powered water
pumps to take up water from the
ground

Figure 2.1c Plant Function: Water Transfer (Up)

III   Contaminant location: within soil, water (groundwater, stormwater or wastewater) or air?

Pollutants can be encountered in many different site locations and may act differently when either found in soil, dissolved in water or found in air. This book is primarily dedicated to covering the basics of plant-based remediation and pollution prevention of soils and groundwater. Since contaminants in water and air act quite differently, when generalizations are made throughout this publication, soil-based pollutant removal should be assumed unless otherwise noted.

A Soil

Pollutants can be found in soils after spills or after long-term accumulation of repeated small releases. For phytotechnologies to be considered for soil remediation, the pollution must be located at a depth that plants can reach. Most herbaceous plant species have a maximum root depth of 2 feet, with tap-rooted tree species maximizing root depth at 10 feet (ITRC, 2009). Soil pollution within 10 feet of the soil surface is about the maximum depth where phytotechnologies should be considered. Soil contamination within the top 3 feet is the most effective zone for phytotechnologies.

B Water

1 Groundwater

Water located below the soil surface in continuous soil-pore spaces is referred to as groundwater. When groundwater is contaminated, plumes of pollutants can migrate within the flow. Since groundwater is naturally recharged by rain, precipitation events can make groundwater and pollutant plumes flow faster. Contamination is especially concerning, since groundwater is often tapped for drinking water and agricultural wells, and eventually emerges into surface water bodies. The depth to groundwater varies greatly from site to site, and can be as shallow as 1 foot or begin hundreds of feet below the surface. Water-loving trees, call phreatophytes, have been shown to tap contaminated groundwater (Negri, 2003). For phytotechnologies targeting groundwater to be effective, the maximum depth to groundwater should be no greater than 20 feet below the surface. Phytotechnologies for groundwater are most effective where the water is shallower and within 10 feet of the surface. Specialized deep-rooting planting can be utilized in these applications to reach the groundwater sooner (Figure 2.16, p. 46), or in some cases water can be pumped up and irrigated onto plants for remediation (see Figure 4.3, p. 207).

2 Stormwater and wastewater

The ability of planted systems and wetlands to filter pollutants out of stormwater and wastewater is well known and highly documented. Once the pollutants are filtered from the water, they either remain in the soil media or can be additionally remediated by plants. The remediation occurring in a wetland can be significantly different from a planted soil system, since reactions often take place in saturated anaerobic (without oxygen) environments, very different from planted aerobic (with oxygen) soil-based systems. For the most part, these systems will not be covered in this book, since they are extensively documented in other publications.

C Air

Air pollution remediation with plants is a topic unto itself and will be only briefly covered in this book, in Chapter 3 (see p. 189).

31

IV   Contaminant type: organic vs inorganic

Phytotechnology treatment techniques are contaminant specific. The first step in deciding if phytotechnology systems may be applicable is to consider which of two categories the targeted pollutant falls into, either organic or inorganic (Figure 2.2).

Organic pollutants are compounds that typically contain bonds of carbon, nitrogen and oxygen, are man-made and foreign to living organisms (Pilon-Smits, 2005) (Figure 2.3). Since these pollutants are compounds, if phytotechnology is an applicable solution, many can be degraded, breaking them down into smaller, less toxic components. Organic contaminants may be degraded outside the plant in the root zone, taken into a plant, bound to the plant tissues, degraded to form non-toxic metabolites or released to the

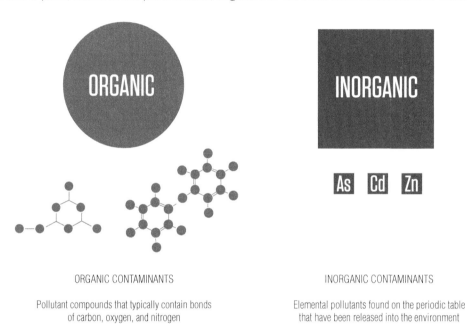

ORGANIC CONTAMINANTS

Pollutant compounds that typically contain bonds
of carbon, oxygen, and nitrogen

INORGANIC CONTAMINANTS

Elemental pollutants found on the periodic table
that have been released into the environment

Figure 2.2 Organic vs Inorganic Pollutants

Figure 2.3 List of Common Organic Pollutants Successfully Degraded or Volatilized at Field Scale with Phytotechnologies

Pollutant	Typical Sources
Petroleum Hydrocarbons: Oil, Gasoline, Benzene, Toluene, PAHs, gas additive: MTBE: Methyl Tertiary Butyl Ether	Fuel spills, leaky underground or above-ground storage tanks
Chlorinated Solvents: such as TCE: trichloroethylene (most common pollutant of groundwater), Perc	Industry and transportation, dry cleaners
Pesticides: Atrazine, Diazinon, Metolachlor, Temik (to name a few)	Herbicides, insecticides and fungicides from agricultural and landscape applications
Explosives: RDX	Military activities

List of Common Organic Pollutants not Easily Degraded or Volatilized at Field Scale with Phytotechnologies

Pollutant	Typical Sources
Persistent Organic Pollutants: Including DDT, Chlordane, PCBs	Historic use as pesticides or in products such as insulation and caulking
Explosives: TNT	Military activities

atmosphere (Ma and Burken, 2003). Phytotechnology systems for the treatment of organics can be an ideal scenario where the pollutant is degraded and disappears and there is no need to harvest the plants.

Inorganic pollutants are naturally occurring elements on the periodic table such as lead and arsenic (Figure 2.4). Human activities such as burning of fossil fuels, industrial production and extraction mining create releases of inorganic pollutants into the environment, causing toxicity (Figure 2.5). These are elements, so they cannot be degraded and destroyed; however, in some instances they can be taken up and extracted by plants. If extraction is possible, the plants must be cut down and harvested to remove the pollutant from a site (Chaney et al., 2010). In addition, inorganic elements can exist in many forms: anions, cations, oxidized states, or as solids, liquids or gases. If extraction for remediation is not a

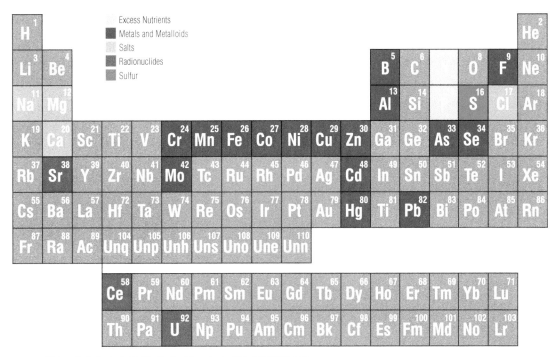

Figure 2.4 Periodic Table: Typical Inorganic Site Contaminants

Figure 2.5 List of Common Inorganic Pollutants Successfully Extracted (and Harvested) or Volatilized with Phytotechnologies

Pollutant	Typical Source
Plant Macronutrients: Nitrogen and Phosphorus	Wastewater, landfills, agriculture and landscape practices
Metals: Arsenic, Nickel, Selenium (shorter time frame) Cadmium and Zinc (longer time frame)	Mining, industry, emissions, automobiles and agriculture

List of Common Inorganic Pollutants not easily Extracted or Volatilized with Phytotechnologies

Pollutant	Typical Source
Metals: Boron (B), Cobalt (Co), Copper (Cu), Chromium (Cr), Iron (Fe), Manganese (Mn), Molybdenum (Mo), Lead (Pb), Fluorine (F) Lead (Pb), Mercury (Hg), Aluminum (Al)	Mining, industry, emissions, automobiles, agriculture, and lead paint
Salt: Sodium chloride, Magnesium chloride	Road de-icing, gas fracking and oil drilling, fertilizers, herbicides
Radioactive Isotopes: Cesium, Strontium, Uranium	Military and energy production activities

possibility (which is true for most inorganic contaminants), plants and their associated microbes are sometimes able to stabilize or change the state of an inorganic contaminant to reduce exposure risk and danger to humans and the environment.

For pollutants found in soils, plant-based treatment technologies have been best utilized for the treatment of organic contaminants and nitrogen (Dickinson et al., 2009). Inorganic-contaminant phytoremediation for removal of pollutants from a site has been less successful. This generalization refers to soil-based contamination and does not apply to remediation of pollutants within water, since inorganics in water can often be filtered out by various types of wetlands and held within the soil matrix (Kadlec and Wallace, 2009).

A site may be heterogeneous, with many different contaminants at different concentrations in the soil, and each one must be considered individually in a treatment system, since each contaminant may require different remediation technologies in a particular sequence. The system may require a 'treatment train' where individual components of a contamination 'cocktail' are targeted one by one in a particular sequence. In addition, the reactions between contaminants must be considered, since the presence of some chemical compounds may greatly influence the reaction of others in the system.

34   V   **Phytotechnology mechanisms: how plants assist in remediating contaminants**

The plant processes involved in both organic and inorganic contaminant transformations have been simplified here into seven phytotechnology mechanisms. Each mechanism describes a particular way in which a pollutant can be modified by plants. 'Phyto' precedes many of the mechanism words, for example phytodegradation, phytovolatilization, phytoextraction. All these terms can lead very quickly to phytoconfusion (White, 2010)! The following section provides a simple way to explain the most common mechanisms encountered. A short phrase for each mechanism is presented to help decipher the scientific terms. Multiple mechanisms may be at work at the same time in any given phytotechnology installation.

A Organic pollutant mechanisms

The following mechanisms are utilized in phytotechnology systems for organic contaminants only.

1 Phytodegradation (also called phytotransformation)

REMINDER: Plant destroys it.

This mechanism is the process in which a contaminant is taken up by the plant and broken down into smaller parts (Figure 2.6). In most cases the smaller parts, called metabolites, are non-toxic. The plant often uses the byproduct metabolites in its growth process, so little contamination remains. The degradation occurs during photosynthesis or by internal enzymes and/or microorganisms living within the plant.

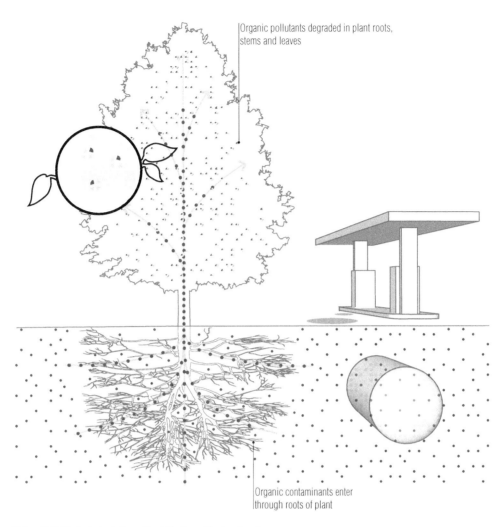

Organic pollutants degraded in plant roots, stems and leaves

Organic contaminants enter through roots of plant

35

Figure 2.6 Phytodegradation: Plant Destroys Contaminant

 2 Rhizodegradation (also called phytostimulation, rhizosphere biodegradation, or plant assisted bioremediation/ degradation)

REMINDER: Microbes in the soil destroy it.

When rhizodegradation is at work, the root exudates released by the plant and/or the soil microbiology around the roots break down the contaminant (Figure 2.7). While the soil microbes are doing the breakdown, the plant is still a critical part of this process because it releases phytochemicals and sugars that create the environment for the microbes to thrive (Reynolds et al., 1999). The plant essentially provides a reactor for the contaminant to be broken down by helping to increase numbers of microorganisms and sometimes encouraging the growth of specific degrading communities of microbes (White and Newman, 2011). Microorganisms readily metabolize many simple compounds (Reynolds et al., 1999). "Environmental contaminants are more complex compounds and generally are metabolized by a smaller percentage of the soil microbial population. However, if the soil microbial population is robust and simple carbon sources become depleted, the soil microbial community can often adapt and use the contaminants as a carbon source" (Reynolds et al., 1999, p. 167).

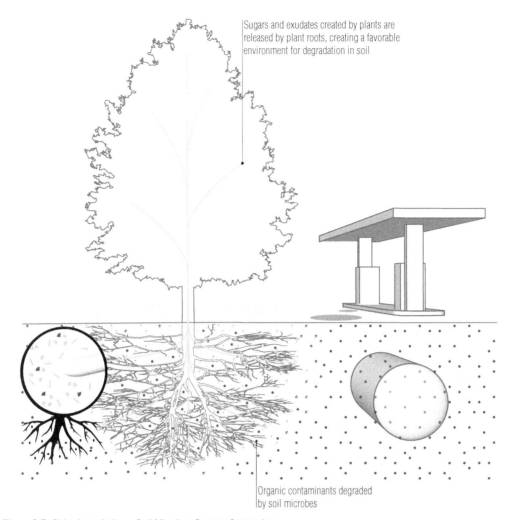

Sugars and exudates created by plants are
released by plant roots, creating a favorable
environment for degradation in soil

Organic contaminants degraded
by soil microbes

Figure 2.7 Rhizodegradation – Soil Microbes Destroy Contaminant

Both the phytodegradation and rhizodegradation mechanisms are the best-case scenarios in phytotechnologies, since the original contaminant is degraded and no plant harvesting is needed. Other phytotechnology mechanisms are listed below.

B Organic and inorganic pollutant mechanisms

The following mechanisms are utilized in phytotechnology systems for both organic and inorganic contaminants.

1 Phytovolatilization

REMINDER: Plant releases it as a gas.

Contaminants can exist in several forms, for example as a solid, liquid and a gas. In this mechanism, the plant takes up the pollutant in either form and transpires it to the atmosphere as a gas, thus removing it from the site (Figure 2.8). The gas is usually released slowly enough that the surrounding air quality is not significantly impacted. The net benefit of removing the

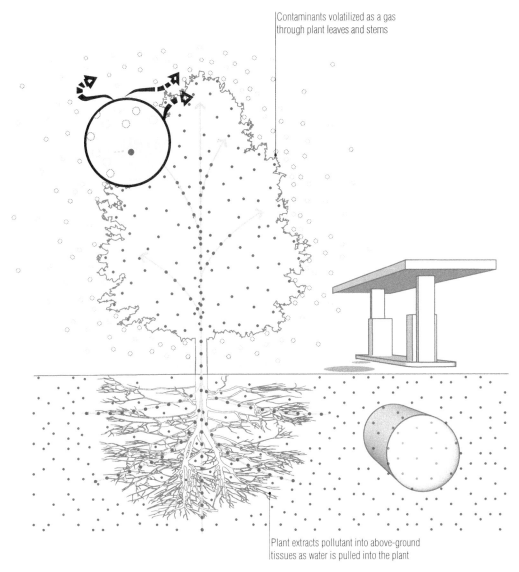

Contaminants volatilized as a gas through plant leaves and stems

Plant extracts pollutant into above-ground tissues as water is pulled into the plant

37

Figure 2.8 Phytovolatilization: Plant Extracts and then Releases Contaminant as a Gas

contaminant from the ground is typically better than any effect of releasing the pollutant into the atmosphere. In some cases, a breakdown product derived from the previous mechanisms of rhizodegradation or phytodegradation may be volatilized (ITRC, 2009).

2 Phytometabolism (also called phytotransformation)

REMINDER: Plant uses it in growth, incorporates it into biomass.

For plants to grow, they need nutrients as building blocks for photosynthesis and biomass creation. Phytometabolism is the process in which the nutrients needed by plants (inorganic elements such as N, P, K) are processed and turned into plant parts (Figure 2.9). In addition, once organic contaminants have been broken down by a plant (phytodegradation), the metabolites that are left over from the process are often phytometabolized and incorporated into the plant's biomass.

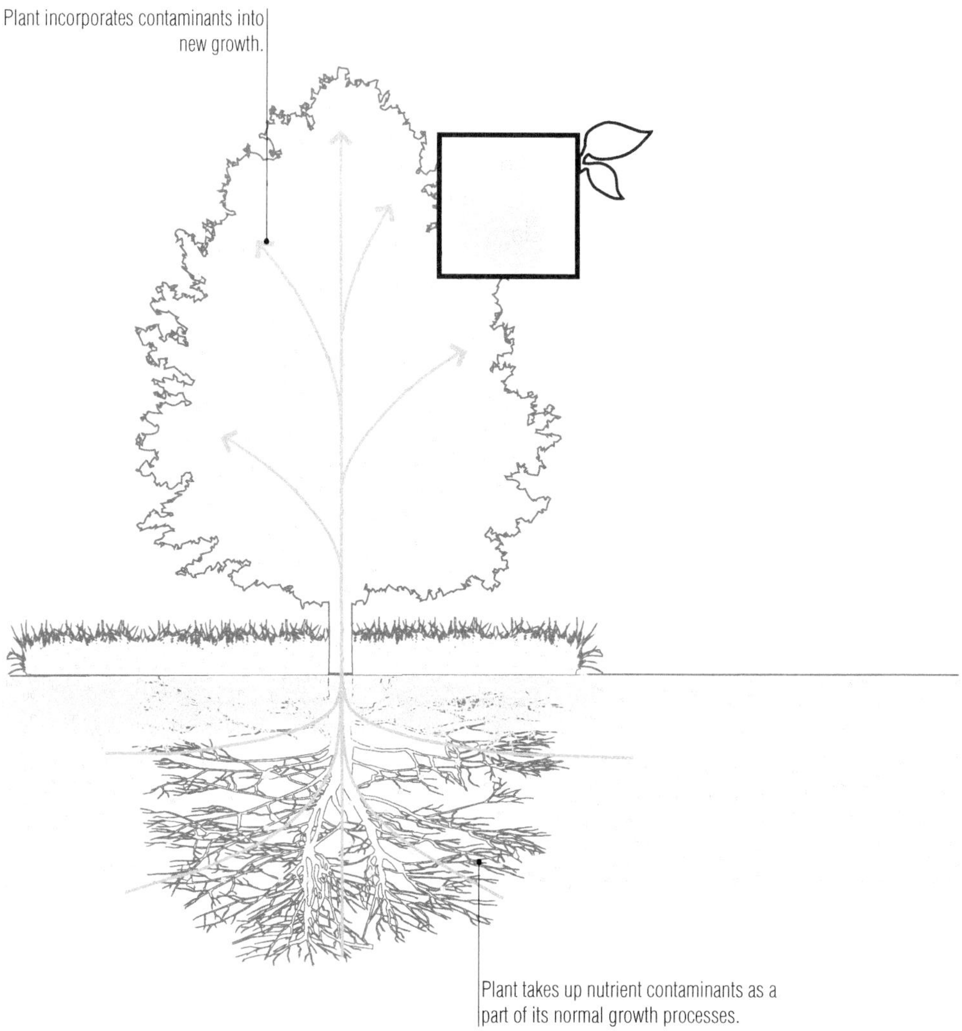

Plant incorporates contaminants into
new growth.

Plant takes up nutrient contaminants as a
part of its normal growth processes.

Figure 2.9 Phytometabolism: Plant Incorporates Nutrient Contaminants into Growing New Biomass

38

3 Phytoextraction

REMINDER: Plant extracts it, and for inorganics it is stored and must be harvested for removal.

Phytoextraction is the ability of the plant to take up a pollutant from soils and water and move it into plant parts (Figure 2.10). When phytoextraction is coupled with phytodegradation for organics, the contaminant essentially disappears from site. However, since inorganics are elements on the periodic table, they cannot be degraded and broken down into smaller parts. Instead, the plant stores away the extracted inorganic pollutant in the shoots and leaves. For the pollutant to be removed from the site, the plant must be harvested before the leaves drop or the plant dies back. The harvested plant material can be burned, followed by disposal in a landfill, reused for biomass (fuel, hardwoods and pulp) or burned and smelted into ore to collect valuable metals (called phytomining) (Chaney et al., 2007).

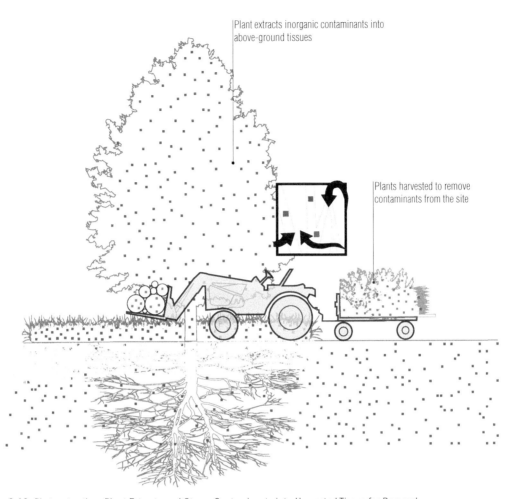

Plant extracts inorganic contaminants into above-ground tissues

Plants harvested to remove contaminants from the site

Figure 2.10 Phytoextraction: Plant Extracts and Stores Contaminants into Harvested Tissue for Removal

4 Phytohydraulics

REMINDER: Plant pulls up water, and the contaminant may come with it.

Plants need water, and the pull created as water is brought into the roots is referred to as phytohydraulics (Figure 2.11). The pull can be so great that groundwater can be drawn towards a plant, and masses of plants can actually change the direction or stop the flow of groundwater. If the groundwater is contaminated, phytohydraulics may be able to stop migrating plumes. In addition, the plant will often use one of the other mechanisms, such as phytodegradation or phytovolatilization, to eliminate the pollutant.

5 Phytostabilization (also called phytosequestration, phytoaccumulation, rhizofiltration)

REMINDER: Plant holds it in place.

The plant holds the contaminant in place so that it does not move off site (Figure 2.12). This occurs because vegetation is physically covering the contamination and the plant may also release phytochemicals into the soil that bind contaminants and make them less bioavailable. In addition, phytoaccumulation refers to the collection of airborne pollutants onto leaf surfaces, physically filtering contaminants out of the air and holding them in place.

Figure 2.11 Phytohydraulics: Plants Change Groundwater Hydrology, Take up Water and Contaminants

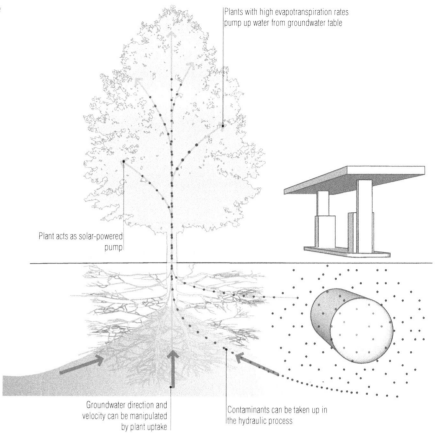

Plants with high evapotranspiration rates pump up water from groundwater table

Plant acts as solar-powered pump

Groundwater direction and velocity can be manipulated by plant uptake

Contaminants can be taken up in the hydraulic process

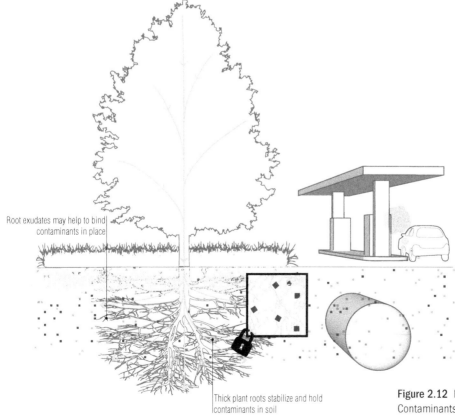

Root exudates may help to bind contaminants in place

Thick plant roots stabilize and hold contaminants in soil

Figure 2.12 Phytostabilization: Plant Holds Contaminants in Place and Prevents Mobilization

6 Rhizofiltration

REMINDER: Roots and soil filter water.

In constructed wetlands and stormwater filters, the roots of plants filter out pollutants from the water. The plants add oxygen and organic matter to the soil to maintain binding sites for contaminant filtration and storage.

C Summary of mechanisms

The eight primary mechanisms described previously are summarized in Figures 2.13a and 2.13b.

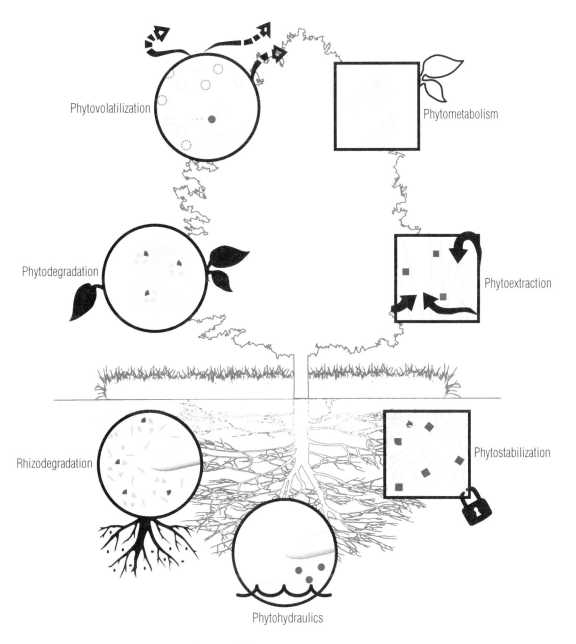

Figure 2.13a Phytomechanisms: Summary Diagram

Figure 2.13b Phytotechnology Mechanism Summary Table

Icon	Name	Description	Contaminant type addressed: organic ○ or inorganic ☐
	Phytodegradation	Plant destroys it	○
	Rhizodegradation	Soil biology destroys it	○
	Phytovolatilization	Plant turns it into a gas	○ ☐
	Phytometabolism	Plant uses it in growth, incorporates it into biomass	○ ☐
	Phytoextraction	Plant takes it up, stores it and is harvested	○ ☐
	Phytohydraulics	Plant draws it close and contains it with water	○ ☐
	Phytostabilization/ Phytosequestration	Plant caps and holds it in place	○ ☐
	Rhizofiltration	Contaminant is filtered from water by roots and soil	○ ☐

D Natural attenuation

Where low levels of contamination exist, spontaneous vegetation may utilize these mechanisms to start to remove pollutants without any human intervention. This process, called natural attenuation, may also involve the breakdown of contaminants by microbes in unvegetated soils as well.

VI Phyto plant characteristics and installation considerations

A Tolerant to pollution and competitive

Choosing plant species that will live in contaminated soil is the first critical selection criterion. If plants cannot grow on a site, it is impossible for a phytotechnology system to be successful. Many contaminants will be toxic to plants and inhibit plant growth. When selecting species, the very first qualifier to consider is whether it will tolerate the encountered concentrations of pollutants. In addition, plants that are hardy perennials, adapted to the local climate and will aggressively out-compete weeds and other plants are preferred. Once these selection criteria have been applied, the characteristics below can be evaluated before extraction, degradation or stabilization capabilities are considered.

B Root depth and structure

Phytotechnologies are limited by root depth, as the plant must be able to reach the pollutant. Most wetland species have a root depth of less than 1 foot, herbaceous species have a maximum root depth of 2 feet and tap-root trees have a maximum root depth of 10 feet (Figure 2.14). However, drought-tolerant species and phreatophytes (see below) may reach even greater root depths and are selected for phytotechnology applications.

1 Drought-tolerant species

Species from arid climates typically have longer, more developed root zones, so these plants are often good experimental choices for phytoremediation (Negri, 2003). For example, grass species from the North American prairie tend to have long, deep root zones, reaching 10–15 feet in some instances. These prairie grass species have been used successfully in phytoremediation applications. However, 70–80% of the root structure will be in the first 2 feet of soil, therefore the top 2 feet is often the most effective zone for phytotechnologies (ITRC, 2009).

43

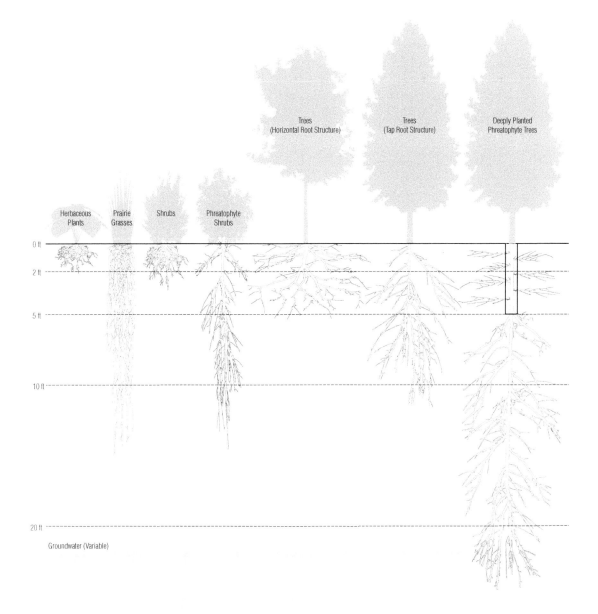

Figure 2.14 Typical Plant Root Depths

2 Phreatophytes

A group of plants called phreatophytes are very deep rooted and usually have at least a part of their root system constantly in touch with water (Figure 2.15). Many use groundwater as their water source and live either in arid environments tapped into groundwater, or in wetlands and river edges in standing water. (The Greek root of the word phreatophyte translates to 'well plant'.) These plants send long root systems in search of water and can reach depths of up to 30 feet or more (Negri, 2003). Poplar and willow species are both phreatophytes, which is one of the reasons these species are used in phytotechnology applications, especially where cleanup of groundwater is required.

3 Deep-root planting

Another way to overcome root-depth restrictions is by drilling deep boreholes or trenches and planting species at the bottom of the hole to allow the roots to penetrate to greater depths (Figure 2.16). This detail is standard practice for many commercial phytotechnology firms. Planting depths by boreholes and trenches can reach 15 feet deep during installation, and typically bare-root plants or dormant cuttings are inserted into the holes. The species must be able to withstand deep-root planting for this technique to be a success. With deep-root planting techniques, the maximum rooting depth the plants typically reach is about 25 feet (Tsao, 2003). Local soil conditions will ultimately dictate the root depth that a plant will reach.

4 Fibrous root zones

When contamination is near the surface rather than deep, species with fibrous root zones are able to come into closer contact with contamination than are tap-root species, because of the number of small, dense roots dispersed through the soil. Fibrous roots provide more surface area for colonization by microorganisms and allow close interaction between the contaminant and the microbiology associated with roots. These species are therefore preferred for soil contamination within 5 feet of the finish grade (Kaimi et al., 2007).

5 High biomass-producing plants

Plants that will grow fast and produce a lot of biomass are often utilized for phytotechnology applications. If degradation is the objective, fast-growing plants tend to release more sugars and exudates at the root zone, creating an environment for enhanced degradation (Robson, 2003). If extraction is the objective, fast-growing plants may take up and store contaminants faster and in larger amounts than the average plant. A list of high biomass-producing plants that have been utilized in remediation plantings is provided in Figure 2.17. The plant families *Salix*, *Populus*, *Vetiveria* and *Brassicacae* are often utilized because of the significant amount of biomass produced (Dickinson et al., 2009). Many nitrogen-fixing pioneer species are currently being tested for phytoremediation capabilities since they are fast growing, produce high biomass and are hardy species adapted to growing in harsh environments in a range of climates (Dutton and Humphreys, 2005).

44

Figure 2.15 Representative Phreatophyte Plant Species (Robinson, 1958) and (McCutcheon and Schnoor, 2003)

Latin	Common	Vegetation Type	USDA Hardiness Zone	Native to
Acacia greggii	Catclaw	Shrub	7–11	Southwestern USA and Mexico
Acer negundo	Boxelder	Tree/Shrub	3–8	North America
Acer rubrum	Red Maple	Tree	3–9	Eastern North America
Alnus spp.	Alder	Tree/Shrub	varies	varies
Amelanchier Canadensis	Serviceberry	Tree	4–8	Eastern North America
Atriplex canescens	Fourwing Saltbush	Shrub	7+	Western USA
Baccharis emoryi	Emory Baccharis	Shrub	5+	Southwestern USA and Mexico
Baccharis glutinosa	Seepwillow	Shrub	7-10	Southwestern USA and Mexico
Baccharis sarothroides	Desert Broom	Shrub	9–10	Southwestern USA and Mexico
Baccharis sergiloides	Squaw Baccharis Waterweed	Shrub	7–10	Southwestern USA and Mexico
Baccharis viminea	Mulefat	Shrub	6–10	Southwestern USA and Mexico
Celtis reticulata	Netleaf Hackberry	Tree	5+	Western USA
Cercidium floridum	Blue Palo Verde	Tree	9–11	Southwestern USA and Mexico
Chilopsis linearis	Desert-Willow	Shrub	7–11	Southwestern USA and Mexico
Chrysothamnus pumilus	Rabbitbrush	Shrub	4+	Western USA
Cornus amomum	Silky Dogwood	Shrub	5–8	Eastern North America
Eucalyptus spp.	Eucalyptus	Tree	varies	varies
Fraxinus velutina	Velvet Ash	Tree	7+	Southwestern USA and Mexico
Hymenoclea monogyra	Burrobush	Shrub		Southwestern USA and Mexico
Juglans microcarpa	Texas Walnut	Tree	5+	Southwestern USA and Mexico
Larrea tridentata	Creosote Bush	Shrub	7+	Southwestern USA and Mexico
Magnolia virginiana	Sweetbay Magnolia	Tree	5–10	Eastern USA
Platanus wrightii	Arizona Sycamore	Tree	7+	Southwestern USA and Mexico
Populus spp.	Poplar	Tree	varies	varies
Populus deltoides	Cottonwood	Tree	2–9	North America
Populus tremuloides	Quaking Aspen	Tree	1+	North America
Prosopis juliflora and *Prosopis pubescens*	Mesquite	Shrub/Tree	6+	Southwestern USA, Central and South America
Prosopis velutina	Velvet Mesquite	Shrub/Tree	7+	Southwestern USA and Mexico
Purshia stansburiana	Vanadium Bush	Shrub	4–9	Southwestern USA and Mexico
Quercus alba	White Oak	Tree	3–8	Eastern USA
Quercus agrifolia	California Live Oak	Tree	7–10	California
Quercus lobata	Roble Oak *Valley Oak*	Tree	5–9	California
Salix spp.	Willow	Tree/Shrub	varies	varies
Sambucus spp.	Elderberry	Tree/Shrub	varies	varies
Sarcobatus vermiculatus	Greasewood	Shrub	3+	Southwestern USA and Mexico
Tamarix spp.	Tamarisk Athel Pine Athel Tree	Tree/Shrub	8–11	Africa, Asia
Taxodium spp.	Cypress	Tree/Shrub	varies	varies

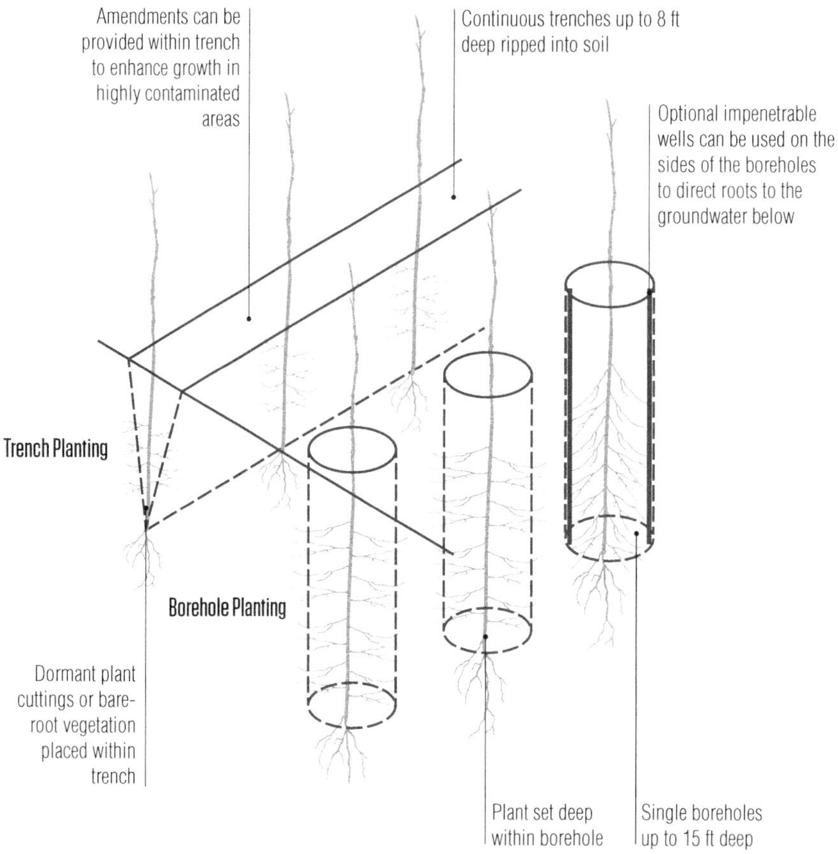

Amendments can be provided within trench to enhance growth in highly contaminated areas

Continuous trenches up to 8 ft deep ripped into soil

Optional impenetrable wells can be used on the sides of the boreholes to direct roots to the groundwater below

Trench Planting

Borehole Planting

Dormant plant cuttings or bare-root vegetation placed within trench

Plant set deep within borehole

Single boreholes up to 15 ft deep

Bare-root vegetation and cuttings can be planted deeply within trenches and boreholes to give plants a head start to reach groundwater.

Figure 2.16 Deep-Root Planting Techniques

Figure 2.17 High-Biomass Species Commonly Used in Phytotechnology Applications

Not intended to be a complete list, but rather a representative list of high-biomass species commonly used

Latin	Common	Vegetation Type	USDA Hardiness Zone	Native to
Bambuseae	Bamboo	Herbaceous	varies	Asia
Brassica juncea	Indian Mustard	Herbaceous	9–11	Russia to Central Asia
Brassica napus	Rapeseed	Herbaceous	7+	Mediterranean
Cannabis sativa	Hemp	Herbaceous	4+	Asia
Chrysopogon zizanioides	Vetiver Grass	Herbaceous	9–11	India
Helianthus annuus	Sunflower	Herbaceous	Grown as annual	North and South America
Linum usitatissimum	Flax	Herbaceous	4+	Asia
Miscanthus giganteus	Giant Chinese Silver Grass	Herbaceous	5–9	China, Japan
Panicum virgatum	Switchgrass	Herbaceous	2–9	North America
Populus spp.	Poplar	Tree	varies	varies
Salix spp.	Willow	Tree/Shrub	varies	varies
Sorghum bicolor	Sorghum	Herbaceous	8+	Africa
Zea mays	Corn	Herbaceous	Grown as annual	North and South America

6 High evapotranspiration-rate species

Species that have higher evapotranspiration rates move more water from the soil to the atmosphere and therefore can better capture contaminants in water than can other species. This is especially important when the contaminant is mobilized in water, such as in stormwater or groundwater. High evapotranspiration-rate plants can take up a lot of water, and may be installed in masses to prevent contaminants from migrating in groundwater plumes. However, these plants do need a lot of water for survival and are usually not drought tolerant. Supplemental irrigation may be required during periods of drought. These plants often have large leaves and surface area for evaporation and, unlike drought-tolerant species like succulents, these high evapotranspiration-rate species have evolved to use a significant amount of water in their growth cycles.

At contaminated groundwater sites, high evapotranspiration-rate species can be utilized in phytohydraulics ⬭ to modify groundwater levels, flow direction and speed (Landmeyer, 2001). The plants' potential influence can be estimated by a hydrologic engineer using a water-balance calculation. Water balance, as it relates to phyto, involves a calculation of the water to be used by the trees and analyses how much groundwater will be taken up, factoring in anticipated precipitation, climate, irrigation, length of growing season and so forth.

Water used by the trees will be extracted either directly from groundwater or from infiltrating soil moisture following precipitation events, or both simultaneously. Many phyto installations targeting groundwater pollution minimize the amount of precipitation and soil moisture available to the trees so that they will search for the groundwater. The trees may also be installed in impervious tubes to help guide their roots downward (Gatliff, 2012). The infiltration of precipitation through soils may also be limited on these sites to prevent vertical migration of additional soil pollutants into groundwater, and also to minimize the recharge rate and groundwater migration speed.

Lists of high evapotranspiration-rate species typically utilized in remediation plantings are provided in Figures 2.18a and 2.18b.

7 Hybrid species

Plant species and various cultivars must be carefully selected, since hybrids and crosses of similar species may produce very different results from the parents or relatives. For example the genus *Salix* (willow) is often widely utilized for remediation of organics, groundwater plume control, precipitation infiltration control, uptake of some inorganics and stabilization and exclusion of some inorganics. There is a big variation in uptake, translocation to the shoot and tolerance to many metals between various hybrid clones of *Salix*. This makes *Salix* unique because it can be possible to choose a certain clone for a certain phytoremediation purpose, such as phytostabilization ⬭ vs phytoextraction ⬭. For example, some hybrids of the species *Salix viminalis* tend to be excluders of heavy metals and good for stabilization, rather than extractors and accumulators of metals, which is more common in *Salix* (Gawronski et al., 2011).

Hybrid poplars are continually seen in phytoremediation research and field applications for many reasons. First, they have many of the desirable plant properties described earlier

47

Figure 2.18a High Evapotranspiration-Rate Woody Plant Species

Not intended to be a complete list, but rather a representative list of high-evapotranspiration rate species commonly used in phytotechnologies (ITRC 2009)

Latin	Common	Vegetation Type	USDA Hardiness Zone	Native to
Alnus spp.	Alder	Tree/Shrub	varies	varies
Betula nigra	River Birch	Tree	4–9	Eastern USA
Eucalyptus spp.	Eucalyptus	Tree	varies	varies
Fraxinus spp.	Ash	Tree	varies	varies
Populus spp.	Hybrid Poplar	Tree	varies	varies
Populus deltoides	Eastern Cottonwood	Tree	2–9	North America
Populus tremuloides	Aspen	Tree	1+	North America
Prosopis glandulosa	Mesquite	Tree/Shrub	6–9	North America
Salix spp.	Willow	Tree/Shrub	varies	varies
Sarcobatus vermiculatus	Greasewood	Shrub	5+	North America
Tamarisk gallica	Salt Cedar	Tree	5–9	Southeast Europe and Central Asia
Taxodium distichum	Bald Cypress	Tree	6–10	North America

48

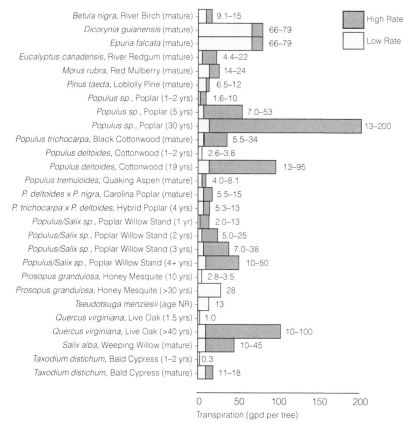

Rates of transpiration of various woody species (redrawn from ITRC, 2009)

Figure 2.18b Plant Species with High Evapotranspiration Rates

in this chapter. They are fast growers, producing a lot of biomass, have some of the highest evapotranspiration rates of any species and are phreatophytes with long, deep tap-root systems. In addition, they are very hardy and virtually impossible to kill, unless enough water is not supplied. Many different cultivars are available that will grow in many different climates.

Hybrid poplars (crosses between two species) are typically utilized rather than the straight species of origin. Two parents are crossed and the hybrids have traits that can exceed those of either parent. For example, *Populus deltoides* (Eastern Cottonwood) and *Populus trichocarpa* (Black Cottonwood) are commonly crossed because their offspring have much larger leaves (Landmeyer, 2012). Hundreds if not thousands of poplar clones have been developed, originally for the wood-product and biomass industries. The Pacific Northwest Research Station in Oregon, run by the US Forest Service, has developed and tested many of the clones used in the US.

Another advantage of poplars in the scientific research work is that the genome of the species is mapped, making it easier for scientists to study plant function and mechanisms in a laboratory environment. In addition, there is a high similarity between the 'lab rat' of the plant world, *Arabidoposis* (a plant which propagates easily and has a short life cycle), and poplar, allowing poplar to directly benefit from research performed on this other plant species.

8 Contaminant concentration and soil amendments

Phytotechnology systems are often best suited for sites with low to moderate contamination, where contaminants will not be toxic enough to inhibit plant growth. For conditions of higher concentration, plants that will tolerate these conditions must be selected. A phytoremediation expert should be consulted to determine if the contamination on a particular site falls within concentrations that can be treated with plant-based systems.

Agronomic soil tests should always be conducted in the planning stage of phytoremediation projects. Amendments can be added to create better conditions for plant growth; the contamination is often not the only reason plants do not survive. Poor soil porosity, lack of nutrients and incorrect pH can also contribute to failures of plant growth. However, it is also important to ensure that the addition of soil amendments does not mobilize existing contaminants. For example, by the addition of amendments, pollutants may become more water soluble and could leach into groundwater and migrate. In addition, as amendments are being added and soils are moved around, there is potential for wind erosion and for pollutant particles to mobilize, creating an additional risk.

9 Winter dormancy and climate

Evapotranspiration and photosynthesis essentially stop during the winter period, so some phytotechnology systems will become dormant. However, if rhizodegradation is the primary mechanism at work, soil biology often still functions, but likely at a reduced rate, due to lower soil temperatures. Dormant conditions and timing have to be considered in any phytotechnology installation.

10 Plant spacing

Preliminary recommendations for plant spacing in phytoremediation systems are the following (ITRC, 2009).

- Trees: A general rule is to provide at least 75 square feet of space per tree. Phreatophyte poplars are typically installed at 10 feet or 12 feet on center spacing, with the same amount of distance between rows (Figure 2.19).
- Biomass plants: Sometimes poplars are planted closer, at 6 feet on center with 12 feet spacing between rows if biomass is an economic product. Any plants that will be harvested for biomass should use the preferred spacing for cutting equipment to determine plant spacing.
- Common grass seeds (Rye Grass, Fescue, etc.): Typically installed at around 400 lbs per acre
- US Prairie Grass seed: These warm-season, native grasses are usually installed at about 10 lbs per acre.
- Shrubs and perennials: Can be installed per the standard practice of the specific species being installed.

VII Principles of organic and inorganic phytotechnologies

A Organic contaminants: basic phytotechnology principles

50

Many organic contaminants can be degraded by plants and their associated root microbes into non-toxic parts. This can be performed by the associated microbes living in the soil, the plant itself or even microbes living within the plant (including endophyte bacteria). In addition, the plant can also

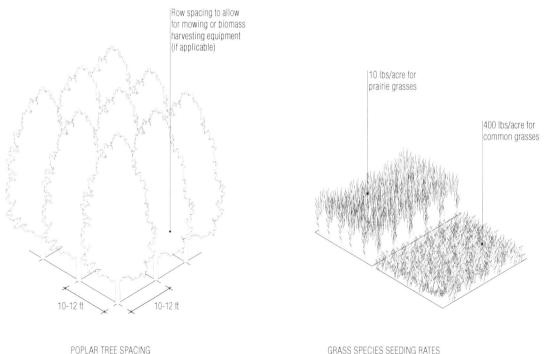

POPLAR TREE SPACING   GRASS SPECIES SEEDING RATES

Figure 2.19 Common Plant Spacing and Seeding Rates

help transform the state of some organic pollutants from liquid form to a gas, releasing it into the atmosphere. "Organic pollutants are relatively less toxic to plants, because they are less reactive and do not accumulate" (Cherian and Oliveira, 2005, p. 3978). The following phyto principles apply to organic contaminants only, and not inorganic pollutants.

1 Log K_{ow}

One of the good predictors to determine if an organic pollutant can be removed from soils by plants is the value of the log K_{ow} (octanol-water partition coefficient). Since organic pollutants are typically man-made and foreign to plants, there are no transporters for uptake and the usual mechanism for uptake is passive diffusion into the plant (Cherian and Oliveira, 2005). Typically, the higher the log K_{ow}, the more unlikely it will be that a plant-based system will be able to take up the contaminant. Log K_{ow} is a measure of a pollutant's hydrophobicity (aversion to water) (Trapp and McFarlane, 1995). The higher the log K_{ow}, the more likely it is that the pollutant will bind to soil particles and not dissolve in water in pore spaces between soil particles. This means that the pollutant is attracted to the soil particles to such a degree that it is unavailable to plants for uptake. With a lower log K_{ow}, the pollutant will often dissolve in water in the pore spaces between the soil particles and the plant systems can access it for degradation (Pilon-Smits, 2005).

A log K_{ow} value can be looked up for every organic compound (Figure 2.20b). Contaminants with a log K_{ow} between 0.5 and 3.5 can likely be taken up and translocated into a plant (Figure 2.20a), with good opportunities for degradation, release into the atmosphere or sequestration into plant parts (Briggs et al., 1982; Burken and Schnoor, 1998; ITRC, 2009). For hydrophobic compounds with a log K_{ow} over 3.5, it is extremely rare for plants to take up these contaminants. For this reason, if there is an organic contaminant on a site, determining if the log K_{ow} is in the range $0.5 < \log K_{ow} < 3.5$ is a good first initial step to see if plants might be useful in remediation.

Organic pollutants such as PCBs and DDT have very high log K_{ow} values, are categorized as 'recalcitrant' and tend to bind tightly to soil particles (Pilon-Smits, 2005). They can persist in

51

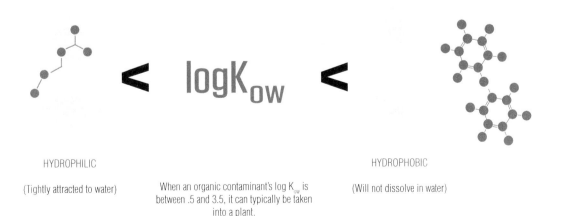

HYDROPHILIC

(Tightly attracted to water)

When an organic contaminant's log K_{ow} is between .5 and 3.5, it can typically be taken into a plant.

HYDROPHOBIC

(Will not dissolve in water)

Figure 2.20a log K_{ow} – Octonal-Water Partition Coefficient

Figure 2.20b Log K_{ow} List of Common Organic Contaminants

Icon	Contaminant	Log K_{ow}
●	PCBs (McCutcheon and Schnoor, 2003)	5.02–7.44
●	PAHs (McCutcheon and Schnoor, 2003)	3.37–7.23
●	Persistent Organic Pollutants (White and Newman, 2011)	3.0–8.3
●	Toluene	2.73
●	Xylenes	3.12–3.20
●	Ethylbenzene	3.15
●	MTBE	0.94
●	Benzene	2.13
●	PCE (Perchloroethylene)	3.4
●	TCE (Trichloroethylene)	2.42
●	RDX	0.87–0.90
●	HMX	0.17
●	TNT	1.73
●	DDT	6.36
●	Chlordane	6.22
●	Lindane	3.55
●	Atrazine	2.61

Source: Sangster Research LOGKOW Databank (except where noted). Recommended values indicated. Available at: http://logkow.cisti.nrc.ca/logkow/index.jsp.

soils for decades. If any remediation of high log K_{ow} pollutants does take place in plant-based systems, this usually can be attributed to microbial degradation in the soil, rather than uptake and degradation within the plant itself.

There are some exceptions to using log K_{ow} as a predictor for contaminant uptake. Sometimes log K_{ow} values may be out of the applicable range, yet a plant may still take up the compound. In these cases, other factors such as molecular mass or the number of hydrogen bonds may influence the uptake (Limmer and Burken, 2014). In general, however, using log K_{ow} as a predictor for organic contaminant uptake can be an initial tool for the designer to consider if an organic contaminant may be able to be taken up and potentially degraded by plants.

2 High biomass-producing plants

For organic pollutants within the desired log K_{ow} range, choosing plant species that will live in the contaminated soil and will grow fast and produce a lot of biomass will usually produce the best remediation results. Most plant species will take up organic contaminants in the desired log K_{ow} range, therefore the specific plant family, genus and species selection may be less important than how fast the plant will grow. Some studies have suggested that when plants have a higher growth rate, they can process and degrade contaminants faster than other species (Robson, 2003). In addition, fast-growing plants may release more sugars at the root zone, creating an environment for enhanced degradation in the root zone (see Figure 2.17 for plant list).

3 High levels of oxygen-degrading enzymes and specific exudate compounds

Another favorable plant property for degradation of organics is selecting species that release high levels of oxygen-degrading enzymes or specific root exudates. For example, Mulberry trees have been reported to release specific compounds that stimulate the growth of microbes involved in the degradation of persistent organics (Fletcher and Hegde, 1995). In addition, some plant enzymes can affect a pollutant's water solubility, oxidization state or other chemical factors that aid in degradation (Volkering et al., 1998).

The root exudate profile is most important when the contaminants have a high log K_{ow} value (above 3.5) and are recalcitrant (persistent and difficult to degrade). These larger organic compounds are often tightly bound to soils and cannot be translocated into plants. Many of these compounds are complex ring structures that are unable to be brought into the plant, but may possibly be broken down in the root zone by microbes. Microbial activity in soils fluctuates not only from plant effects, but also with temperature, fertility and soil moisture (Reynolds et al., 1999).

4 Phreatophytes

Another factor that makes for a good organic-pollutant phytoremediator is if the roots can survive both in aerobic soils with oxygen and in flooded, anaerobic conditions, such as in groundwater without oxygen. If roots can survive in both these conditions, there is a greater diversity of microbes and reactions, both with and without oxygen, that can take place at the root zone, and therefore a greater chance for degradation of organic contaminants (Licht, 2012). Phreatophytes are plants that have at least a part of their root system constantly in touch with water, and tend to be tap-root species. Their roots typically thrive in both flooded (anaerobic) and dry (aerobic) conditions and they are therefore great starting species for organic-contaminant degradation. Two of the most common groundwater contaminants are organics – petroleum products and chlorinated solvents – so phreatophytes can be good selections for both contaminated soils and groundwater, as long as water is readily available (see Figure 2.15 for plant list).

5 Will it vaporize?

Some pollutants, when they are drawn out of soils and water by plants, they vaporize through the plant and are released into the atmosphere as a gas. To see if a pollutant will release into the atmosphere in a volatile form, the dimensionless Henry's law constant (H_i) of a particular contaminant can be looked up. Henry's law constant (H_i) is a measure of a compound's tendency to move into air relative to water (Figure 2.21) (Davis et al., 2003). When the dimensionless Henry's law constant of a pollutant is $H_i > 10^{-3}$ the compound can volatilize through the plant into the air (Pilon-Smits, 2005; Tsao, 2014). The plant basically acts like a wick, drawing up volatile contaminants from the soil and diffusing them into the air. When the H_i is below 10^{-3} these compounds tend to partition in water, and other phyto mechanisms such as phytodegradation or phytosequestration must be considered (Pilon-Smits, 2005; Tsao, 2014).

53

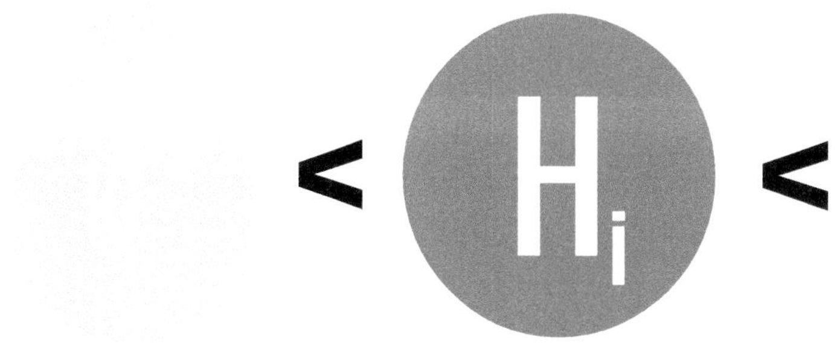

AQUEOUS STATE	Contaminant is mobile in both air and water when H_i is in this range, and plants can likely be used to volatilize the pollutant	GASEOUS STATE
Contaminant moves predominantly in water		Contaminant moves predominantly in air space between soil particles

Figure 2.21 Henry's Law Constant (Dimensionless) – Contaminant Phytovolatilization Potential

6 Soil amendments: organic compounds

Organics tend to be attracted to other organics. When soils have a lot of organic content, organic pollutants may tend to stick to those organic soil particles, making them less bioavailable to plants for uptake and degradation. In addition, when rhizodegradation is utilized, organic soil amendments "may provide an easy-to-digest carbon source that microbes may prefer to use instead of the organic pollutant" (Pilon-Smits, 2005, p. 23). In general, adding amendments with high organic content is usually avoided when remediation of organic pollutants is desired. However, in some cases organics are added to cause microbes to use up all the oxygen, resulting in anaerobic conditions, which can be preferable for certain degradation pathways.

7 Organic-pollutant phyto-plant characteristics: summary

The field of phytoremediation for treatment of organics is promising. As one expert in the field notes:

> Originally, plant uptake and translocation of organic contaminant compounds such as herbicides, pesticides, other petroleum hydrocarbons, and chlorinated solvents was seen as a potential vector for increased exposure risk to wildlife and human populations. The extent of this risk was unclear, because the potential for bioaccumulation of these contaminants in plants, especially those used as food crops, was unknown. Today, however, plants are actively being added to many contaminated sites around the country to reduce environmental risk, because certain plants have been found to take up, sequester, transform some organic compounds into innocuous end products. (Landmeyer, 2011)

Specific plant species shown to degrade certain categories of organic contaminants are listed by contaminant in Chapter 3. To summarize, below are the generalized rules for selecting plants for organic-pollutant remediation.

- If groundwater is being targeted, deep-rooted phreatophytes with high evapotranspiration rates should be used.
- In soils, if log K_{ow} is between 0.5 and 3.5, the contaminant will likely be taken up by plants and possibly degraded. All plants will likely have some effect and it is most important to select plants that will grow fast and produce a lot of biomass. Designers may consider high-biomass plants that have not previously been used in phyto studies as potential candidates for remediation.
- In soils, if log K_{ow} is greater than 3.5 the pollutant is likely too hydrophobic to get into a plant. Instead, remediation tools other than phytotechnologies should be explored.

B Inorganic pollutants

Inorganic contaminants cannot be degraded in the plant or root zone; in fact, they cannot be degraded at all. The target mechanism is either extraction into the plant, where it can be collected, stored and harvested; transformation into gas to be released in the atmosphere; or stabilization, where plants help to cap the inorganic contaminants on site. Extraction of inorganics in phytoremediation is not very practical for field application at this time (Dickenson, 2009), except in a few limited cases covered in more detail in Chapter 3. For this reason, most phytotechnology applications for inorganics in soils involve stabilization – holding pollutants on site with the assistance of vegetation to prevent exposure to risk. However, in wetland systems, inorganics can be filtered out of water and bound into the soils of the wetland.

In a few cases where extraction may be possible, two approaches have been used: either (1) hyperaccumulating species or (2) accumulator, fast-growing, high biomass species are installed.

1 Hyperaccumulators

Some plant species will take up certain elements at concentrations 10–100 times greater than normal plants. These plants, called hyperaccumulators, can translocate elements from soils into the above-ground plant tissues at unusually high concentrations (Van der Ent et al., 2013).

A specific pathway must be present in a species to allow the inorganic contaminant to enter the plant. Pathways exist in all plants to transport all of the essential nutrients required by plants, which are all inorganic elements. These include the following:

- primary macronutrients: Nitrogen (N), Phosphorus (P) and Potassium (K)
- secondary macronutrients: Calcium (Ca), Magnesium (Mg) and Sulfur (S)
- micronutrients, required in trace quantities: Boron (B), Chloride (Cl), Copper (Cu), Iron (Fe), Manganese (Mn), Molybdenum (Mo), Nickel (Ni) and Zinc (Zn).

In addition to the essential trace elements, there are beneficial elements which promote plant growth in many species, but are not absolutely necessary for plant growth: Silicon (Si), Sodium (Na), Cobalt (Co) and Selenium (Se).

Plants will sometimes take up inorganic elements not on this list. The plant has a pathway for each of the nutrients above, and some pollutants are similar in chemical structure to the essential nutrients and the pollutant can be taken up in the same way. In addition, many of the hyperaccumulators have developed a specific pathway for taking up a particular element. Often, these plant traits maybe have developed over time as natural defense mechanisms to make the plant toxic to insects or other predators that might eat it (Hanson et al., 2004).

About 500 plants have been cited in the literature as hyperaccumulators – only a small fraction of 300,000 recognized plant species (Van der Ent et al., 2013). To complicate plant selection, the uptake rate of elements within the same species of plant can vary widely between populations and different cultivars (Van der Ent et al., 2013). In addition, accumulation is a term of relativity – a hyperaccumulator is a plant that takes in more metals than other plants do, which still might not be very much (Rock, 2014). For this reason, utilizing hyperaccumulators for phytoextraction 🔗 must be approached cautiously, and detailed species trial studies must first be conducted to determine removal rates before any large-scale extraction projects are even considered.

"Hyperaccumulation of nickel, zinc, cadmium, arsenic and selenium have been confirmed without a doubt in a range of plant species. Hyperaccumulation of lead, copper, cobalt, chromium and other metals have not (yet) been demonstrated beyond a doubt" (Van der Ent et al., 2013). Many older studies can be found where species have been studied and named as 'hyperaccumulators' for contaminants other than the five indicated here. However, many of these studies have since been disproven or doubted and they must be approached with caution.

2 High-biomass plants

In addition, some highly productive, non-hyperaccumulator plants (sometimes referred to as 'accumulators') may also take up contaminant concentrations at higher ranges than those found in most plants, due to their fast growth rate. The high-biomass, high-yield plants can be utilized in phytoextraction 🔗 applications in addition to hyperaccumulator species. This alternative can be most viable where productive crop cultivation is desired (such as harvesting to make biofuels, hardwood products or pulp) or where hyperaccumulator species' biomass yield is quite low (Dickinson et al., 2009). See Figure 2.17 for a list of typical high biomass-producing plants used in phyto projects.

3 Bioavailability

The "bioavailable fraction" is often described as the amount of contaminant accessible for uptake by organisms (Alexander, 2000). If an inorganic contaminant is in the soil, it may not necessarily be bioavailable to a plant even if it seems that they should be a good match. Pollutants can be chemically and physically attached to other clay or organic soil particles (sorption) (Alexander,

1994) or be attracted and bound to soils with opposite charges. Inorganics are usually present as charged cations (+ positive charge) or anions (- negative charge) and can stick to soils of the opposite charge. Low-pH, acidic soils, have a lot of H+ ions and are positively charged, therefore inorganics that are negatively charged tend not to be bioavailable. The opposite is true for high-pH, alkaline soils, which have more OH- ions and are negatively charged, attracting positively charged cations. The bioavailability of an inorganic can also be influenced by how much oxygen is in the system. In soils, oxygen is usually available and elements tend to exist in their most oxidized form (selenate, arsenate, etc.). In flooded soils or wetlands, reducing conditions exist with little oxygen, which favor reduced elemental forms (selenite, arsenite, etc.). The form of the element can greatly affect its bioavailability to the plant (Pilon-Smits, 2005). Some soil tests are unavailable to determine the bioavailable fraction of a certain contaminant rather than the total amount of that contaminant in the sample.

In summary, differences in inorganic contaminant availability can depend on the following factors (Van der Ent et al., 2013):

- the form of the pollutant and presence in different phases or chemical forms
- the charge of the soil and pH
- the presence and concentration of other soil elements
- physical factors such as local climate, soil porosity and the addition or subtraction of organic matter or other amendments
- the total concentration of the pollutant in the soil.

Extraction of inorganics is complicated and several of the above factors may be at work on any given site. For this reason, extraction of inorganics is discouraged in most cases.

4 Chelants

In the past, several studies have reported plants that extracted or even 'hyperaccumulated' contaminants because a chemical was added to the soil. These chelants, such as EDTA (ethylene diamine tetra-acetic acid), oxalic acid and citric acid, have been shown to speed up accumulation in plants (Evangelou et al., 2007). When these chemical-extraction methods are used, the plants used should not be considered hyperaccumulators (Van der Ent et al., 2013). For the purposes of this publication, only plants considered hyperaccumulators under natural conditions without chemical additives have been listed as such.

There are also a lot of inherent risks in using chelants in phytoapplications, so they are not recommended. The same chelator chemicals are well known to increase leaching of metals from soils, which may be unavoidable, thus potentially mobilizing metals toward groundwater. The overall efficacy of such treatments is likely to be compromised by cost, leaching risk and the lesser-known impact of the chelating agents on soil biota and related functional processes in the soil. Recent reviews of the use of chelating agents have voiced the concern that no solution that is effective in preventing the leaching of metals has been found, and it has been argued that phytoextraction should distance itself from chelate-assisted phytoextraction (Chaney et al., 2007; Evangelou et al., 2007; Dickinson et al., 2009, p. 101).

57

5 Harvesting

Unlike phyto for organic contaminants, annual harvesting of the plant's biomass above ground, followed by proper disposal, is required for most inorganic extraction projects. Longer-lived, high-biomass species (which can accumulate but not hyperaccumulate inorganics) may be harvested less frequently. Harvesting can be a labor-intensive and expensive process. If high concentrations of removal are predicted, harvested material must be tested to see if it needs to be disposed of in a hazardous waste facility or can be disposed of in a municipal landfill.

6 Excluders and stabilization

Many plants can tolerate only low concentrations of bioavailable inorganics in soil before they die, due to phytotoxicity. Some plants can ignore these metals and grow in a wide range of toxic soils even though they can't take them up (Van der Ent et al., 2013). These plants are called excluders (Figure 2.22) and can often be used for phytostabilization, essentially capping metals on site with vegetation so that they do not pose a risk of exposure. Stabilization with plants may be the best-recommended treatment for inorganic contaminants in cases where phytotechnology is being considered. For lists of excluders and stabilization plants by contaminant, see Chapter 3.

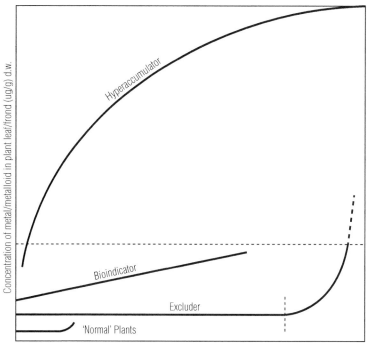

Hyperaccumulators translocate 100–1000x more of a particular metal into their above ground parts compared to normal plants. Excluders tend to take up even less metals than normal plants as concentrations in soil increase.

(Redrawn from Van der Ent et al., 2013. Hyperaccumulators of metal and metalloid trace elements: facts and fiction. *Plant and Soil*, 362 (1–2), pp. 319–334.)

Figure 2.22 Metals Hyperaccumulator vs Excluder Plants – Species Characteristics

VIII   Field application and challenges

Phytoremediation requires more effort than simply planting vegetation and, with minimal maintenance, assuming that the contaminant will disappear. It requires an understanding of the processes that need to occur, the plants selected, and what needs to be done to ensure plant growth. Research is needed upon the identification of suitable plant species that can colonize the polluted area and remove, degrade or immobilize the contaminant of interest. (US EPA, 2001)

Despite our understanding of the mechanisms of remediation, and the success of studies in the laboratory and greenhouse, efforts to translate phytoremediation research to the field have proven challenging. Although there have been many encouraging results in the past decade, there have also been numerous inconclusive and unsuccessful attempts at phytoremediation in the field. There is a need to critically assess why remediation in the field is not satisfactory, before negative perceptions undermine the progress that has been made with this promising remedial strategy. Two general themes have emerged in the literature: (1) Plant stress factors not present in laboratory and greenhouse studies can result in significant challenges for field applications. (2) Current methods of assessing phytoremediation may not be adequate to show that contaminant concentrations are decreasing, although in many cases active remediation may be occurring. If phytoremediation is to become an effective and viable remedial strategy, there is a need to mitigate plant stress in contaminated soils. There is also a need to establish reliable monitoring methods and evaluation criteria for remediation in the field. (Gerhardt et al., 2009, p. 20)

For the above reasons, it is encouraged that landscape architects and designers work closely with a team of experienced phytotechnology scientists on any given project. There are many applications where phyto is a good fit and other conditions where it is not very applicable. Landscape architects can assist in moving the field forward with designed experiments and research incorporated into landscape projects.

One of the best uses for phytotechnologies may be as a holding strategy for sites while they remain vacant, awaiting other uses. Many of these sites do not yet have regulatory requirements; phyto systems can be considered voluntary, and performance standards are not required. The deployment of phytotechnologies during this holding time may significantly reduce or eliminate the need for a comprehensive 'dig and haul' effort in the future. In these circumstances, installations can be considered 'safe to fail'; if the low-cost phyto installation does not work, the risk involved is minimal. When starting a phytotechnology application, smaller pilot projects should be considered so as to test applicability before full-scale field installations are completed. Chapter 3 will delve into specific contaminants and case studies, providing plant lists of species used in research applications.

59

3: Contaminant classifications and plant selection

This chapter provides a summary of the current status of phytotechnology research by contaminant type, to prompt landscape architects to consider integrating these systems into site-design projects. In order to identify potential plant-based systems for a particular site, the chemical composition and concentration of existing and potential site contaminants must first be investigated. The types of pollutants found in the landscape are presented one by one through this chapter, so the site designer can easily look up information by contaminant. Case studies are provided to illustrate results and implementation techniques. Lists of potentially applicable plants are provided for each pollutant type. The goal is to promote integration of phytoremediation concepts into day-to-day landscape design practice and to initiate the practice of buffering typical land uses from pollutant releases before sites become contaminated. In this way phytotechnology can become part of a projective approach to site design and sustainable landscape practice over time. For sites with existing contamination requiring remediation, the research presented should be used as a starting point only. Plant-based systems for remediation should not be applied without the help of an experienced professional in this field. Individual project-site conditions, such as concentrations and mix of contaminants, hydrology, soil properties and climate, will greatly affect the selection of potential phytoremediation systems. It is critical to involve a phytotechnology specialist at the outset of a project to determine if remediation using plants is a valid option and to engage in selecting species and modifying soil conditions and hydrology. Agronomic challenges and the variability inherent in natural systems are significant hurdles in addressing site pollutants. Failed projects can not only be locally dangerous, but can also damage the reputation and regulatory acceptance of the entire field. It cannot be reiterated strongly enough that professionals experienced in these technologies, and specifically in the encountered contaminant, must be engaged to create successful remediation installations. The process for meeting regulatory acceptance must be carefully considered as well. Continual monitoring systems and risk assessment, as required by the regulating agency, must be designed with the planting approach so as to track progress.

A Introduction

The plant species listed in this book are the most common species encountered in the research studies surveyed. The field of phytotechnology is ever evolving and lists of applicable plants will change in future as new research is completed and updated. The plant lists are not intended to be complete lists of possibilities, but rather an initial compilation of plants utilized in the systems surveyed. Most of the plants included in this publication prefer temperate climates. This is not because tropical species do not perform remediation functions, but rather that this book has been authored in the northeastern United States and the research has focused on temperate climate systems. The principles included in this chapter can be applied in other climates, and it is encouraged that peer-reviewed literature be reviewed to select applicable plant species.

Much of the peer-reviewed literature in this field has been produced by agricultural institutions with grant funding and considers food and agricultural crop safety; therefore many of the species included in vegetation lists are agricultural crops. This does not mean that ornamental landscape plants are not valid for the particular use in question. In many cases, comparison species may be able to be tested and added to the plants listed here. It is suggested that practitioners strongly consider using their projects as testing grounds to evaluate potential new species as opportunities arise.

B How to use this chapter

Contaminant groups found in the landscape are presented one at a time through this chapter; first organic contaminants are covered, then inorganic contaminants. Before you use this chapter and reference a particular contaminant of interest, it is suggested that Chapter 2 first be reviewed. Understanding the difference between organic and inorganic contaminants, as covered in Chapter 2, along with the specific contaminant groups covered in this chapter, will provide insight for species selection and application.

Basic information and scientific case studies for each contaminant group are included in this chapter, as well as introductory plant lists of species that have been shown in studies to be useful for the particular pollutant. The plant lists can be used in conjunction with the planting types included in Chapter 4 and site programs illustrated in Chapter 5 for landscape design application.

C Contaminant chart

The summary diagram of site contaminants in Figure 3.1 has been developed to provide a clear, graphic understanding of the following:

- The most significant kinds of contaminants found in the landscape. Contaminants are organized in general groupings by chemical composition. Each group is assigned a color in the chart and the color is used as a key throughout the book. Organic contaminants are shown as circles and inorganic contaminants are represented as squares. Each contaminant group is discussed in a separate section within this chapter.

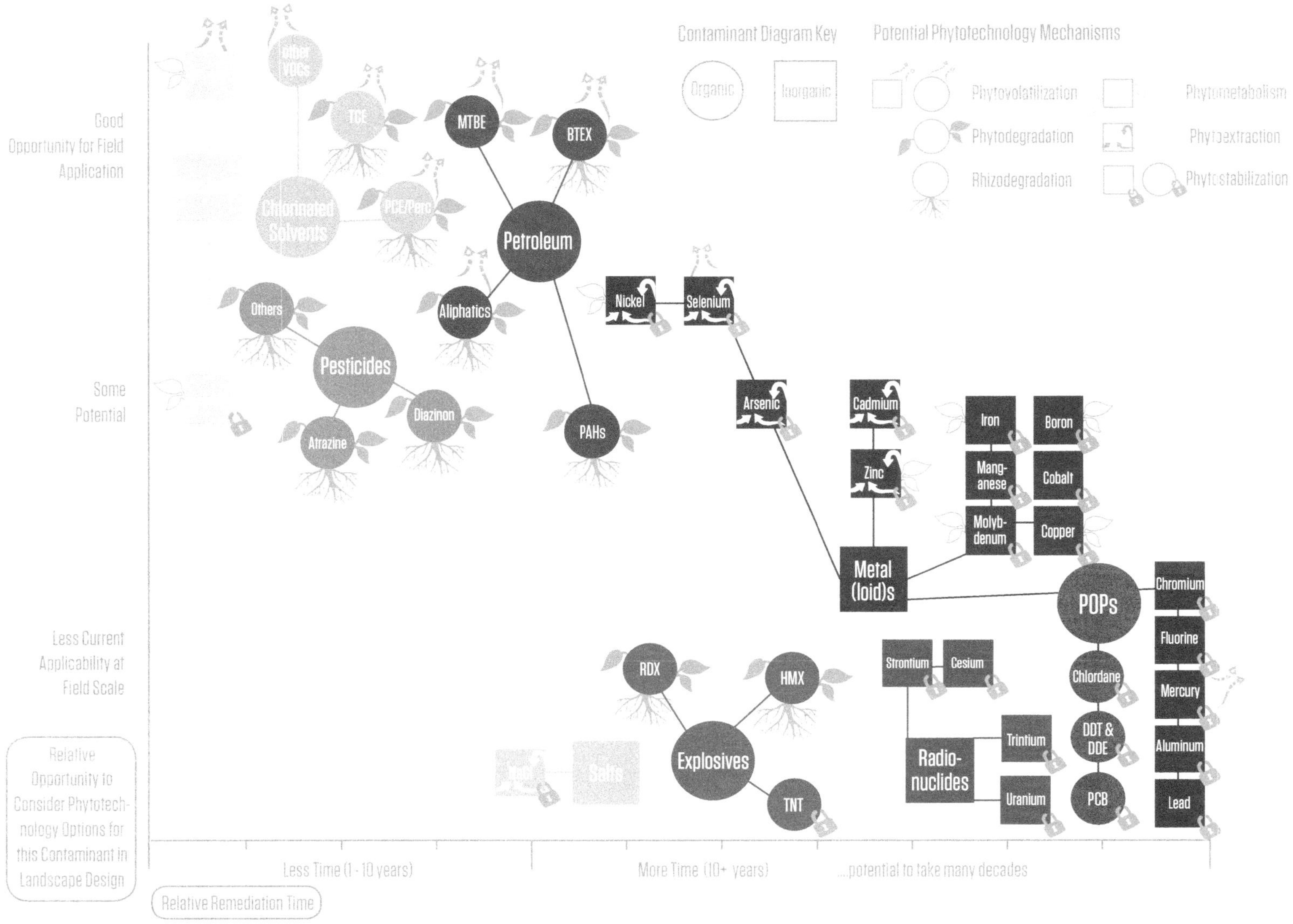

Figure 3.1 Site Contaminants

Figure 3.2 Contaminant Groupings and Typical Sources of Pollutants

Organic Pollutants	
Contaminant Group*	**Typical Source of Pollutants in this Category**
Petroleum: Oil, Gasoline, Benzene, Toluene, PAHs and additives such as MTBE	Fuel spills, petroleum extraction, leaky storage tanks, industrial uses, railway corridors
Chlorinated Solvents: TCE, PCE and organic compounds with a chlorine component	Dry cleaners, military activities, industrial uses
Explosives: RDX, TNT, HMX	Military activities, munitions manufacturing and storage
Pesticides: Herbicides, Insecticides and Fungicides	Agricultural and landscape applications, railway and transportation corridors, residential spraying for termites and pests
Persistent Organic Pollutants (POPs): DDT, DDE, PCBs, Aldrin, Chlordane	Agricultural and landscape applications of historic pesticides, former industry, atmospheric deposition
Other Organic Contaminants of Concern: Ethylene and Propylene Glycols, Formaldehyde, Pharmaceuticals	Aircraft de-icing fluids, embalming fluids, wastewater
Inorganic Pollutants	
Contaminant Group	**Typical Source of Pollutants in this category**
Plant Macronutrients: Nitrogen and Phosphorus	Wastewater, stormwater, agriculture and landscape applications, landfill leachate
Metals: Arsenic, Cadmium, Selenium, Nickel (to name a few)	Mining, industrial uses, agricultural applications, roadways, landfill leachate, pigments, lead paint, emissions
Salt: Sodium, Chloride, Magnesium, Calcium	Agricultural activities, roadways, mining, industrial uses
Radioactive Isotopes: Cesium 127 and Strontium 90	Military activities, energy production

- The vertical position of each contaminant group in Figure 3.1 indicates the relative viability of *in situ* phytotechnology systems for contaminant *removal in soil and groundwater*. Groups near the top of the chart are more viable for implementation of contaminant removal in design projects, as shown by field-scale tests. This chart specifically applies to *removing* the contaminant from a site, and not to the viability of using phytotechnologies for stabilization (keeping the contaminant on the site and eliminating exposure to risk). In addition, the chart specifically addresses the viability of plant-based remediation for *soil and groundwater* cleansing, and not for cleansing the contaminant from air, wastewater or stormwater. The viability of using phytotechnologies for cleansing from air and water vectors will be presented as each contaminant group is discussed, but the focus of this book is on using phytotechnologies for contaminants found in soil and groundwater.

- The horizontal position of each contaminant group on the chart illustrates the relative time frame estimated for contaminant removal from the soil using plant-based remediation systems. Actual removal times on sites will vary significantly, based on concentrations and site-specific factors. The time frames are provided only to give a relative understanding of the number of years it might take to remove a contaminant from soil on the site, so that an initial evaluation of the potential applicability of plant-based remediation can be made.

The chart illustrates three main aspects of the applicability of plant-based systems for the removal of contaminants from soil.

- The most promising utilization of plant-based systems for contaminant removal in soil and groundwater is for: nitrogen ⬛, VOCs, including chlorinated solvents ●, petroleum products ● and some pesticides ●.
- There is research supporting the use of plant-based systems for extraction of arsenic ⬛, nickel ⬛ and selenium ⬛ from soils with moderate to low levels of contamination. In addition, cadmium ⬛ and zinc ⬛ removal may be possible over very long time frames.
- Research is still evaluating the potential for plants to remove explosives ●, metals (other than the ones listed above) ⬛, persistent organic pollutants ● and radionuclides ⬛ from soil. At this time, the applicability of *in situ* phytoremediation for these groups of contaminants for removal from soil is limited but some areas are promising for future development. However, exceptions do exist and plant-based methods for mitigating risk via stabilization on site and hydraulic control are often effectively utilized. In addition, if these contaminants are in water rather than soil, it is possible to filter them from water using constructed wetlands, trapping the contaminant in the soil matrix. These methods are discussed in detail as each contaminant group is presented throughout this chapter.

| Organic contaminant classifications

65

● Petroleum compounds (also known as petrochemicals)

Specific contaminants in this category: Petroleum products contain hundreds of hydrocarbon compounds, which are included in this category. Some of the more typical petroleum contaminants include: oil, gasoline, TPH (Total Petroleum Hydrocarbons – a measure of the many types of hydrocarbons on a site), coal tar and creosote (both sticky black materials, usually from industrial processing of coal); subgroups of petroleum including: PAHs (Polycyclic Aromatic Hydrocarbons), GRO (Gasoline Range Organics), DRO (Diesel Range Organics), BTEX (Benzene, Toluene, Ethyl benzene, and Xylene: VOCs found in gasoline), MOH (Motor Oil Hydrocarbons); and specific petroleum compounds such as MTBE (Methyl Tertiary Butyl Ether – an additive in gasoline that prevents knocking) and ethanol.

Typical sources of petroleum contamination: Releases of petroleum products include: leaking underground storage tanks (LUSTs) containing fuel or oils; gasoline and fuel spills; exhaust from vehicles, machines and trains; byproducts of coal burning and processing; and creosote-treated wood, such as railroad ties.

Typical land uses with potential petroleum contamination: Gas stations and machine-repair shops, maintenance facilities, roadsides, rail yards and tracks, industrial facilities, oil and gas refineries, fracking and drilling facilities, oil and gas and transfer facilities, former manufactured-gas plants and any use that has a fuel storage tank, including residences with current or historic fuel tanks for home heating.

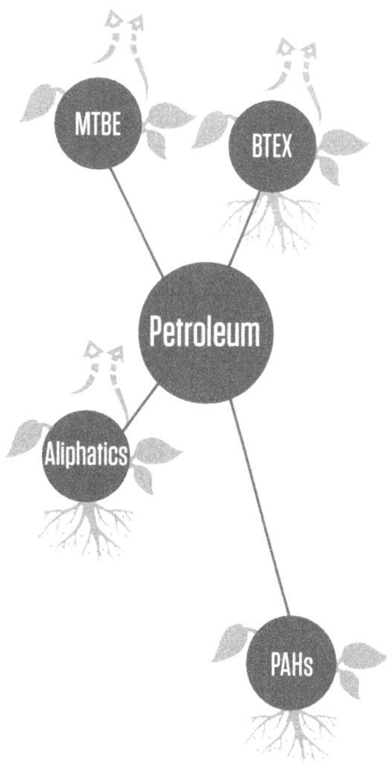

Figure 3.3 Petroleum Compounds

Why these contaminants are a danger: Petroleum hydrocarbons are some of the most common contaminants found in sediments and soils globally (Stroud et al., 2007). Over 500,000 instances of fuel leaks from LUSTs were reported in the US between 1984 and 2011 (OUST 2011). Many hydrocarbons have been found to be potential carcinogens, and benzene and benzo(a)pyrene are known carcinogens (Mueller et al., 1996).

Summary

Petroleum hydrocarbons can generally be placed into two categories.

1 Easily degradable: with log K_{ow} between 0.5 and 3.5*

   These include the 'lighter' fractions of petroleum that are often single chains of molecules. Some volatilize easily, such as BTEX and MTBE (these have Henry's law constant (H_i) values over 10^{-3} – see Chapter 2, p. 53). When hydrocarbons volatilize easily, they are considered VOCs (Volatile Organic Compounds). These molecules often create the odor of gasoline and oil.

2 Persistent, more difficult to degrade: with log K_{ow} over 3.5*

   These include the 'heavier' fractions of petroleum that are often multiple ring structures that are difficult to break apart, such as PAHs, coal tar, crude oil, heating oil and creosote.

* *Note*: Log K_{ow} (octanol-water partition coefficient) is a measure of hydrophobicity, where a higher value means lower water solubility (Pilon-Smits, 2005). Values for specific contaminants can be looked up in charts – see Chapter 2, p. 57 for more information.

Petroleum hydrocarbons in both of these categories have been successfully remediated from soil, groundwater, wastewater and stormwater with plant-based systems. These organic pollutants can be completely degraded by plants and associated microbes without the need for harvesting the plants.

Petroleum remediation has accounted for many of the success stories within the phytoremediation field.

Petroleum hydrocarbons – lighter fractions, easiest to treat: MTBE, BTEX, gasoline, diesel fuel and other aliphatic (straight chain, no ring) compounds of carbon and hydrogen.

Mechanisms utilized: Rhizodegradation ⚲, Phytohydraulics ◌, Phytovolatilization ◌, Phytodegradation ◌.

This is one of the most promising contaminant categories to treat with plant-based systems. Typically, if hydrocarbons in this category were released on site, natural attenuation (natural soil microbial activity as well as exposure to sun, wind and humidity) would likely volatilize or degrade them. However, if plants are introduced they can help to speed up this process and also help to access, contain and treat the fractions that have mobilized into the groundwater. Plants are utilized in two ways.

1 To speed up the natural attenuation process

Plants introduce oxygen, sugars, enzymes and other root exudates into the soils, which promote soil biology in the rhizosphere of the plant. The soil biology in this situation will provide more degradation than soil alone without the plant (Reynolds et al., 1997). This process, called rhizodegradation ⚲, enhances the speed of the natural attenuation process. In addition, plants improve degradation by modifying the oxygen status of the environment. While roots respire and use oxygen, they also remove water and allow more oxygen to diffuse into the soil from the atmosphere. Plants also physically create oxygen conduits as the roots break through the soil (Doucette, 2014). Generally, plants that can tolerate the pollutant, grow fast and produce a large amount of biomass are the best to use for this application. Since these compounds are quite easy to break down, the most important factor is introducing fast-growing, high biomass-producing plants that can tolerate growing in the contaminated soil in specific contexts (Robson, 2003).

2 To control, degrade and volatilize hydrocarbons in groundwater

Light fractions of hydrocarbons dissolve easily in water and can leach very quickly into groundwater. The flow of groundwater can then spread the contaminant into a plume, which may migrate off site. By installing masses of specific tree species that tap the groundwater, the trees can pump up the groundwater and release the water through evapotranspiration. In the process, the light hydrocarbon is pulled into the plant and degraded within the plant, or volatilized as a gas into the air (Hong et al., 2001).

Scientists have collected and measured off-gassing of leaves using gas-collection equipment placed over tree canopies, to ascertain that there is no detrimental effect on human health. Many VOCs can break down quite quickly once they enter the air (Atkinson, 1989). VOCs released by plants can combine

67

with other molecules in the air to create ozone and other compounds, so this potential outcome needs to be considered (US EPA, 2013b).

Petroleum hydrocarbons – heavier fractions, more difficult to treat: PAHs, coal tar, crude oil, heating oil.

Mechanisms utilized: Rhizodegradation .

Petroleum hydrocarbons with a log K_{ow} over 3.5 can be degraded with plant-based systems, but the process is more challenging (Lee et al., 2008). The time frame for degradation is longer and plant establishment can be difficult. These compounds tend to be ring structures consisting of three or more fused benzene rings of carbon and hydrogen. Usually, the more rings in the petroleum compound, the more difficult it is to degrade (White and Newman, 2011; Shuttleworth and Cerniglia, 1995). These complex molecules tend to stick to soil particles, especially organic matter, and they typically are not translocated into a plant (Wattiau, 2002). They have limited water solubility (Reynolds et al., 1997) and tend to persist on sites and not leach into groundwater. The contaminants are not typically degraded through natural attenuation in soils alone. The degradation instead can occur in the rhizosphere of certain plant species as a result of increased oxygen levels provided by the plant roots and root exudates released by the plant, which encourage degradation by soil microbes. The objective of phytotechnologies with these 'heavy' petroleum products is not increased uptake by the plants, but increased microbial activity (Reynolds et al., 1999) and increased oxygen and root exudates in the below-ground environment (Doucette, 2014). When microbes are at work, petroleum can be used as a carbon and energy source and the pollutant is broken down in the process (Mueller et al., 1996). It has been shown that by introducing tolerant plant species into these deposits, the plant species can bring oxygen and root exudates to the subsurface and promote the development of microbiology that can degrade some recalcitrant hydrocarbons slowly over time (Schwab and Banks, 1994).

Plants are likely to contribute to the rhizosphere degradation in two ways.

1 By creating an environment where microorganisms can thrive

The plants deliver oxygen and sugars to enhance microbial activity within the hydrocarbon deposit (Reynolds et al., 1999). In addition, the process of planting, including the addition of soil amendments, also helps microbial growth. Unlike conventional bioremediation, where microbes are introduced into the soil and injection might need to be carried out frequently, plant-based systems use the power of the sun to keep the system active. The need for repeat injections is minimized, since the plants are continually releasing root exudates into the soil. Plants are passive solar-powered systems, which is an advantage over traditional bioremediation practices (Huang et al., 2004). Plant species that will tolerate high levels of the contamination must be established. Many species will simply die in this kind of environment, but several species have been shown to establish better in highly toxic soils. In general, better growth rate and fibrous root production correlates with better degradation in the soil (Robson, 2003).

2 By creating an environment favorable for degradation

Different varieties of plants release different root exudates. The presence of these root exudates can attract specific kinds of microbes that target degradation of a particular type of compound (Pilon-Smits, 2005). Some plants even release weak acid root exudates and particular enzymes that also aid in degradation (White and Newman, 2011). For example, *Robinia pseudoaccacia*, Black Locust, exudes high amounts of flavonoids, chemicals with six carbon rings with great similarity to PAHs, into the soil. This stimulates the development of a specific rhizobial microorganism community which can use the carbon bonds of the flavonoids or the PAHs as an energy source, thereby degrading the contaminant (Gawronski et al., 2011). In past studies, a direct correlation was found between biomass production and a plant's ability to produce remediation benefits (Robson, 2003). However, for the heavier fractions of organic compounds, it is now reasonably well established that effectiveness may be more closely related to the soil environment, root-exudate profile and plant-community interactions. Therefore specific species selection is more important. Research in this area is developing, and until mechanisms are more clearly defined, species with proven, peer-reviewed degradation capabilities for PAH degradation should be utilized.

Typical petroleum-contaminated sites often have both heavier and lighter fractions of hydrocarbons. The lighter fractions of the hydrocarbons can be mobilized in groundwater, and this often happens in the early post-release years. As the site ages, the lighter fractions (i.e. aliphatic hydrocarbons) may be removed from the system with natural attenuation, leaving only the larger, recalcitrant hydrocarbon molecules (i.e. PAHs) sorbed to the soil on aged sites (Schwab and Banks, 1994).

Petroleum-contaminated soils are often tested for TPH. This is a combined measurement of the hundreds of kinds of petroleum hydrocarbons that may be found in a given sample. For both light and heavy categories of petroleum, TPH reduction rates in phytotechnology systems are often initially rapid, followed by a period of slower losses (Schwab and Banks, 1994). Initial losses may be due to chemical and physical processes, such as newly created contact with the air when planting, and volatilization (Loehr and Webster, 1996). TPH remediation rates tend to be faster in sites polluted with gasoline and diesel products, and slower in heavy, oil petroleum-polluted sites. (Hall et al., 2011; Reynolds et al., 1999).

Planting specifics

Grasses vs trees and shrubs

Because of the ability of lighter hydrocarbons to mobilize quickly, trees with long roots that can access groundwater have typically being chosen for remediation of light hydrocarbon fractions, while grasses have been more commonly used for remediation of PAHs and TPH fractions that tend to stick to soils (Cook and Hesterberg, 2013). Most of the literature to date has favored using deep-rooted grass species for degradation of the heavier PAH compounds (Kaimi

69

et al., 2007). The use of tree and shrub species for PAH degradation has had mixed success and some studies have suggested that the larger plants might compete with microbes for available nitrogen, lowering degradation (Hall et al., 2011). A recent study has compared 52 research studies using grasses with those using trees and shrubs, and finds little difference between grasses and trees with respect to the average reduction of hydrocarbons (Cook and Hesterberg, 2013). Until additional research is published and the mechanisms for removal have been clearly identified, species that have shown proven PAH degradation results in several studies should be utilized. Lists of degradation species for which results have been reported in more than one study are provided in Figure 3.5.

Fertilizer

Several studies have found that fertilizer application may speed up hydrocarbon degradation. This is likely because of the excessive carbon content in hydrocarbon-contaminated soils, which upsets the ratio to nitrogen and phosphorus necessary for proper plant and microbial growth (Hall et al., 2011 and Reynolds et al., 1997).

However, fertilizer application can also affect species succession over the long term, which can affect remediation. A study originally started by the US Army Cold Regions Research and Engineering Laboratory (Reynolds et al., 1997) and continued by University of Alaska Fairbanks looked at petroleum degradation in Alaska over 15 years on both easier-to-degrade diesel-contaminated soils and more difficult-to-degrade crude oil-contaminated soils seeded with readily available commercial grasses. Some experimental plantings were fertilized and others were not. Within one year, the diesel in plots planted with and without fertilizer was 31–76% removed. However, the more difficult-to-degrade crude oil was only 26–36%

On a petroleum phytoremediation site that was originally seeded with grasses and left to natural succession for 15 years, plots that were not originally fertilized (back right) had greater species diversity, more native and woody plants and more diverse microbial communities than plots where fertilizer had been originally added (front left) 15 years prior.

Figure 3.4 Case Study: Fairbanks Alaska, US Army Cold Weather Research Site

removed with plants and 36–40% removed with plants and fertilizer. In summary, in this situation, fertilizer did speed up the degradation of the harder-to-degrade hydrocarbons. The plots were then left for 15 years. No additional inputs were added and natural succession took over. When the plots were resurveyed after 15 years, the contamination had reached cleanup levels on all sites, meaning that the petroleum levels had dropped below the cleanup limit set by the Alaska Department of Environmental Conservation, and the grasses originally planted on the plots were no long present. However, new species had moved in and it was found that more native plants and woody vegetation were present in the non-fertilized plots. In addition, TPH levels were lowest on plots with higher numbers of woody plants. The results indicate that the addition of plants and fertilizer sent the plots on different successional trajectories, resulting in different plant community assemblages and associated microbial communities (Leigh, 2014; Leewis et al., 2013). This has a significant implication for landscape architects, illustrating that not only plant but initial soil and fertilizer inputs specified can greatly impact long-term succession.

Legumes

The addition of legumes to phytotechnology systems can also be considered to help in the degradation of hydrocarbons. The efficacy of many grass species for petroleum remediation is widely proven and the majority of studies have used them as a candidate because of their fast growth and dense fibrous root systems (Kaimi et al., 2007). It has been recently proposed, however, that legumes may add an important component to the degradation of more persistent PAH compounds. Legumes can add nitrogen in these deficient soils and their tap-root systems have also been shown to provide increased pore space in the soil structure, allowing for more oxygen, the presence of which is critical for hydrocarbon degradation (Hall et al., 2011).

71

Planting types

For petroleum contamination, the following planting typologies (described in detail in Chapter 4) can be considered in conjunction with the plant lists provided in Figure 3.5.

In soil: Natural attenuation will often degrade lighter fractions of hydrocarbons on their own, as long as enough oxygen and carbon is present in the soils. However, adding plants can speed up this process. For these lighter fractions, any introduction of vegetation has been shown to have a beneficial effect. Species does not matter as much as the growth rate and biomass production; the greater these are, the faster the remediation (Robson, 2003). However, for the heavier fractions, the pollutant tends to be bound to soil. With the introduction of plants, they can often be slowly degraded over time. The most challenging concern is to find plant species that will both tolerate the pollutant concentrations and also degrade the contaminant. Several studies have shown that certain plant species can survive in very high concentrations of petroleum and the plant roots will penetrate from clean soils into contaminated soils, even if the plants were started in unpolluted soil media (Rogers et al., 1996; Reynolds et al., 1999). The following planting typologies can be utilized in conjunction with the plant lists in Figure 3.5.

Degradation typologies for petroleum in soil: Time frame 0–5 years for lighter fractions, 5–20+ years for heavier fractions.

- Degradation Cover: Chapter 4, see p. 222
- Degradation Hedge: Chapter 4, see p. 220
- Degradation Living Fence: Chapter 4, see p. 220
- Degradation Bosque: Chapter 4, see p. 218

In groundwater: Since lighter fractions of petroleum dissolve in water, they can quickly move into the groundwater and migrate off site. To stop the contamination from migrating and to degrade or volatilize it, the following planting typologies can be used in conjunction with the plant lists in Figure 3.5.

Typologies for petroleum in groundwater: Time frame 3–10 years or more (highly dependent on how contaminated the groundwater is, flowrate, depth and volume of plume).

- Interception Hedgerow: Chapter 4, see p. 216
- Groundwater Migration Tree Stand: Chapter 4, see p. 213
- Phytoirrigation: Chapter 4, see p. 207

In stormwater and wastewater: Hydrocarbons are very common contaminants in run-off from roads and impervious surfaces, as well as in industrial wastewater. When hydrocarbons have a log K_{ow} between 0.5 and 3.5, that is an indicator that they may easily be degraded in Stormwater Filters and Constructed Wetlands. As long as there are plants in the system that are thriving, the degradation should occur. However, plants that grow faster and produce more biomass will usually create better degradation systems (Robson, 2003). Hydrocarbons with log K_{ow} values >3.5 don't often mobilize into water. They tend to stay on site, bound to soil. However, in stormwater systems, PAHs can be captured when particulate matter sloughs off from roadway surfaces (also made of petroleum) and the stormwater carries away particulates. These are best treated by first settling the particulate matter out of the water through engineering practices such as sedimentation basins and tanks, and then treating the sediment with species that target the degradation of recalcitrant hydrocarbons. In addition, biofilter and wetland systems with greater diversity of plant species tend to perform better than monocultures, likely due to the increased diversity of associated microbes (Coleman et al., 2001).

The following planting typologies have been effectively utilized. Species for these systems have not been included in this book, since they have been widely documented elsewhere.

Stormwater and wastewater typologies: From the time a pollutant enters the system, treatment may be possible in 0–5 years for lighter fractions, 5–20+ years for heavier fractions.

- Stormwater Filters: Chapter 4, see p. 235
- Surface-Flow Constructed Wetland: Chapter 4, see p. 238
- Subsurface Gravel Wetland: Chapter 4, see p. 241

The plant list in Figure 3.5 is an initial list of species that have been shown in more than one study to be useful for degrading hydrocarbons.

Plant species with maximum hydrocarbon spill tolerance

In 2003, the global energy company BP completed forward-thinking research considering which species could be planted around gas stations to tolerate spills and potentially provide remediation benefits (Tsao and Tsao, 2003; Fiorenza (BP) and Thomas (Phytofarms), 2004). A series of experiments were conducted to test which typical US ornamental horticulture species might have the best tolerance to straight doses of gasoline. Although scientific studies document many species for gasoline remediation (see the plant list in Figure 3.5), the authors found that many of the plants may not be suitable as landscape plants for aesthetic and maintenance reasons. In addition, the documented phytoremediation plants tend to be grasses and trees, while the dominant retail landscape plant type is a shrub. The landscape plants tested were identified from plans developed by local landscape architects designing BP retail stations in the US. Potted plants were dosed with gasoline, and tolerance was used as an indicator of the plant's potential ability for phytoremediation. There were only limited attempts to identify degradation mechanisms and characterize the fate and transport of the added gasoline in this study. However, this important research, provided by Dr. David Tsao, is an initial list of ornamental plants for phytoscaping (term coined by D. Tsao) with hydrocarbon tolerance that may also prove to be successful degradation species in the future. Out of a total of 113 species tested, 53 demonstrated some tolerance (Figure 3.6). Intolerant plants (Figure 3.7) can also be considered as indicator species, i.e. they could be planted in high-risk areas and help identify leaks and spills early after their occurrence. Since gasoline causes toxic responses in these species, plant decline and death could signify a leak.

73

Figure 3.5 Petroleum Degradation Plant List

Latin	Common	Petroleum Category Targeted	Contaminant	Vegetation Type	USDA Hardiness Zone	Native to	Reference
Acer platanoides	Norway Maple	Easy	BTEX	Tree	3–7	Europe	Cook and Hesterberg, 2012 Fagiolo and Ferro, 2004
Agropyron cristatum	Crested Wheatgrass	Both	TPH	Herbaceous	3+	Asia	Cook and Hesterberg, 2012 Muratova et al., 2008
Alnus glutinosa	Black Alder	Both	MOH	Tree/Shrub	3–7	Europe, Africa	Tischer and Hubner, 2002
Andropogon gerardii	Big Bluestem	Hard	PAH	Herbaceous	4–9	North America	Aprill and Sims, 1990 Balcom and Crowley, 2009 Cook and Hesterberg, 2012 Euliss, 2004 Olson et al., 2007 Rugh, 2006
Avena sativa	Oat	Both	TPH	Herbaceous	5–10	Europe	Cook and Hesterberg, 2012 Muratova et al., 2008
Axonopus compressus	Carpet Grass	Both	TPH	Herbaceous	7–10	North America, South America	Efe and Okpali, 2012
Betula pendula	European White Birch	Hard	PAH	Tree	3–6	Europe	Cook and Hesterberg, 2012 Rezek et al., 2009
Bouteloua curtipendula	Side Oats Grass	Both	TPH PAH	Herbaceous	3–9	North America, South America	Aprill and Sims, 1990 Cook and Hesterberg, 2012
Bouteloua dactyloides	Buffalo Grass	Both	PAH TPH	Herbaceous	3–9	North America	McCutcheon and Schnoor, 2003 Qiu et al., 1997
Bouteloua gracilis	Blue Grama	Hard	PAH	Herbaceous	3–9	North America	Aprill and Sims, 1990 Cook and Hesterberg, 2012
Brachiaria decumbens	Signal Grass	Both	TPH	Herbaceous	Not available	Africa	Cook and Hesterberg, 2012 Gaskin and Bentham, 2010
Brachiaria serrata	Velvet Signal Grass	Both	TPH	Herbaceous	Not available	Africa	Maila and Randima, 2005
Brassica juncea	Indian Mustard	Both	PAH	Herbaceous	Grown as annual	Asia, Europe, Africa	Roy et al., 2005
Bromus inermis	Smooth Brome	Both	TPH	Herbaceous	3–9	Europe, Asia	Cook and Hesterberg, 2012 Muratova et al., 2008
Canna × generalis	Canna	Easy	BTEX	Herbaceous	8–12	Central and South America, Southern USA	Boonsaner et al., 2011

Latin	Common	Petroleum Category Targeted	Contaminant	Vegetation Type	USDA Hardiness Zone	Native to	Reference
Carex cephalophora	Ovalhead Sedge	Hard	PAH	Herbaceous	3–8	Eastern USA	Cook and Hesterberg, 2012 Euliss, 2004
Carex stricta	Sedge	Both	TPH	Herbaceous	5–8	North America	Euliss et al., 2008
Celtis occidentalis	Hackberry	Both	BTEX TPH PAH	Tree	2–9	North America	Cook and Hesterberg, 2012 Fagiolo and Ferro, 2004 Kulakow, 2006b
Cercis canadansis	Eastern Redbud	Hard	PAH	Herbaceous	4–9	North America	Ferro et al., 1999
Chrysopogon zizanioides	Vetiver Grass	Hard	PAH	Tree/ Herbaceous	9–11	India	Cook and Hesterberg, 2012 Paquin et al., 2002
Conocarpus lancifolius	Axlewood	Both	TPH	Tree	Not available	Africa	Cook and Hesterberg, 2012 Yateem et al., 2008
Cordia subcordata	Kou	Both	TPH PAH	Tree	Not available	Hawaii, Pacific, Africa	Tang et al., 2004
Cymbopogon citrullus	Lemon-Scented Grass	Both	TPH	Herbaceous	10–11	India	Cook and Hesterberg, 2012 Gaskin and Bentham, 2010
Cynodon dactylon	Bermuda Grass	Both	Fluoranthene Phenanthrene Pyrene TPH PAH	Herbaceous	7–10	Africa	Banks, 2006 Banks and Schwab, 1998 Cook and Hesterberg, 2012 Flathman and Lanza, 1999 Hutchinson et al., 2001 Kulakow, 2006e Olson and Fletcher, 2000 White et al., 2006
Cyperus brevifolius	Sedge rottb.	Both	TPH	Herbaceous	8+	Australia	Basumatary et al., 2013
Cyperus rotundus	Purple Nutsedge	Both	TPH	Herbaceous	8+	India	Basumatary et al., 2013 Efe and Okpali, 2012
Dactylis glomerata	Orchardgrass	Both	TPH PAH	Herbaceous	3+	Europe	Cook and Hesterberg, 2012 Kulakow, 2006b
Eleusine coracana	African Millet	Both	TPH	Herbaceous	Grown as annual	Africa	Maila and Randima, 2005
Elymus canadensis	Canada Wild-Rye	Hard	TPH PAH	Herbaceous	3–9	North America	Aprill and Sims, 1990 Cook and Hesterberg, 2012
Elymus hystrix	Bottlebrush Grass	Hard	PAH	Herbaceous	4–9	North America	Cook and Hesterberg, 2012 Rugh, 2006
Elytrigia repens	Couch Grass	Both	TPH	Herbaceous	3–9	Europe, Asia	Cook and Hesterberg, 2012 Muratova et al., 2008

Figure 3.5 *(continued)*

Latin	Common	Petroleum Category Targeted	Contaminant	Vegetation Type	USDA Hardiness Zone	Native to	Reference
Eucalyptus spp.	Eucalyptus	Easy	BTEX	Varies	Varies	Australia	Coltrain, 2004 Cook and Hesterberg, 2012
Fabaceae	Legumes	Both	TPH PAH	Varies	Varies	Worldwide	Cook and Hesterberg, 2012 Kulakow, 2006a Kulakow, 2006b Kulakow, 2006e Liu et al., 2010 Tsao, 2006a Tsao, 2006b
Festuca spp.	Fescue	Both	TPH PAH BTEX	Herbaceous	Varies	Worldwide	Banks, 2006 Cook and Hesterberg, 2012 Kulakow, 2006b Kulakow, 2006e Tsao, 2006a Tsao, 2006b White et al., 2006
Festuca arundinacea	Tall Fescue	Both	Anthracene Ethylene glycol Fluoranthene Phenanthrene Pyrene TPH PAH PAE	Herbaceous	3–8	Europe	Banks and Schwab, 1998 Batty and Anslow, 2008 Chen and Banks, 2004 Cook and Hesterberg, 2012 Flathman and Lanza, 1998 Hutchinson et al., 2001 ITRC PHYTO 3 Karthikeyen et al., 2012 Kulakow, 2006d Liu et al., 2010 Ma et al., 2013 Olson et al., 2007 Parrish et al., 2004 Reilley et al., 1996 Reilley et al., 1993 Rice et al., 1996a Robinson et al., 2003 Roy et al., 2005 Schwab and Banks, 1994 Siciliano et al., 2003 Sun et al., 2011

76

Latin	Common	Petroleum Category Targeted	Contaminant	Vegetation Type	USDA Hardiness Zone	Native to	Reference
Festuca pratensis	Meadow Fescue	Both	TPH	Herbaceous	3–9	Europe, Asia	Cook and Hesterberg, 2012 Muratova et al., 2008
Festuca rubra	Red Fescue	Both	TPH PAH	Herbaceous	4–10	North America, Europe	Cook and Hesterberg, 2012 Kulakow, 2006c Palmroth et al., 2006
Ficus infectoria	Wavy Leaf Fig Tree	Both	TPH	Tree	Not available	India	Cook and Hesterberg, 2012 Yateem et al., 2008
Fraxinus pennsylvanica	Green Ash	Hard	PAH	Tree	2–9	Eastern USA	Cook and Hesterberg, 2012 Spriggs et al., 2005
Geranium viscosissimum	Sticky Geranium	Hard	PAH	Herbaceous	2+	Western North America	Olson et al., 2007
Gleditsia triacanthos	Honey Locust	Easy	BTEX	Tree	3–9	North America	Cook and Hesterberg, 2012 Fagiolo and Ferro, 2004
Helianthus annuus	Sunflower	Hard	PAH	Herbaceous	Grown as annual	North America, South America	Cook and Hesterberg, 2012 Euliss, 2004
Hibiscus tiliaceus	Dwarf Hau	Hard	PAH	Tree/ Herbaceous	10–12	Australia	Cook and Hesterberg, 2012 Paquin et al., 2002
Hordeum vulgare	Barley	Both	TPH Pyrene	Herbaceous	Grown as annual	Asia, North Africa	Cook and Hesterberg, 2012 Muratova et al., 2008 White and Newman, 2011
Juncus effusus	Common Rush	Hard	PAH	Wetland	2–9	Worldwide	Cook and Hesterberg, 2012 Euliss, 2004
Juniperus virginiana	Eastern Red Cedar	Easy	BTEX	Tree	2–9	Eastern USA	Cook and Hesterberg, 2012 Fagiolo and Ferro, 2004
Kochia scoparia	Burningbush	Both	TPH	Herbaceous	9–11	Europe, Asia	Zand et al. 2010
Leymus angustus	Altai Wildrye	Both	TPH	Herbaceous	2–8	Europe, Asia	Cook and Hesterberg, 2012 Phillips et al., 2009
Linum usitatissumum L.	Flax	Both	TPH	Herbaceous	4–11	Europe, Asia	Zand et al. 2010

77

Figure 3.5 (continued)

Latin	Common	Petroleum Category Targeted	Contaminant	Vegetation Type	USDA Hardiness Zone	Native to	Reference
Lolium multiflorum	Annual Rye	Both	TPH PAH	Herbaceous	5+	Europe	Cook and Hesterberg, 2012 Flathman and Lanza, 1998 ITRC PHYTO 3 Lalande et al., 2003 Parrish et al., 2004
Lolium perenne	Herbaceous Ryegrass	Both	Acenaphthene Benzo(a)anthracene Benzo(a)pyrene Benzo(b)fluoranthene Benzo(ghi)perylene, Benzo(k)fluoranthene Chrysene Dibenzo(ah)anthracene Fluoranthene Indeno(123cd)pyrene Naphthalene Pyrene TPH PAH BTEX PAE	Herbaceous	3-9	Europe, Asia	Binet et al., 2000 Cook and Hesterberg, 2012 Ferro et al., 1999 Ferro et al., 1997 Fu et al., 2012 Gunther et al., 1996 ITRC PHYTO 3 Johnson et al., 2005 Kulakow, 2006a Kulakow, 2006c Ma et al., 2013 Olson et al., 2007 Palmroth et al., 2006 Reynolds, 2006a Reynolds, 2006b Reynolds, 2006c Reynolds, 2006d Reynolds, 2006e Rezek et al., 2009 Yateem, 2013
Lolium spp.	Ryegrass	Both	TPH PAH	Herbaceous	Varies	Europe, Asia, North Africa	Banks, 2006 Kulakow, 2006e Muratova et al., 2008 Nedunuri et al., 2000 Tsao, 2006a Tsao, 2006b White et al., 2006
Lotus corniculatus	Birdsfoot Trefoil	Both	TPH PAH	Herbaceous	5+	Europe	Karthikeyen et al., 2012 (5) Smith et al., 2006

Latin	Common	Petroleum Category Targeted	Contaminant	Vegetation Type	USDA Hardiness Zone	Native to	Reference
Medicago sativa *Medicago sativa Mesa var. Cimarron VR*	Alfalfa	Both	Anthracene Ethylene glycol MTBE Phenol PAH (total priority) Pyrene Toluene TPH PAH Benzene PAE	Herbaceous	3–11	Middle East	Cook and Hesterberg, 2012 Davis et al., 1994 Ferro et al., 1997 ITRC PHYTO 3 Komisar and Park, 1997 Liu et al., 2010 Ma et al., 2013 Muralidharan et al., 1993 Muratova et al., 2008 Phillips et al., 2009 Pradhan et al., 1998 Reilley, Banks and Schwab, 1993 Rice et al., 1996a Schwab and Banks, 1994 Sun et al., 2011 Tossell, 2006 Tsao, 2006a Tsao, 2006b Yateem, 2013
Melilotus officinalis	Sweet Clover	Both	TPH PAH	Herbaceous	4–8	Europe, Asia	Cook and Hesterberg, 2012 Karthikeyen et al., 2012 Kulakow, 2006d
Microlaena stipoides	Weeping Grass	Both	TPH	Herbaceous	9–11	Australia	Cook and Hesterberg, 2012 Gaskin and Bentham, 2010
Miscanthus × giganteus	Giant Maiden Grass	Hard	PAH	Herbaceous	4–9	Japan	Techer et al., 2012
Morus alba	White Mulberry	Hard	PAH	Tree	3–9	China	Cook and Hesterberg, 2012 Euliss, 2004
Morus rubra	Red Mulberry	Hard	PAH	Tree	5–10	Eastern USA	Cook and Hesterberg, 2012 Euliss, 2004 Rezek et al., 2009
Myoporum sandwicense	False Sandalwood	Both	TPH PAH	Tree	10–11	Hawaii	Tang et al., 2004
Onobrychis viciifolia	Sainfoin	Both	TPH	Herbaceous	3–10	Europe	Cook and Hesterberg, 2012 Muratova et al., 2008

79

Figure 3.5 *(continued)*

Latin	Common	Petroleum Category Targeted	Contaminant	Vegetation Type	USDA Hardiness Zone	Native to	Reference
Panicum coloratum	Klinegrass	Hard	PAH	Herbaceous	10–12	Africa	Balcom and Crowley, 2009 Olson et al., 2007 Qiu et al., 1994, 1997
Panicum virgatum	Switchgrass	Both	Anthracene PAH (total priority) Pyrene TPH PAH	Herbaceous	2–9	North America	Aprill and Sims, 1990 Cook and Hesterberg, 2012 Euliss et al., 2008 Kulakow, 2006d Pradhan et al., 1998 Reilley et al., 1996 Reilley et al., 1993 Schwab and Banks, 1994 Wilste et al., 1998
Pascopyrum smithii (syn. *Agropyron smithii*)	Western Wheatgrass	Both	TPH PAH	Herbaceous	4–9	North America	Aprill and Sims, 1990 Cook and Hesterberg, 2012 Karthikeyen et al., 2012 Kulakow, 2006d Olson et al., 2007
Paulownia tomentosa	Empress Tree Princess Tree	Both	PAH	Tree	7–10	China	Macci et al., 2012
Pennisetum glaucum	Millet	Both	TPH	Herbaceous	Grown as annual	Africa, Asia	Cook and Hesterberg, 2012 Muratova et al., 2008
Phalaris arundinacea	Reed Canary Grass	Hard	PAH	Herbaceous	4–9	Europe	McCutcheon and Schnoor, 2003
Phragmites australis	Common Reed	Both	Benzene Biphenyl Ethylbenzene Toluene p-Xylene TPH MTBE	Wetland	4–10	Europe, Asia	Anderson et al., 1993 Reiche and Borsdorf, 2010 Ribeiro et al., 2013 Unterbrunner et al., 2007
Picea glauca var. densata	Black Hills Spruce	Easy	BTEX	Tree	2–6	North Dakota	Cook and Hesterberg, 2012 Fagiolo and Ferro, 2004
Pinus banksiana	Jack Pine	Easy	BTEX	Tree	3–8	North America	Cook and Hesterberg, 2012 Fagiolo and Ferro, 2004
Pinus spp.	Conifers	Both	MTBE TBA (Tert-butyl Alcohol)	Tree	Varies	Worldwide	Arnold et al., 2007

Latin	Common	Petroleum Category Targeted	Contaminant	Vegetation Type	USDA Hardiness Zone	Native to	Reference
Pinus sylvestris	Scots Pine	Both	TPH	Tree	3–8	Europe, Asia	Cook and Hesterberg, 2012 Palmroth et al., 2006
Pinus taeda	Loblolly Pine	Both	Dioxene, BTEX, TPH	Tree	6–9	North America	Ferro et al., 2013 Guthrie Nichols et al., 2014
Pinus thunburgii	Japanese Pine	Both	Dioxene	Tree	5–10	Japan	Ferro et al., 2013
Pinus virginiana	Virginia Pine	Both	Dioxene	Tree	5–8	North America	Ferro et al., 2013
Poa pratensis	Kentucky Bluegrass	Both	TPH PAH	Herbaceous	3–8	Europe	Kulakow, 2006c Palmroth et al., 2006
Poaceae	Grasses	Both	TPH PAH BTEX	Herbaceous	Varies	Worldwide	Luce, 2006 Tossell, 2006
Populus nigra var. italica	Black Poplar Lombardy Poplar	Both	PAH	Tree	4–9	Italy	Macci et al., 2012
Populus spp. Populus deltoides Populus deltoides × Populus nigra Populus deltoides × nigra DN34 Populus trichocarpa × deltoides 'Hoogvorst' Populus trichocarpa × deltoides 'Hazendans'	Poplar species and hybrids	Both	Aniline Benzene Ethylbenzene Phenol Toluene m-Xylene PAH BTEX MTBE DRO TPH	Tree	varies	varies	Applied Natural Sciences, Inc., 1997 Barac et al., 2009 Burken and Schnoor, 1997a Coltrain, 2004 Cook et al., 2010 Cook and Hesterberg, 2012 El-Gendy et al., 2009 Euliss et al., 2008 Euliss, 2004 Fagiolo and Ferro, 2004 Ferro et al., 2013 Ferro, 2006 ITRC PHYTO 3 Kulakow, 2006b Kulakow, 2006 Luce, 2006 Ma et al., 2004 Olderbak and Erickson, 2004 Palmroth et al., 2006 Spriggs et al., 2005 Tossell, 2006 Unterbrunner et al., 2007 Weishaar et al., 2009 Widdowson et al., 2005

81

82

Figure 3.5 (continued)

Latin	Common	Petroleum Category Targeted	Contaminant	Vegetation Type	USDA Hardiness Zone	Native to	Reference
Quercus macrocarpa	Bur Oak	Easy	BTEX	Tree	3–8	North America	Cook and Hesterberg, 2012 Fagiolo and Ferro, 2004
Quercus phellos	Willow Oak	Easy	Dioxin	Tree	6–9	North America	Ferro et al., 2013
Robinia pseudoacacia	Black Locust	Both	PAH MOH	Tree	4–9	North America	Gawronski et al., 2011 Tischer and Hubner, 2002
Sagittaria latifolia	Arrowhead	Both	TPH	Herbaceous	5–10	North America, South America	Cook and Hesterberg, 2012 Euliss et al., 2008
Salix alaxensis	Felt-Leaf Willow	Both	TPH	Tree/Shrub	2–8	Alaska, Canada	Cook and Hesterberg, 2012 Soderlund, 2006
Salix alba	White Willow	Easy	BTEX	Tree	2–8	Europe, Asia	Cook and Hesterberg, 2012 Fagiolo and Ferro, 2004 Ferro et al., 2013
Salix babylonica L.	Weeping Willow	Both	MTBE TBA (Tert-butyl Alcohol)	Tree	6–9	China	Yu and Gu, 2006
Salix nigra	Black Willow	Both	PAH, BTEX, TPH	Tree/Shrub	2–8	Eastern USA	Spriggs et al., 2005 Guthrie Nichols et al., 2014
Salix spp. Salix interior Salix exigua	Willow	Both	DRO TPH BTEX PAH	Tree/Shrub	Varies	Worldwide	Applied Natural Sciences, Inc., 1997 Carman et al., 1997, 1998 Coltrain, 2004 Cook et al., 2010 Cook and Hesterberg, 2012 Euliss et al., 2008 ITRC PHYTO 3 Kulakow, 2006b Kulakow, 2006c
Salix viminalis	Basket Willow	Hard	PAH	Shrub	4–10	Europe, Asia	Cook and Hesterberg, 2012 Hultgren et al., 2010 Hultgren et al., 2009 Roy et al., 2005
Schizachyrium scoparium	Little Bluestem	Hard	PAH	Herbaceous	2–7	Eastern USA	Aprill and Sims, 1990 Cook and Hesterberg, 2012 Pradhan et al., 1998 Rugh, 2006

Latin	Common	Petroleum Category Targeted	Contaminant	Vegetation Type	USDA Hardiness Zone	Native to	Reference
Schoenoplectus lacustris	Bulrush	Both	Phenol	Wetland	5–10	North America	ITRC PHYTO 3 Kadlec and Knight, 1996
Scirpus atrovirens	Green Bulrush	Hard	PAH	Wetland	4–8	North America	Thomas et al., 2012
Scirpus maritimus	Alkali Bulrush	Both	TPH	Wetland		North America	Couto et al., 2012
Scirpus spp.	Bulrush	Both	Phenol Biological oxygen demand Chemical oxygen demand Oil and gasoline Phenol Total suspended solids	Wetland	Varies	Worldwide	ITRC PHYTO 3 Kadlec and Knight, 1996
Secale cereale	Winter Rye	Both	Pyrene TPH PAH	Herbaceous	3+	Asia	Cook and Hesterberg, 2012 ITRC PHYTO 3 Kulakow, 2006a Kulakow, 2006b Muratova et al., 2008 Reynolds et al., 1998
Senna obtusifolia	Coffee Weed	Hard	PAH	Herbaceous	7+	North America, South America	Cook and Hesterberg, 2012 Euliss, 2004
Solidago spp.	Goldenrod	Both	TPH PAH	Herbaceous	Varies	North and South America, Europe, Asia	Cook and Hesterberg, 2012 Kulakow, 2006b
Sorghastrum nutans	Indiangrass	Both	TPH PAH	Herbaceous	2–9	North America	Aprill and Sims, 1990 Cook and Hesterberg, 2012
Sorghum bicolor Sorghum bicolor subsp. Drummondii	Sorghum	Both	TPH Anthracene Pyrene	Herbaceous	8+	Africa	Cook and Hesterberg, 2012 Flathman and Lanza, 1998 ITRC PHYTO 3 Liu et al. 2010 Muratova et al., 2008 Nedunuri et al., 2000 Reilley et al., 1996 Reilley et al., 1993 Schwab and Banks, 1994
Sorghum vulgare	Sudan Grass	Hard	PAH	Herbaceous	8+	Africa	Reilley et al., 1996

84

Figure 3.5 *(continued)*

Latin	Common	Petroleum Category Targeted	Contaminant	Vegetation Type	USDA Hardiness Zone	Native to	Reference
Spartina pectinata	Prairie Cordgrass	Hard	PAH	Herbaceous	5+	North America	Cook and Hesterberg, 2012 Rugh, 2006
Stenotaphrum secundatum	St. Augustine Grass	Both	TPH PAH	Herbaceous	8–10	North America, South America	Cook and Hesterberg, 2012 Flathman and Lanza, 1998 ITRC PHYTO 3 Nedunuri et al., 2000
Thespesia populnea	Milo	Both	TPH PAH	Tree	8–10	Hawaii	Tang et al., 2004
Thinopyrum ponticum	Tall Wheatgrass	Both	TPH	Herbaceous	3–8	Mediterranean, Asia	Cook and Hesterberg, 2012 Phillips et al., 2009
Trifolium hirtum	Rose Clover	Both	TPH	Herbaceous	8–11	Europe, Asia	Cook and Hesterberg, 2012 Siciliano et al., 2003
Trifolium pratense	Red Clover	Both	TPH	Herbaceous	3+	Europe	Cook and Hesterberg, 2012 Karthikeyen et al., 2012 Muratova et al., 2008
Trifolium repens	White Clover	Both	Fluoranthene Phenanthrene Pyrene TPH PAH	Herbaceous	3+	Europe	Banks and Schwab, 1998 Cook and Hesterberg, 2012 Flathman and Lanza, 1998 ITRC PHYTO 3 Johnson et al., 2005 Kulakow, 2006c
Trifolium spp.	Clover	Both	TPH PAH BTEX	Herbaceous	Varies	Worldwide	Banks, 2006 Cook and Hesterberg, 2012 Parrish et al., 2004 Reynolds, 2006a Reynolds, 2006b Reynolds, 2006c Reynolds, 2006d Reynolds, 2006e
Triglochin striata	Three-Rib Arrowgrass	Both	TPH	Wetland	5–9	North America, Europe	Ribeiro et al., 2013

Latin	Common	Petroleum Category Targeted	Contaminant	Vegetation Type	USDA Hardiness Zone	Native to	Reference
Tripsacum dactyloides	Eastern Gamagrass	Both	TPH PAH	Herbaceous	4–9	Eastern USA	Cook and Hesterberg, 2012 Euliss, 2004 Euliss et al., 2008
Triticum spp.	Wheat	Both	TPH	Herbaceous	Varies	Asia	Muratova et al., 2008
Typha spp.	Cattail	Both	DRO Oil and gasoline Phenol Total suspended solids Biological oxygen demand Chemical oxygen demand	Wetland	3–10	North America, Europe, Asia	ITRC PHYTO 3 Kadlec and Knight, 1996 Kadlec and Knight, 1998
Ulmus parvifolia	Chinese Elm	Both	Dioxene	Tree	5–9	Asia	Ferro et al., 2013
Vetiveria zizanioides	Vetiver Grass	Both	TPH	Herbaceous	8b-10	India	Danh et al., 2009
Vicia faba	Broad Bean	Both	TPH	Herbaceous	Grown as annual	Africa, Asia	Radwan et al., 2005 Yateem, 2013
Vulpia microstachys (Nutt.) Munro	Small Fescue	Both	TPH PAH	Herbaceous	Not available	Western USA	Cook and Hesterberg, 2012 Kulakow, 2006a
Zea mays	Corn	Both	TPH	Herbaceous	Grown as annual	North America, Central America	Cook and Hesterberg, 2012 Muratova et al., 2008

Figure 3.6 Petroleum-Tolerant Plants from BP Study

Latin	Common	Variety evaluated	Vegetation Type	USDA Hardiness Zone
Agapanthus africanus	Lily-of-the-Nile		Perennial	8–11
Arbutus unedo 'compacta'	Compact Strawberry Bush		Shrub	7–9
Bulbine frutescens	Snake Flower	Orange	Groundcover	8–11
Bulbine frutescens	Snake Flower	Yellow	Groundcover	8–11
Cassia corymbosa	Senna		Shrub	8–11
Cercis canadensis	Eastern Redbud	Oklahoma	Tree	4–9
Cistus × purpureus	Purple Rock Rose	Firescaping Plant	Shrub	8–11
Clytostoma callistegioides	Lavender Trumpet Vine		Vine	8–11
Dietes irioides	Fortnight lily	(Moraea) bicolor	Shrub	8–11
Euonymus coloratus	Purple Leaf Wintercreeper		Groundcover	4–9
Ficus pumila	Creeping/Climbing Fig		Vine	9–11
Fraxinus pennsylvanica	Green Ash	Patmore	Tree	2–9
Hedera helix	English Ivy		Groundcover	5–9
Hemerocallis hybrid	Daylily, dwarf yellow	Happy Returns	Perennial	3–10
Hemerocallis hybrid	Daylily	Scarlet Orbit	Perennial	3–10
Ilex cornuta	Dwarf Burford Holly	Burfordii Nana	Shrub	7–9
Ilex cornuta	Carisa Holly	Carisa	Shrub	7–9
Ilex vomitoria	Yaupon Holly	Nana	Shrub	7–9
Juniperus procumbens	Juniper	Green Mound	Shrub	4–9
Lagerstroemia indica	Dwarf Crape Myrtle	Tightwad Red	Shrub	7–9
Lantana montevidensis	Creeping Lantana	New Gold	Groundcover	8–10
Ligustrum japonicum	Waxleaf Ligustrum	Texana	Shrub	7–10
Liriope muscari	Aztec Grass	Ophiopogon	Groundcover	6–10
Liriope muscari	Lily Turf	Giant	Groundcover	6–10
Macfadyena unguis-cati	Yellow Trumpet Vine		Vine	9–11
Millettia reticulata	Evergreen Wisteria		Vine	8+
Moraea bicolor	Fortnight Lily		Shrub	8–11
Moraea iridioides (D.iridioides)	African Iris		Shrub	8–11
Nandina domestica	Heavenly Bamboo	Jaytee Harbor Belle	Shrub	6–10
Nerium oleander	Oleander		Shrub	9–11
Phormium tenax	New Zealand Flax	Wings of Gold	Shrub	8–11
Photinia fraseri	Red Tip Photinia		Shrub	7–9
Picea pungens	Dwarf Globe Blue Spruce		Shrub	2–8
Pinus mugo pumilo	Dwarf Mugo Pine		Shrub	2–8
Pistacia chinensis	Chinese Pistachio		Tree	6–9
Pittosporum tobira	Pittosporum	Variegata	Shrub	8–10
Podranea riscasoliana	Pink Trumpet Vine	Sprague	Vine	9+
Pyrus calleryana	Bradford Flowering Pear	Holmford	Tree	5–9
Raphiolepis indica	Indian Hawthorne	Snow	Shrub	8–11
Rumohra adiantiformis	Leather Leaf Fern		Perennial	9–11
Sabel minor	Palmetto Bush		Shrub	7–11
Sedum mexicana	Sedum		Groundcover	7–10
Spiraea spp.	Neon Flash		Shrub	4–9

Latin	Common	Variety evaluated	Vegetation Type	USDA Hardiness Zone
Strelitzia reginae	Bird of Paradise		Shrub	9+
Tecomaria capensis	Cape Honeysuckle	Orange	Shrub	9–11
Thuja occidentalis	Rheingold Arborvitae		Shrub	2–7
Trachelospermum asiaticum	Asian Jasmine		Groundcover	7–11
Tulbaghia violacea	Society Garlic		Perennial	7–10
Veronica spicata	Spiked Speedwell	Sunny Blue Border	Perennial	4–9
Viburnum obovatum dentata	Compact Walter's Viburnum		Shrub	6–9
Viburnum odoratissimum	Sweet Viburnum	Spring Bouquet	Shrub	8–10
Washingtonia filifera	California Fan Palm		Tree	8+
Yucca hesperaloe parvifolia	Red Yucca		Shrub	5–11
Yucca recurvifolia	Yucca	Soft Leaf Yucca	Shrub	7–9

Source: Tsao and Tsao, 2003

Figure 3.7 Petroleum-Intolerant Plants from BP Study

Latin	Common	Variety Evaluated	Vegetation Type	USDA Hardiness Zone
Abelia × grandiflora	Glossy Abelia		Shrub	5–9
Abelia mosanensis	Bridal Bouquet Abelia	Monia	Shrub	5–9
Abutilon hybridum	Pink Flowering Maple	Roseus	Shrub	9–10
Acer rubrum	Red Sunset Maple	Frank's Red	Tree	3–9
Arecastrum romanzoffianum	Queen Palm		Tree	9–11
Artemisia spp.	Wormwood	Powis Castle	Shrub	4+
Asparagus densiflorus	Asparagus Fern	Sprengeri	Perennial	9–11
Aspidistra elatior	Cast Iron Plant		Perennial	7–11
Berberis thunbergii	Barberry	Crimson Pygmy	Shrub	4–9
Bougainvillea cvs.	Bougainvillea		Vine	9+
Buxus microphylla	Boxwood	Winter Gem	Shrub	6–10
Carissa macrocarpa	Natal Plum	Fancy	Shrub	9–11
Cassia splendida	Senna	Golden Wonder	Tree	9–11
Cedrus deodara	Golden Deodar Cedar	Aurea	Tree	7–9
Chamaecyparis pisifera filifera	Dwarf Gold Thread Cypress		Shrub	4–8
Cornus kousa chinensis	Chinese Dogwood		Tree	5–8
Cotoneaster apiculatus	Cranberry Cotoneaster		Shrub	4–7
Cuphea hyssophyla	Mexican Heather		Perennial	8–11
Cycas revoluta	King Sago Palm		Tree	8–11
Delosperma cooperi	Yellow Ice Plant	Aurea	Groundcover	5–11
Distictis buccinatoria	Scarlet Trumpet Vine		Vine	9–11
Eleagnus × ebbingei	Ebbinge's Silverberry		Shrub	7–9
Escallonia × exomiensis	Pink Princess Escallonia	Frades	Shrub	7–9
Euryops pectinatus	Green-Leaved Euryops	Viridis	Shrub	8–11
Gardenia jasminoides	Gardenia	August Beauty	Shrub	7–10

Figure 3.7 *(continued)*

Latin	Common	Variety Evaluated	Vegetation Type	USDA Hardiness Zone
Gelsemium sempervirens	Carolina Jessamine		Vine	6–9
Grevillea × Noell	Noell Grevillea		Shrub	8–11
Hemerocallis hybrid	Daylily	Stella d'Oro	Perennial	3–10
Jacaranda mimosifolia	Jacaranda		Tree	9–11
Juniperus chinensis	Chinese Juniper	Spartan	Shrub	5–11
Juniperus chinensis	Chinese Juniper	Sea Green	Shrub	5–11
Juniperus communis	Alpine Carpet Juniper	Mondap	Shrub	2–6
Juniperus horizontalis	Andorra Juniper	Youngstown	Shrub	3–9
Juniperus scopulorum	Gray Green Juniper	Gray Gleam	Shrub	4–9
Lagerstroemia indica	Crape Myrtle	Raspberry Sundae	Tree	7–10
Lavendula dentata	Toothed Lavender	Goodwin Creek Gray	Shrub	5–9
Leucophyllum frutescens	Texas Sage	Compactum	Shrub	7–10
Limonium perezii	Sea Lavender		Perennial	10–11
Loropetalum chinensis	Chinese Loropetalum	Rubrum 'Purple Majesty'	Shrub	7–9
Macfadyena unguis-cati	Yellow Trumpet Vine or Cat's Claw		Vine	9–11
Magnolia grandiflora	Southern Magnolia	Little Gem	Tree	6–10
Mahonia aquifolium	Oregon Grape Holly	Compacta	Shrub	5–9
Mahonia bealei	Leatherleaf Mahonia		Shrub	7–9
Nandina domestica	Heavenly Bamboo, dwarf	Firepower	Shrub	6–10
Olea europaea 'Mowber'	Fruitless Olive	Majestic Beauty	Tree	8+
Osmanthus fragrans	Sweet Olive		Shrub	7–9
Perovskia atriplicifolia	Russian Sage		Shrub	4–9
Picea abies	Nest Spruce	Nidiformis	Shrub	2–7
Plumbago auriculata	Plumbago	Imperial Blue	Shrub	9–11
Prunus cerasifera	Purple Leaf Plum		Tree	4–9
Rosa banksiae	White Banksian Rose	Alba Plena	Vine	6–9
Rosmarinus officinalis	Rosemary	Benenden Blue	Shrub	8–11
Rosmarinus officinalis	Huntington Carpet Rosemary		Shrub	8–11
Salvia leucantha	Mexican Bush Sage	Santa Barbara	Shrub	9–11
Spiraea cantoniensis	Spiraea	Double Bridal Wreath	Shrub	5–9
Syringa meyeri	Dwarf Korean Lilac	Palibin	Shrub	3–7
Taxus × Media	Spreading Japanese Yew		Shrub	4–7
Thuja occidentalis	Emerald Arborvitae		Shrub	2–7
Trachelospermum jasminoides	Star or Confederate Jasmine		Vine	8–11
Viburnum rhytidophylloides	Allegheny Viburnum		Shrub	5–8
Washingtonia robusta	Mexican Fan Palm		Tree	8+

Source: Tsao and Tsao, 2003

Case studies

Following, two case studies are presented which illustrate successful phytotechnology installations at petroleum-impacted sites.

Project name: US Coast Guard Former Fuel Storage Facility (Guthrie Nichols et al., 2014; Cook et al., 2010)

Location: Elizabeth City, NC

Scientists: Elizabeth Guthrie Nichols (a); Rachel L. Cook (a); James E. Landmeyer (b); Brad Atkinson (c); Jean-Pierre Messier (d)

Institutions: (a) Department of Forestry and Environmental Resources, North Carolina State University, Raleigh, NC

(b) US Geological Survey, SC Water Science Center, Columbia, SC

(c) North Carolina Department of Environment and Natural Resources – Division of Waste Management, Raleigh, NC

(d) US Coast Guard, Elizabeth City, NC

Date installed: 2006–2007

Number of trees/species installed

- Hybrid poplar – *Populus deltoides* Bartram ex Marsh. *x nigra* L. Clones OP-367, DN-34, 15–29 and 49–177
- Willow – *Salix nigra* "Marsh," *Salix interior* "Rowle's" and *Salix exigua* "Nutt"
- Loblolly Pine – *Pinus taeda*

2006: 112 bare-root poplars, 1.2 meter (4 foot) height, and 403 unrooted poplar and willow cuttings were installed. Plantings were placed in 8 centimeter (3 inch) diameter boreholes, 1.2 meters (4 feet) deep and backfilled with unamended on-site soil. Trees were mulched. All plantings were spaced 3 meters (10 feet) apart. Mortality averaged 28% because the contaminant concentration was too high for plant survival in some areas.

2007: 2,176 new trees (2,123 poplars, 43 willows, and 10 trial Loblolly Pines) were planted in 23 centimeter (9 inch) diameter boreholes, 1.2 meters (4 feet) deep. Boreholes were backfilled with clean topsoil from off site. Trees were mulched. All plantings were spaced 2 meters (6.5 feet) apart. Mortality decreased and averaged 13% because the borehole width was increased and boreholes were backfilled with clean soil to aid in plant establishment.

2007: 65 poplars and 208 willows were planted using a dibble tool to create 15–30 centimeter (6–12 inch) deep holes just wide enough for each cutting. No backfill was used. All plantings used a 2 meter (6.5 foot) spacing. Mortality averaged 89% because soil was too contaminated for establishment.

Contaminant: Petroleum compounds ●: TPH, BTEX, MTBE, PAH. Estimates from monitoring wells show 567,000–756,000 liters (150,000–200,000 US gallons) of gasoline,

89

diesel and aviation jet fuel at the site in the groundwater. Up to 85 centimeters (33 inches) of petroleum product was floating on top of the water table at the site.

Target media and depth: Soil and groundwater with depth ranging from 0.9 to 1.2 meters (3–4 feet) below ground surface. The groundwater table fluctuates from 1.2 to 2.7 meters (4 to 9 feet) below surface except after major precipitation events.

A 5-acre former US Coast Guard fuel farm had leaked large amounts of fuel since World War II and it was migrating towards the Pasquotank River, 150 meters (490 feet) from the site. A viable tree community (2,984 trees) was established to prevent further groundwater discharge of fuel to the nearby river. Coastal plain surface waters are particularly vulnerable to LUST contamination, due to shallow water tables, porous soils and proximity of LUSTs to natural surface water bodies.

A phytoremediation system was installed from 2006 to 2008 (Cook et al., 2010) using *Salix*, *Pinus taeda* and four hybrid poplar clones (*Populus*). During the tree installation process (April 2006–April 2008), hydrocarbon contaminants were monitored using groundwater samples and soil gas analyses. Tree plantings began in 2006, but high contamination concentrations on part of the site resulted in a tree mortality of 26%. Fuel contamination on the site was very well characterized and documented, but groundwater data did not indicate the high levels of contamination found once the trees were planted. In the next round of planting, in 2007, trees were backfilled with clean soil and survival rates were much higher. Within 2 years, soil gas analyses of total petroleum hydrocarbons and BTEX in soil showed reductions of 95% and 99%, respectively. For the contaminants in the groundwater, as of 2013, only one monitoring well remained above regulatory NCAC 2L standards for benzene, but three wells were above

Poplar trees in 2010 after 3 growing seasons

On-site signage informs public of tree planting intent

Cadets training in newly planted poplar field in 2007. Tree area was not fenced and is open to the public.

Soil-gas monitors installed within planting

Figure 3.9 Case Study: US Coast Guard, Elizabeth City NC

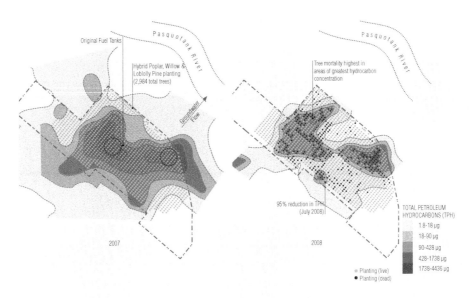

Figure 3.8a Case Study: US Coast Guard, Elizabeth City NC – TPH Degradation in Soil Gas
Scale: 1 in = 250 ft

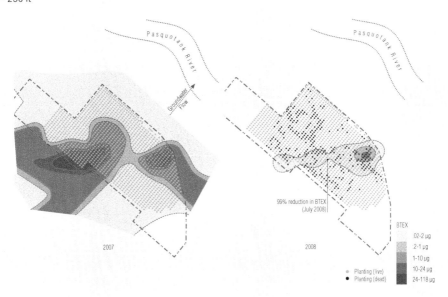

Figure 3.8b Case Study: US Coast Guard, Elizabeth City NC – BTEX Degradation in Soil Gas
Scale: 1 in = 250 ft

MTBE NCAC 2L standards, due to hydraulic control downgradient of the site. Removal of not only the lighter, easier-to-degrade hydrocarbon fractions took place, but heavier polycyclic aromatic hydrocarbons (PAHs) removal was significant, based on analyses of soil transects in 2008, 2010 and 2012 (Guthrie Nichols, 2013).

Additional lessons learned

Tree height and growth rate as measured in 2010 correlated significantly with the TPH levels in the soil below. The lower the contamination, the higher the tree height, with more polluted areas having lower tree heights. The plants are acting as visual indicators of the contaminant concentrations below ground.

Project name: Ford Motor Company Factory (Barac et al., 1999)

Location: Genk, Belgium

Scientists: Tanja Barac (a); Nele Weyens (a); Licy Oeyen (a); Safiyh Taghavi (b); Daniel van der Lelie (b); Dirk Dubin (c); Marco Spliet (d); Jaco Vangronsveld (a)

Institutions: (a) Environmental Biology, Hasselt University, Diepenbeek, Belgium

(b) Biology Department, Brookhaven National Laboratory, Upton, NY, USA

(c) Ford Motor Company, Genk, Belgium

(d) Ingenieurbetrieb Dr.-Ing. W. PützBrühl, Germany

Date installed: 1999

Number of trees/species installed: 275 *Populus trichocarpa x deltoides* cv. "Hoogvorst" and "Hazendans"

Amendments: Soil mixed with compost and backfilled into boreholes.

Contaminant: Hydrocarbons ●: BTEX, TPH

Target media and depth: Groundwater at 4–5 meter (13–16.5 foot) depth.

Underground storage tanks at a Ford factory in Genk, Belgium were found to have been leaking solvent and fuel into the ground since the 1980s. Contamination resulted in 4–5 meter (13–16.5 foot) deep groundwater laden with BTEX and fuel oil as well as nickel and zinc. The leaking tanks were removed and above-ground replacements were constructed. A conventional pump-and-treat system was installed at the core of the plume and ran 23 hours a day. However, the plume was still migrating. A phytoremediation scheme was installed to halt further migration of the contamination plume by providing hydraulic containment. ◯

The phytoremediation installation began in April 1999 and consisted of 275 hybrid poplar trees (*Populus trichocarpa x deltoides* cv. "Hoogvorst" and "Hazendans"). Poplars were chosen to target the migrating groundwater, based on their high evapotranspiration capacity, and the individual cultivars were selected based on their resistance to fungal disease. The poplars were planted in a 2 hectare (5 acre) zone (75 × 270 meters or 246 × 846 feet), perpendicular to the flow of the contaminated groundwater. Four meter (13 foot) tall cuttings were placed in 80 centimeter (32 inch) deep boreholes. Soil in the boreholes was amended with compost to supply cuttings with sufficient nutrients, and backfilled. The cuttings were planted 7 meters (23 feet) on center in 9 rows of 30 trees.

By May 2000, 13 months after planting, roots from the cuttings had not reached the groundwater and the contaminant plume. However, after 42 months, in October 2002, the BTEX plume had been "cut off" by the planting installation, whereas previously it had extended beyond the factory area and under a nearby highway. In June 2003, 55 months after planting, the plume concentration had declined by 50–90% in the planting zone. Measurements taken in 2003, after five growing seasons, indicated that the plume had been entirely eliminated from the planted area.

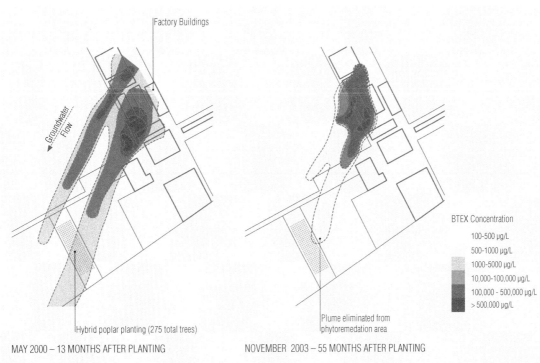

Factory Buildings

Groundwater Flow

Hybrid poplar planting (275 total trees)

Plume eliminated from phytoremedation area

BTEX Concentration

100-500 µg/L
500-1000 µg/L
1000-5000 µg/L
10,000-100,000 µg/L
100,000 - 500,000 µg/L
> 500,000 µg/L

MAY 2000 – 13 MONTHS AFTER PLANTING

NOVEMBER 2003 – 55 MONTHS AFTER PLANTING

Figure 3.10 Case Study: Ford Factory, Genk, Belgium
Scale: 1 in = 900 ft

93

Poplar trees planted 7 meters apart perpendicular to the plume.

Some bore holes were drilled directly into asphalt paving areas. Poplar cuttings were not affected.

Existing poor sandy soils were amended in the bore hole to promote establishment.

Poplar buffer planted between parking lot areas after nine years of growth

New poplar and willow buffers were planted around above-ground storage tanks in 2006 to create a preventative remediation buffer for any potential future releases.

Figure 3.11 Case Study: Ford Factory, Genk, Belgium (Photographs from 2008)

Additional lessons learned

- Because the remediation strategy was successful, preventative phytoremediation buffer plantings of poplar and willows were later planted around above-ground storage tanks and around parking lots on the factory campus, to treat any potential future releases.

- This study also included an examination of endophytic bacteria (living in the plant roots, stems and leaves) and rhizosphere bacteria (living in the root zone) in the phytoremediation treatment zone. The presence of the BTEX plume increased the number of bacteria capable of degrading toluene around the poplar planting. Laboratory cultures indicated that the bacteria were using toluene as their sole carbon source. Tests outside the treatment zone found fewer numbers of these bacteria. Testing in 2006, after the plume had been completely degraded, found no toluene-degrading bacteria inside or outside the treatment zone, indicating that once toluene had been lost as a carbon source, its capacity to be degraded was lost as well. Testing also showed horizontal gene transfer, indicating that DNA coding for toluene-degrading abilities was being shared within the microbial population and could be built up and increased over time.

Chlorinated solvents

Specific contaminants in this category: Trichloroethylene (TCE), and perchloroethylene (Perc or PCE), Vinyl Chloride (VCM), Carbon tetrachloride (Freon)

Typical sources of chlorinated solvent and alcohol contamination: Degreasers, solvents, rocket propellants, cleaners, refrigerants and fire retardants.

Typical land uses with potential chlorinated solvent contamination: Defense sites, dry cleaners, industrial sites, rail maintenance yards, older automobile repair shops.

Why these contaminants are a danger: Exposure to chlorinated solvents is associated with several types of cancers in humans (US EPA, 2014a). Contamination with chlorinated solvents is very prevalent in the US because of their widespread use. TCE is one of the most common pollutants of groundwater in the US (Newman et al., 1997b). Chlorinated solvents are all clear, sweet-smelling liquids that volatilize to a gas upon contact with air. When TCE plumes are under buildings, they often evaporate from the groundwater, enter air spaces between soil particles and migrate through building foundations into the building's indoor air. This vapor is highly toxic to humans and called 'soil vapor intrusion.' In addition to air-quality problems created by plumes, another significant problem arises when these contaminants are deep below the surface, not exposed to air, and move into groundwater

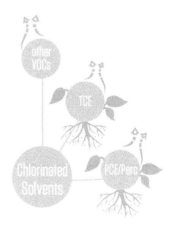

Figure 3.12 Chlorinated Solvents

94

sources used for drinking supply. When they exist in a separated, pure product (DNAPL) state, they can sink deep, since DNAPL is denser than water, and can contaminate large areas of drinking-water aquifers. Vinyl Chloride (VCM) is used to make PVC (polyvinyl chloride), a type of popular plastic, and is one of the top 20 most-produced chemicals in the world. Carbon tetrachloride, commonly known as Freon–10®, is used as a refrigerant, a fire suppressant, an industrial degreaser and in the cleaning industry.

Summary

Many chlorinated solvents, especially TCE and Perc or PCE (also called tetrachloroethene) can be effectively removed, degraded and volatilized by certain plants in a relatively short time frame. These pollutants are very mobile and dense, and may quickly leach to very deep groundwater levels. Once they are in groundwater they can disperse quickly, meaning it may take decades or even centuries to remove them with conventional pump-and-treat systems (Newman et al., 1997). Conventional *in situ* treatment typically requires energy-intensive groundwater pumping (Newman et al., 1997). Phytoremediation offers a potentially effective control and degradation alternative for both soil and groundwater contamination (when the pollutant hasn't sunk too deep as DNAPL), and is now a commonly accepted remediation strategy used by the US EPA (US EPA, 2005a; US EPA 2005b). Plant-based systems for chlorinated solvent removal can be considered, due to their widespread use and many success stories.

Mechanisms utilized: Phytohydraulics , Phytovolatilization , Phytodegradation , Rhizodegradation .

Most successful phytoremediation projects targeting chlorinated solvents have used deep tap-rooted trees with high evapotranspiration rates to access the groundwater and pump the water up through the tree. During the process of pumping, the contaminant is either degraded in the roots, stems and leaves of the plant or volatilized with the plant transpiration into the atmosphere.

Studies have documented that a percentage of the chlorinated solvents taken up by trees will volatilize and not degrade, and will instead be released through the tree into the atmosphere as a gas. This mechanism is known as phytovolatilization . Research findings, however, are not entirely consistent, as some studies have shown minimal to no above-ground transpiration of chlorinated solvents and others have shown some volatilization (Ma and Burken, 2003). However, it is generally agreed that the human health risk of exposure, even in low doses, to contaminated groundwater can often be much higher than the risk from exposure to volatilized gas. In addition, when TCE is released into the air, it can quickly break down.

Recent research has found that when the contaminants are degraded rather than volatilized, the degradation work occurring within plants may be assisted by endophytic bacteria living within the plant roots, stems and leaves (Weyens et al., 2009). Researchers are conducting experiments to see if these particular bacteria can be isolated and inoculated into other plants to provide maximum degradation, so that the chlorinated solvents are no longer released to the atmosphere. The initial results are promising, showing a 90% reduction of TCE gas release through the leaves with the introduction of beneficial bacteria into the plant (Weyens et al., 2009).

Planting specifics

Since dense chlorinated solvents may move quickly through groundwater to the bottom of the water table, trees with long roots have typically been chosen for remediation. Deep-rooted planting techniques are often utilized for installation (see Chapter 2). Where these contaminants exist in surface soils and shallow groundwater, grass species and some shrubs have also been used. To stop the contaminants from migrating and to also degrade or volatilize them, the following planting typologies have been effectively utilized in conjunction with the plant lists provided in this section.

Groundwater typologies for chlorinated solvents: Hydraulic control time frame 2–10 years or more (highly dependent on how contaminated the groundwater is, speed and volume of plume, climate and evapotranspiration rate of chosen plants).

- Interception Hedgerow: Chapter 4, see p. 216
- Groundwater Migration Tree Stand: Chapter 4, see p. 213
- Phytoirrigation: Chapter 4, see p. 207

Soil degradation typologies for chlorinated solvents: Degradation time frame 1–10 years or more.

- Degradation Cover: Chapter 4, see p. 222
- Degradation Hedge: Chapter 4, see p. 220
- Degradation Living Fence: Chapter 4, see p. 220
- Degradation Bosque: Chapter 4, see p. 218

In stormwater and wastewater: Volatile chlorinated solvents typically do not persist in stormwater and wastewater, since these vectors are often exposed to the atmosphere and the chlorinated solvents freely volatilize into the air.

The plant list in Figure 3.13 is an initial list of species that have been shown in studies to be useful for controlling and/or degrading chlorinated solvents.

Case Study Chlorinated Solvents–1

Project name: Travis Air Force Base (Klein, 2011; Doucette et al., 2013; Parsons, 2010)

Location: Fairfield, CA

Institutions: (a) Utah Water Research Laboratory, Utah State University, 8200 Old Main Hill, Logan, Utah 84322-8200, United States

(b) Parsons, 1700 Broadway, Suite 900, Denver, Colorado 80290, United States

Date installed: 1998–2000

Number of trees/species installed: 480 *Eucalyptus sideroxylon* 'Rosea' (Red Ironbark Eucalyptus)

Figure 3.13 Chlorinated Solvents–Degradation and Hydraulic Control Phytotechnology Plant List

Latin	Common	Contaminant	Vegetation Type	USDA Hardiness Zone	Native to	Reference
Agropyron desertorum cv. Hycrest	Crested Wheatgrass	Pentachlorophenol	Herbaceous	3+	Asia	Ferro et al., 1994 ITRC PHYTO 3
Betula pendula	European White Birch	TCE	Tree	3–6	Europe	Lewis et al., 2013
Eucalyptus sideroxylon 'rosea'	Red Ironbark	TCE	Tree	10–11	Australia	Doucette et al., 2011 Klein, 2011 Parsons, 2010
Glycine max	Soybean	Dodecyl linear alcohol ethoxylate Dodecyl linear alkylbenzene sulfonate Dodecyltrimethyl ammonium chloride TCE	Herbaceous	Grown as annual	Eastern Asia	Anderson and Walton, 1991, 1992 Anderson et al., 1993 ITRC PHYTO 3
Lespedeza cuneata	Bush Clover	TCE	Herbaceous	5–11	Eastern Asia	Anderson and Walton, 1991, 1992 ITRC PHYTO 3
Liquidambar styraciflua	Sweetgum	TCE	Tree	5–10	Eastern/ Southeastern USA	Stanhope et al., 2008 Strycharz and Newman, 2009b
Lolium perenne	Perennial Ryegrass	Pentachlorophenol	Herbaceous	3–9	Europe	Ferro et al., 1997 ITRC PHYTO 3
Paspalum notatum	Bahia Grass	TCE	Herbaceous	7–10	Central and South America	Anderson and Walton, 1992 ITRC PHYTO 3
Phragmites australis	Common Reed	Bromoform Chlorobenzene Chloroform Dichloroethane PCE TCE	Herbaceous	4–10	Europe, Asia	Anderson et al., 1993 ITRC PHYTO 3
Pinus palustris	Longleaf Pine	TCE	Tree	7–10	Southeastern USA	Strycharz and Newman, 2009a
Pinus taeda	Loblolly pine	TCE 1,4 dioxene	Tree	6–9	Southeastern USA	Anderson and Walton, 1991, 1992, 1995 ITRC PHYTO 3 Stanhope et al., 2008 Strycharz and Newman, 2009a

Figure 3.13 (continued)

Latin	Common	Contaminant	Vegetation Type	USDA Hardiness Zone	Native to	Reference
Plantanus occidentalis	Sycamore	TCE	Tree	4–9	Eastern USA	Strycharz and Newman, 2009b, 2010
Populus spp. *Populus deltoides* *Populus deltoides × nigra DN34* *Populus trichocarpa × P. deltoides 50–189* *Populus trichocarpa × P. maximowiczii 289–19*	Poplar species and hybrids	PCE TCE Pentachlorophenol 1,2,4-Trichlorobenzene Carbon tetrachloride 1,4 Dioxane	Tree	varies	varies	Burken and Schnoor, 1997 Gordon et al., 1997 Harvey, 1998 ITRC PHYTO 3 Jones et al., 1999 Miller et al., 2011 Newman et al., 1997a, 1997b Ferro et al., 2013 Orchard et al., 1998 Strycharz and Newman, 2009b, 2010 Wang et al., 1999
Quercus palustris	Pine Oak	TCE, PCE, vinyl chloride	Tree	4–8	Eastern USA	Ferro et al., 2000
Quercus virginiana	Live Oak	TCE	Tree	7–10	Southeastern USA	Hayhurst et al., 1998 ITRC PHYTO 3
Ricinus communis	Castor Bean	TCE	Herbaceous	9–11	Middle East	Hayhurst et al., 1998 ITRC PHYTO 3
Salix spp.	Coyote Willow	PCE, TCE	Shrub/Tree	4–10	Southwestern USA	Stanhope et al., 2008 Landmeyer, 2012
Serenoa repens	Saw Palmetto	TCE	Tree	8–11	Southeastern USA	Hayhurst et al., 1998 ITRC PHYTO 3
Solidago spp.	Goldenrod	PCE, TCE	Herbaceous	4–9	North America	Anderson et al., 1991, 1992 ITRC PHYTO 3
Typha spp.	Cattail	Dodecyl linear alcohol ethoxylate, sulfonate and chloride	Herbaceous	3–10	North America, Europe, Asia	Anderson et al., 1993 ITRC PHYTO 3
Zea mays	Corn	Dodecyl linear alcohol ethoxylate, sulfonate and chloride	Herbaceous	Grown as annual	USA	Anderson, Guthrie and Walton, 1993 ITRC PHYTO 3

Contaminant: TCE ⬤

Target media and depth: Groundwater at 6–12 meters (20–40 feet). Concentrations of TCE in groundwater varied between <500 micrograms/liter and >9000 micrograms/liter

Travis Air Force Base is located near the city of Fairfield, CA. The base contains several contaminated areas, including Building 755, which was once used to test liquid-fueled rocket engines and later used as a battery and electric shop. The shop regularly discharged battery acids and chlorinated solvents into the groundwater until 1978. Since then, the building structure has been removed and the source of the solvent plume excavated. Beginning in 1998, 100 Red Ironbark Eucalyptus trees (*Eucalyptus sideroxylon* 'Rosea') were installed on a 2.24 acre plot southeast of the former building to hydraulically control the solvent plume. Two years later, in 2000, another 380 eucalyptus trees were installed. By 2009 almost 100 trees had died, leaving 388 trees remaining; however, this was still enough to create an effective system.

A phytotechnology system was chosen for this site because of its likely effectiveness and low cost in reducing the amount of TCE in the groundwater. In addition, dry climates with low summer precipitation, as found at this site, help ensure that the trees depend solely on the targeted contaminated groundwater for their water needs. The Travis Air Force Base phytotechnology site is part of a larger network of demonstration projects at six US Air Force bases across the country that were installed in the late 1990s. Phytotechnologies are of significant interest to the DOD because of the frequent past use of TCE as a degreasing solvent in aircraft maintenance, resulting in widespread groundwater contamination at Air Force bases around the world.

In addition to the tree planting downgradient from the source, a bioreactor consisting of a mixture of iron pyrite, gravel and woodchip mixed with vegetable oil was installed near the source area in 2008. Since installation of the bioreactor, TCE contamination directly around the source area has decreased by as much as 94%. Phytovolatilization ⬡ of TCE from the trees and soil surface into the atmosphere and phytodegeneration ⬡ were the primary mechanisms of TCE removal from the groundwater that were evaluated in this study. Using groundwater data collected in 2004 and 2009, calculations based on the field measurements indicate that volatilization from leaves and soil accounts for almost half the 3.75 lbs of TCE removed from the phytoremediation site each year. Transfer of TCE into the atmosphere is not considered a concern at this site, since TCE has a relatively short half-life in the atmosphere (Atkinson, 1989) and air-quality sampling has verified worker health is not at risk (Parsons, 2010). The tree stand has been successful enough that the engineers have recommended expansion onto other areas of the base impacted with TCE groundwater plumes.

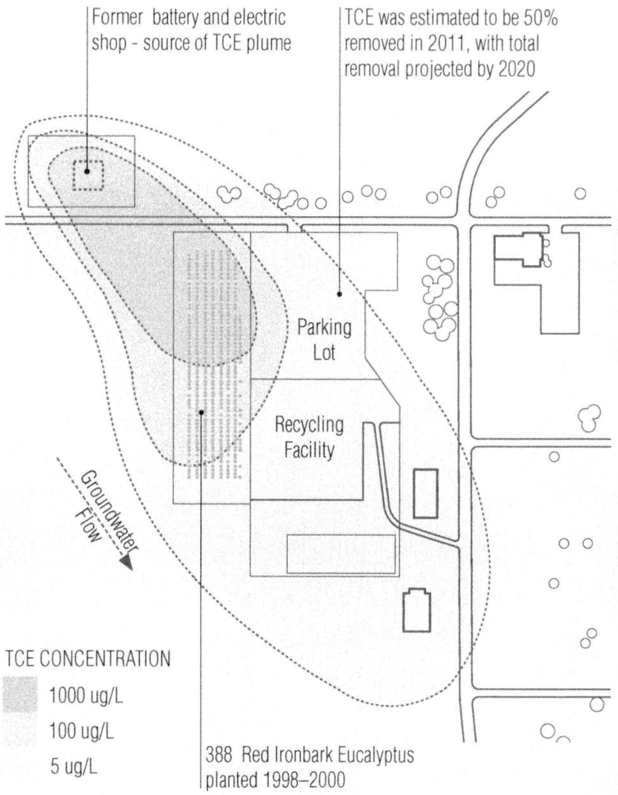

Former battery and electric shop - source of TCE plume

TCE was estimated to be 50% removed in 2011, with total removal projected by 2020

Parking Lot

Recycling Facility

Groundwater Flow

TCE CONCENTRATION

1000 ug/L
100 ug/L
5 ug/L

388 Red Ironbark Eucalyptus planted 1998–2000

100

Figure 3.14 Case Study: Travis Air Force Base, Fairfield, CA
Scale: 1 in = 1500 ft

12-year-old phytoremediation planting of euycalyptus trees at Travis Air Force Base is one of six phytoremediation plantings installed at US Department of Defense sites in the late 1990s, and is expected to have completely remediated the plume by 2020.

Figure 3.15 Case Study: Travis Air Force Base, Fairfield, CA – 2012

Project name: Pinehurst Hotel Dry Cleaners (ATC, 2013; Sand Creek, 2013)

Location: Pinehurst, NC

Consultants/Scientists: Sand Creek Consultants, Wisconsin; ATC Associates, North Carolina

Date installed: 2010

Species installed: Hybrid poplars *(Populus spp.)* and willows *(Salix spp.)* were installed downgradient of the source area to prevent migration of the contaminant plume.

Area A – hybrid poplars planted on 3.2 × 2.4 meter (8 × 6 foot) grid spacing.

Area B – hybrid poplars and willows planted between existing Magnolia trees.

Amendments: Iron (zero valent) was injected into the source area to help break down pollutants in the most contaminated areas. Phytoremediation was not used in the source area; only downgradient to control the migrating contaminant plume. Pine straw and turf grass were utilized for weed control within the phytoremediation planting.

Contaminants: PCE ⬤, TCE ⬤, Benzene ⬤, Xylenes ⬤

Target media and depth: Groundwater at 4 meters (13.5 feet) depth

Pinehurst Hotel Cleaners was a dry cleaning and laundry facility for the former Pinehurst Hotel and operated from the 1930s until the 1970s. Contamination was identified during a due-diligence assessment on an adjacent property. The property owner registered the site with North Carolina's Dry-cleaning Solvent Cleanup Act (DSCA) Program in 2001.

Beginning in 2008, 750 tons of contaminated soil were excavated at the plume source and treated on site using a mobile steam-distillation unit. Additionally, zero valent iron was added near the plume source to help break down contaminants. A phytotechnology installation consisting of hybrid poplar and willow plantings was installed on two sites in 2010 further downgradient on the plume: a deep-planted zone where groundwater was 18 feet below grade, and a shallow-planted zone where groundwater was less than 3 feet below grade.

These phytotechnology plantings provided hydraulic control of the plume and prevented discharge of dry-cleaning solvents to a nearby surface-water stream. The deep-planting species were hybrid poplar, installed between the source area and the stream, and required planting the 20-foot trees into deep boreholes to access groundwater. The shallow plantings were hybrid willow, and were installed as buffers along the banks of the surface-water stream. Phytotechnology approaches were utilized due to their low cost and to comply with stakeholder desires for a green and environmentally friendly remediation solution. Today the plume is stable and surface-water impacts have been successfully mitigated.

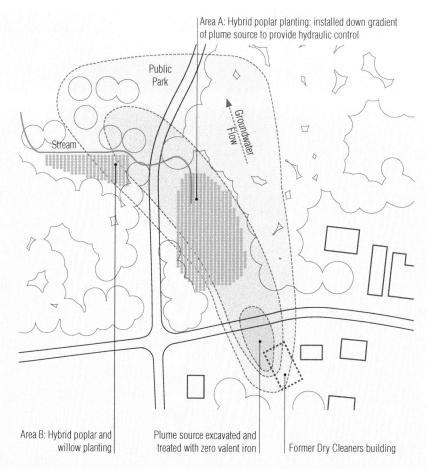

Area A: Hybrid poplar planting: installed down gradient of plume source to provide hydraulic control

Public Park

Groundwater Flow

Stream

Area B: Hybrid poplar and willow planting

Plume source excavated and treated with zero valent iron

Former Dry Cleaners building

Figure 3.16 Case Study: Pinehurst Hotel Cleaners, Pinehurst, NC
Scale: 1 in = 200 ft

Phytoremediation Area A: Before installation

Phytoremediation Area A: 2008 installation – poplar spp.

Phytoremediation Area A: 2010 – 2.5 years of growth

Phytoremediation Area B: Original installation – poplars and willows were installed to protect the surface-water swale visible on the right from contaminated groundwater.

Phytoremediation Area B: 2010 – 2.5 years of growth

Figure 3.17 Case Study: Pinehurst Dry Cleaners, Pinehurst, NC

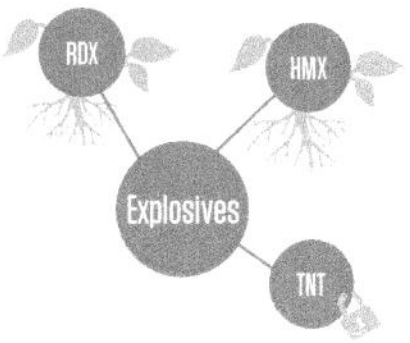

Figure 3.18 Explosives

● Explosives

Specific contaminants in this category

- 1,3,5-trinitroperhydro-1,3,5-triazine (RDX)
- 1,3,5,7-tetranitro-1,3,5,7-tetrazocane (High Melting Explosive or HMX)
  trinitrotoluene (TNT)

Typical sources of explosives contamination: Munitions.

Typical land uses with explosives contamination: Defense sites, explosives manufacturers, stockpiling facilities, weapons-dismantling facilities, other lands where military activities have taken place, mining sites, quarries.

Why these contaminants are a danger: The three most common explosive contaminates are RDX, TNT, and HMX. RDX is the most widely used military explosive, and TNT is commonly mixed with the RDX (Rylott et al., 2011). In recent years, RDX and TNT have been replaced by HMX in numerous military applications (Yoon et al., 2002). The US EPA has classified all of these explosives as priority pollutants and possible human carcinogens, and RDX is especially a danger to humans because it can easily contaminate drinking water supplies. RDX is toxic to humans and mammals; it targets the central nervous system and is known to cause convulsions (Burdette et al., 1988). Unlike other dangerous contaminants that have been banned due to their toxic effects, the constant demand for military explosives means that these explosives will likely continue to be manufactured and pose a significant environmental risk (Rylott et al., 2011).

Summary

RDX, TNT and HMX are very different compounds and act distinctly differently in soils.

RDX

Mechanisms utilized: Phytohydraulics , Rhizodegradation , Phytodegradation , Phytometabolism

RDX leaches easily and can be found in both soil and groundwater. Some bacteria in the root zones and microbes (called endophytes) in the above-ground portions of some plants can degrade RDX (Rylott et al., 2011; Just and Schnoor, 2004). However, RDX can also be toxic to plant growth and plants must be selected that have either evolved or been genetically modified to tolerate soils with RDX. If tolerant plants with deep tap roots and high evapotranspiration rates are selected, they can potentially be used to reach RDX in groundwater and actively pump out the RDX as associated bacteria start to degrade it. With a log K_{ow} of 0.87 in the applicable range, RDX can be mobilized into plants. RDX degradation is possible with the assistance of plants and microbes via both rhizodegradation and phytodegradation .

103

Disturbed soils tend to be poor in nitrogen, and bacteria at sites contaminated with explosives have evolved the ability to tap RDX as a nitrogen source, degrading it as the nitrogen is utilized (Rylott et al., 2011).

The complete mechanism for how degradation pathways work is not completely known, and the metabolite byproducts produced by the phytodegradation process can be toxic. Degradation can occur in environments both with and without oxygen, resulting in the release of toxic and non-toxic metabolites (Rylott et al., 2011). For these reasons, plant-based RDX systems are not fully ready for field application and should be considered only in a scientific experimental setting at this time. In addition, RDX-contamination tends to be mixed with TNT or other pollutants and additional complications can arise (see below). The latest innovation in RDX degradation is a partnership between scientists and the US Army to explore the use of genetically modified grasses that can live on toxic sites and also degrade RDX into metabolites that are less toxic (Rylott, 2012). These grasses are low-growing and can be planted on firing ranges and bases where sight lines are necessary and active military uses are present.

TNT

Mechanisms utilized: Phytostabilization .

TNT is typically tightly bound to soils and toxic to plants, and often persists in soils for centuries, so establishing plants in soils contaminated with TNT is very difficult (Rylott et al., 2010; Thompson et al., 1999). A few select plants in TNT-polluted areas have been shown to develop a tolerance for growing in this toxin. The genes for TNT tolerance in these plants are currently being studied to try to produce genetically modified plants that could tolerate growing in TNT conditions (Rylott et al., 2011). This is important, not for degrading TNT, but for remediating other contaminants that could be degraded on military bases that have a mix of soil contamination including TNT. For example, RDX or petroleum spills may be able to be degraded with plant species, but the plants would have to be tolerant to TNT. The TNT, which is not as large a risk to humans, as it does not typically migrate into groundwater, could be further bound to soils and stabilized on site while other organics are degraded. In some circumstances TNT not bound to soil particles can be extracted into the plant (Thompson et al., 1999), but this condition is unlikely in the field.

HMX

Mechanisms utilized: Phytohydraulics , Phytometabolism , Phytostabilization .

HMX is not phytotoxic, so plants can often grow in soils contaminated with HMX (Yoon et al., 2004). It has a log K_{ow} of 0.19 and can be taken up and stored in the above-ground portions of plants. However, plants cannot degrade it (Yoon et al., 2004). Accumulating it in the leaves creates a pathway for mobilization of this toxic contaminant because HMX can easily leach out of fallen leaves. With bioaccumulation, the risk of insect or animal ingestion is present and therefore phyto is not suggested for field application for HMX remediation at this time (Yoon et al., 2004). If extraction is considered, leaves and other plant parts must be collected to prevent the spread of the contaminant.

Other military contaminants that are not explosives include: petroleum products, which are quite pervasive on military sites (see previous section on petroleum), perchlorate (see section on inorganics –

salts later in this chapter), Perc and TCE (see previous section on chlorinated solvents), metals such as lead, zinc, copper and radionuclides (see following sections on metals and radionuclides).

Since TNT can be quite phytotoxic to plant growth, and explosives contaminants are often mixed on sites, the successful use of phytotechnology systems in soils at field scale has been limited. In addition to phytotoxicity concerns, when the explosives are broken down, they can be metabolized into components that are just as toxic as their original composition. For this reason, plants are currently recommended only for mobilization control of contaminants, and not for degradation or extraction. However, there is significant research being completed in this field and phytotechnologies show considerable promise for future degradation potential. Energy-intensive excavation and incineration practices are most commonly used to remediate explosives-contaminated soils in the US (Subramanian and Shanks, 2003). If research continues and phyto strategies develop further and become viable, they would have tremendous applicability on these often remote, large land-area sites. In addition, they can address environments with existing fragile ecosystems where minimal disturbance is preferred (Reynolds et al., 1999).

When explosives are targeted in water rather than soil as described above, utilizing constructed wetlands for explosives removal and degradation has shown promising results (Kiker et al., 2001).

Planting specifics

The following planting typologies can currently be considered for use in conjunction with the plant lists below.

In soil

To help hold the contaminant on site so it does not move:

- Planted Stabilization Mat: Chapter 4, see p. 202

In water

To help control contaminated groundwater: hydraulic control time frame is 2–10+ years, dependent on how contaminated the groundwater is, speed and volume of plume.

- Groundwater Migration Tree Stand: Chapter 4, see p. 213

 To degrade within water: Some explosives have been successfully removed from water with constructed wetlands.

- Surface-Flow Constructed Wetland: Chapter 4, see p. 238
- Subsurface Gravel Wetland: Chapter 4, see p. 241

The plant list in Figure 3.19 is an initial list of species that have been shown in laboratory studies to be useful for degrading explosives in soils. No plants other than wetland plants for degradation are recommended for field application at this time, since further studies need to be completed before these systems are implemented. Deep-rooted, high evapotranspiration-rate tree species may be considered to prevent groundwater plume migration, but the explosive metabolites that may be translocated to leaves must be carefully considered.

Figure 3.19 Explosives Phytotechnology Plant List

Plants shown in scientific studies to have some extraction and/or degradation potential. Only wetlands should be considered for explosives phytoremediation at this time.

Latin	Common	Contaminant	Vegetation Type	USDA Hardiness Zone	Native to	Reference
Abutilon avicennae	Indian Mallow	TNT	Herbaceous	8–11	India	Chang et al., 2003 Lee et al., 2007
Acorus calamus	Sweet Flag	TNT	Herbaceous	3+	Asia	Best et al., 1999
Aeschynomene indica	Indian Joint Vetch	TNT	Herbaceous		Asia, Africa	Lee et al., 2007
Alisma subcordatum	Water Plantain	RDX	Wetland	3–8	North America	Kiker and Larson, 2001
Arabidopsis thaliana	Arabidopsis (transgenic)	RDX	Herbaceous	1+	Europe, Asia	Rylott et al., 2011 Strand et al., 2009
Carex gracilis	Slim Sedge	TNT	Herbaceous	3–7	North America	Nepovim et al., 2005 Vanek et al., 2006.
Catharanthus roseus	Madagascar Periwinkle	HMX RDX TNT	Herbaceous	10–11	Madagascar	Bhadra et al., 2001 Hughes et al., 1997 Thompson et al., 1999
Ceratophylum demersum	Hornwart	RDX	Wetland	5–11	Worldwide	Kiker and Larson, 2001
Chara	Stonewart	RDX	Wetland	varies	varies	Kiker and Larson, 2001
Cicer arietinum	White Chickpea	TNT	Herbaceous	Grown as annual	Middle East	Adamia et al., 2006
Cyperus esculentus	Yellow Nutsedge	TNT	Herbaceous	3–9	North America, Europe, Asia	Leggett and Palazzo, 1986 Thompson et al., 1999
Dactylis glomerata	Orchardgrass	TNT	Herbaceous	3+	Europe	Duringer et al., 2010.
Echinochloa crus-galli	Barnyard Grass	TNT	Herbaceous	4–8	Europe, Asia	Lee et al., 2007
Festuca arundinacea	Tall Fescue	TNT	Herbaceous	4–8	Europe	Duringer et al., 2010.
Glycine max	Soybean	RDX TNT	Herbaceous	Grown as annual	Eastern Asia	Adamia et al., 2006 Chen et al., 2011 Vila et al., 2007
Helianthus annuus	Sunflower	TNT	Herbaceous	Grown as annual	North America	Lee et al., 2007
Heteranthera dubia	Water Star-Grass	RDX TNT	Wetland	6–11	North and Central America	Best et al., 1997

Latin	Common	Contaminant	Vegetation Type	USDA Hardiness Zone	Native to	Reference
Juncus glaucus	Blue Rush	TNT	Herbaceous	4–9	Europe	Nepovim et al., 2005 Vanek et al., 2006
Lersia oryzoides	Rice Cutgrass	RDX	Wetland	3–8	North America, Europe, Asia	Kiker and Larson, 2001
Lolium multiflorum	Ryegrass	HMX TNT	Herbaceous	5+	Europe	Adamia et al., 2006
Lolium perenne	Perennial Ryegrass	HMX	Herbaceous	3–9	Europe, Asia	Duringer et al., 2010
Myriophyllum aquatica	Parrot Feather	RDX TNT	Wetland	6–10	South America	Bhadra et al., 2001 Hughes et al., 1997 Just and Schnoor, 2000 Just and Schnoor, 2004 Thompson et al., 1999 Wang et al., 2003
Myriophyllum spicaticum	Eurasian Water Milfoil	TNT	Wetland	6–10	Europe, Asia	Hughes et al., 1997 Thompson et. al., 1999
Nicotiana tabacum	Tobacco (transgenic)	RDX TNT	Herbaceous	Grown as annual	North America	French et al., 1999 Hannink et al., 2001 Pieper and Reineke, 2000 Rosser et al., 2001 Van Aken et al., 2004
Oryza sativa	Rice	TNT	Herbaceous	Grown as annual	East Asia	Vila et al., 2007a Vila et al., 2007b Vila et al., 2008
Panicum virgatum	Switchgrass	RDX	Herbaceous	2–9	North America	Brentner et al., 2010
Phalaris arundinacea	Reed Canary Grass	RDX TNT	Herbaceous	4–9	Europe	Best et al., 1997 Best et al., 1999 Kiker and Larson, 2001 Thompson et. al., 1999 Just and Schnoor, 2004
Phaseolus vulgaris	Bush Bean	TNT	Herbaceous	Grown as annual	Central America	Cataldo et al., 1989 Thompson et. al., 1999
Phragmites australis	Common Reed	TNT	Herbaceous	4–10	Europe, Asia	Nepovim et al., 2005 Vanek et al., 2006

Figure 3.19 (continued)

Latin	Common	Contaminant	Vegetation Type	USDA Hardiness Zone	Native to	Reference
Polygonum punctatum	Smartweed	RDX	Herbaceous	3+	North America	Kiker and Larson, 2001
Populus deltoides × nigra DN34 (24, 25) Populus tremula × tremuloides var. Etropole (26)	Populus spp. and hybrids	RDX TNT	Tree	varies	varies	Brentner et al., 2010 Thompson, 1997 Van Dillewijn et al., 2008 Van Akenet al., 2004
Potemogeton spp.	Pondweed	RDX	Wetland			Kiker and Larson, 2001
Sagittaria spp.	Arrowhead	RDX	Wetland	5–10	North and South America	Kiker and Larson, 2001
Scirpus cyperinus	Woolgrass	RDX TNT	Wetland	4–8	North America	Best et al., 1997 Best et al., 1999.
Sorghum sundase	Sorghum	RDX	Herbaceous	8+	Africa	Chen et al., 2011
Triticum aestivum	Wheat	RDX	Herbaceous	Grown as annual	Asia	Chen et al., 2011 Vila et al., 2007a
Typha latifolia	Cattail	TNT	Herbaceous	3–10	North America, Europe, Asia	Nepovim et al., 2005 Vanek et al., 2006
Vetiveria zizanioides	Vetiver Grass	TNT	Herbaceous	9–11	India	Das et al., 2010 Markis et al., 2007a Markis et al., 2007b
Zea mays	Corn	RDX TNT	Herbaceous	Grown as annual	USA	Chen et al., 2011 Vila et al., 2007

Project name: Iowa Army Ammunitions Plant Constructed Wetlands (Kiker et al., 2001; Thompson et al., 1997, 2003)

Location: Middletown, IA

Scientists: Dr. Steve Larson, research chemist with US Army Corps of Engineers ERDC; Randy Sellers, biologist with the US Army Corps of Engineers, Omaha; Jackson Kiker, chemist formerly with the US Army Corps of Engineers, Omaha; Don Moses, civil engineer, US Army Corps of Engineers

Date installed: 1997

Planting description: Constructed wetlands were created and new depressions were lined with soil collected from local lakes. This provided a seedbank for future growth of wetland vegetation: Rice Cutgrass (*Lersia oryzoides*), Smartweed (*Polygonum punctatum*), Reed Canarygrass (*Phalaris arundinacea*), freshwater algae (*Spirogyra spp.*), Barnyard Grass (*Echinochloa crusgalli Michx.*), Pondweed (*Potemogeton sp.*), Water Plantain (*Alisma subcordatum*), Arrowhead (*Sagittaria sp.*), Coontail (*Ceratophylum demersum*), Stonewart (*Chara*).

Contaminants: TNT 2,4,6 Trinitrotoluene in soils ⬤ and RDX Hexahydro-1,3,5-triazine within surface and groundwater ⬤

Target media and depth: Surface and groundwater RDX levels reduced from 778 micrograms/liter to below 2 micrograms/liter from January to July of 1998 and reductions have continued through 2013 to meet regulatory discharge limits. TNT in soils has not been remediated.

The Iowa Army Ammunitions plant is a 19,000-acre facility in southeastern Iowa that has produced ordnance and warheads for the Department of Defense since 1940. Isolated areas within the site were extensively contaminated with explosives related to the disposal of munitions, creating large volumes of wastewater containing TNT and RDX. Contaminated surface waters on the site were often referred to as 'pink water' due to the color change caused by explosives contamination. In 1990, the plant was listed by the US EPA as a Superfund site.

In 1997, two of the most contaminated areas of the Iowa Army Ammunitions plant – the Line 1 Impoundment and Line 800 Lagoon – were excavated to remove most of the contaminated soil. Instead of being backfilled, the areas were reclaimed as engineered wetlands that would treat contaminated surface and groundwater. This was the first known full-scale phytotechnology system designed to treat explosives-contaminated water. The goal was to reduce contamination levels to below the 2 ppb US EPA health advisory lifetime level. Phytotechnology as a remediation option was chosen due to its low cost. The excavated areas were lined with sediment from a nearby lake that acted as a seed source to spur the growth of wetland vegetation of over 50 different species. Line 1 is now a 3.6-acre constructed wetland, while Line 800 is a 5-acre wetland. The wetlands have been successful in explosives reduction to meet regulatory discharge limits, except in winter, when the planting vegetation is dormant. At this time of the year, the water must be held, and discharged when the growing season again is underway.

109

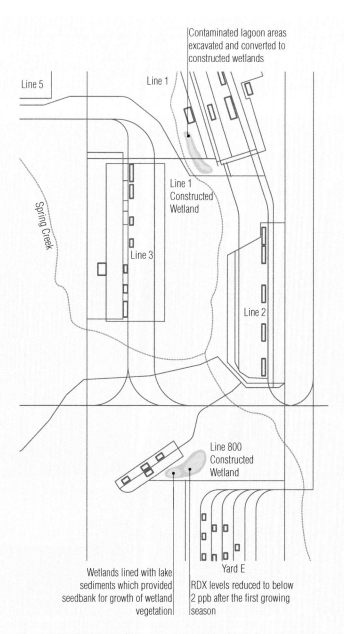

Contaminated lagoon areas excavated and converted to constructed wetlands

Line 5

Line 1

Line 1 Constructed Wetland

Spring Creek

Line 3

Line 2

110

Line 800 Constructed Wetland

Yard E

Wetlands lined with lake sediments which provided seedbank for growth of wetland vegetation

RDX levels reduced to below 2 ppb after the first growing season

Figure 3.20 Case Study: Iowa Army Ammunitions Plant Constructed Wetlands, Middletown, IA
Scale: 1 in = 2500 ft

Line 1 Explosives Constructed Wetland: 2000 – created from excavated explosives impoundment

Line 800 Explosives Constructed Wetland: 2000 – created from excavated area of explosives

Figure 3.21 Case Study: Iowa Army Ammunitions Plant Constructed Wetlands, Middletown, IA

Pesticides

Specific contaminants in this category: Any pesticide with a breakdown half-life of less than one year: Atrazine (herbicide), Picloram (herbicide), Clopyralid (herbicide) and Carbaryl (insecticide), to name a few. Many pesticides are also chlorinated solvents (see p. 94). Pesticides that persist in the environment for more than one year are included in the Persistent Organic Pollutants (POPs) category (see next section, p. 118), and include DDT, DDE, Chlordane and Aldrin.

Typical sources of pesticides: Insecticides, herbicides, fungicides.

Typical land uses with pesticides contamination: Both present and former agriculture or orchards uses, residences (sprayed for termites or other insects), rail corridors, road corridors, utility corridors.

Why these contaminants are a danger: Many pesticides are known endocrine disruptors and may cause cancer in humans. Herbicides such as Atrazine are still widely used in the US but have been banned in the European Union since 2003 because of evidence that they can contaminate groundwater supplies (Sass and Colangelo, 2006).

Mechanisms utilized: Phytohydraulics ⬤, Rhizodegradation ⬤, Phytodegradation ⬤.

Pesticides such as the herbicides Atrazine, Picolram and Clopyralid and the insecticide Carbaryl are in use today and can easily dissolve in water and migrate into surface and groundwater. They are most harmful once mobilized in water, since they break down more slowly once they are in anaerobic water environments. They can contaminate drinking water and be ingested by humans. Many are known carcinogens.

Degradation of these pesticides can be enhanced with planted systems and usually occurs via rhizodegradation ⬤. Since these contaminants are most often mobilized into water, riparian buffers, stormwater filters and constructed wetlands are most often utilized to capture water and break down the pesticides. Even though most of these contaminants have been developed within the past 50 years, microbiology has evolved to use these new nitrogen and carbon sources, facilitating their breakdown.

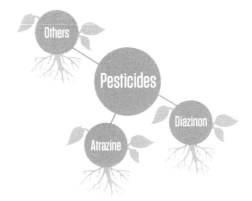

Figure 3.22 Pesticides

Planting specifics

Pesticides, especially herbicides, with a half-life of less than one year, may be present in concentrations that prevent plant growth. However, when mobilized in water, they are often diluted and then can potentially be degraded in constructed wetlands or bioremediation systems without plants.

In soil

Many pesticides are formulated to break down on their own, over time. However, this process can be speeded up with the introduction of plants that encourage microbial degradation in the soil. For these easier-to-degrade pesticides, the following typologies can be considered.

 Soil degradation typologies: Time frame for degradation: 0–3 years

- Degradation Cover: Chapter 4, see p. 222
- Degradation Hedge: Chapter 4, see p. 220
- Degradation Living Fence: Chapter 4, see p. 220
- Degradation Bosque: Chapter 4, see p. 218

In water

Groundwater control typologies: The time frame is dependent on how contaminated the groundwater is, and the speed and volume of the plume.

- Groundwater Migration Tree Stand: Chapter 4, see p. 213

 Stormwater or groundwater degradation: Ongoing, as pollutants continually enter the system. From the time a pollutant enters the system, time frame for degradation may be 0–3 years.

- Stormwater Filters: Chapter 4, see p. 235
- Multi-Mechanism Buffers: Chapter 4, see p. 234
- Surface-Flow Constructed Wetland: Chapter 4, see p. 238
- Subsurface Gravel Wetland: Chapter 4, see p. 241

The plant list in Figure 3.23 is an initial list of species that have been shown in studies to be useful for breaking down easier-to-degrade pesticides in soil and water.

Figure 3.23 Pesticides Phytotechnology Degradation and/or Hydraulic Control Plant List

Latin	Common	Contaminant	Vegetation Type	USDA Hardiness Zone	Native to	Reference
Acorus calamus	Sweet Flag	Atrazine	Wetland	3+	Asia	Marecik et al., 2011 Wang et al., 2012
Andropogon geradi Andropogon geradii var. Pawne	Big Bluestem	Chlorpyrifos Chlorothalonil Pendimethalin Propiconazole Altrazine Pendimethalin	Herbaceous	4–9	North America	Henderson et al., 2006 Smith et al., 2008
Brassica campestris	Mustard	Endosulfan	Herbaceous	7+	Europe	Mukherjee and Kumar, 2012
Brassica napus	Canola	Chlorpyrifos	Herbaceous	7+	Mediterranean	White and Newman, 2007
Cabomba aquatica	Fanwort	Copper sulfate Dimethomorph Flazasulfron	Wetland	9+	South America	Olette et al., 2008
Ceratophyllum demersum	Coontail	Metolachor	Herbaceous	4–10	North America	ITRC PHYTO 3 Rice et al., 1996b
Elodea candensis	Pondweed	Copper sulfate Dimethomorph Flazasulfron	Wetland	4+	North America	Olette et al., 2008
Gossypium spp.	Cotton	Temik	Herbaceous	8–11	Asia	Anderson et al., 1993 ITRC PHYTO 3
Iris pseudacorus	Yellow Flag	Atrazine	Wetland	4–9	Europe, Asia, Africa	Wang et al., 2012
Iris spp.	Iris	Atrazine	Herbaceous	varies	varies	Burken and Schnoor, 1997
Iris versicolor	Blue Flag Iris	Chlorpyrifos Chlorothalonil Pendimethalin Propiconazole	Herbaceous	4–7	Northern North America	Smith et al., 2008
Kochia spp.		Atrazine Metolachlor Trifluralin	Herbaceous	varies	Europe, Asia	Anderson et al., 1994 ITRC PHYTO 3
Lemna minor	Duckweed	Demeton-8-methyl Malathion Metolachlor Copper sulfate Dimethomorph Flazasulfron Isoproturon Glyphosate	Wetland	3+	Worldwide	Dosnon-Olette et al., 2011 Gao et al., 1998 ITRC PHYTO 3 Olette et al., 2008 Rice et al., 1996b

114

Figure 3.23 (continued)

Latin	Common	Contaminant	Vegetation Type	USDA Hardiness Zone	Native to	Reference
Linum spp.	Flax	2,4-D	Herbaceous	5–9	varies	Anderson et al., 1993 ITRC PHYTO 3
Lythrum salicaria	Purple Loosestrife	Atrazine	Wetland	3+	Europe, Asia	Wang et al., 2012
Myriophyllum aquaticum	Parrot feather	Demeton-8-methyl Malathion Ruelene Atrazine Trifluralin Terbutryn Cycloxidin	Wetland	6–10	South America	Gao et al., 1998 ITRC PHYTO 3 Turgut, 2005
Oryza sativa	Rice	Benthiocarb Parathion Propanil Atrazine Lambda-cyhalothrin Diazinon Fipronil	Wetland	Grown as annual	East Asia	Anderson et al., 1993 Hoagland et al., 1994 ITRC PHYTO 3 Moore and Kroeger, 2010 Reddy and Sethunathan, 1983
Panicum virgatum(9) Panicum virgatum var. Pathfinder (10)	Switchgrass	Atrazine Pendimethalin	Herbaceous	2–9	North America	Albright and Coats, 2014 Burken and Schnoor, 1997 Henderson et al., 2006 Murphy and Coats, 2011
Phaseolus vulgaris	Bush bean	Diazinon Parathion Temik	Herbaceous	Grown as annual	Central America	Anderson et al., 1993 Hsu and Bartha, 1979 ITRC PHYTO 3
Pinus ponderosa	Ponderosa Pine	Atrazine	Tree	3–7	North America	Burken and Schnoor, 1997
Pisum sativum	Peas	Diazinon	Herbaceous	Grown as annual	Europe, Asia	Anderson, Guthrie and Walton, 1993 ITRC PHYTO 3
Plantago major	Broadleaf Plantain	Imidacloprid	Herbaceous	3+	Europe, Asia	Romeh, 2009
Populus spp. Populus deltoides × nigra DN34 Populus spp. ' Imperial Carolina' Populus deltoides I-69/55	Poplar Species and Hybrids	Alachlor Dinoseb Atrazine Dioxane Metolachlor Metribuzin Chlorpyrifos	Tree	varies	varies	Applied Natural Sciences, Inc., 1997 Bin et al., 2009 Black, 1995 Burken and Schnoor, 1997b ITRC PHYTO 3 Lee et al., 2012 Nair et al., 1992 Paterson and Schnoor, 1992 Sand Creek, 2013 Schnoor et al., 1997 Schnoor, 1997

Latin	Common	Contaminant	Vegetation Type	USDA Hardiness Zone	Native to	Reference
Saccharum spp.	Sugarcane	2,4-D	Herbaceous	9+	Central and South America	Anderson et al., 1993; ITRC PHYTO 3
Salix alba L. 'Britzensis'	Coralbark Willow	Trifluralin, Metalaxyl	Shrub	2+	Europe, Asia	Warsaw et al., 2012
Salix nigra	Black Willow	Bentazone	Tree/Shrub	2–8	Eastern USA	Conger, 2003; Conger and Portier, 2006
Salix spp.	Willow	Chlorpyrifos	Tree/Shrub	varies	varies	Lee et al., 2012
Sambucus nigra L. 'Aurea'	Elderberry	Trifluralin, Metalaxyl	Shrub	4+	Europe, Asia, Africa	Warsaw et al., 2012
Sorghastrum nutans var. Holt	Yellow Indiangrass	Altrazine, Pendimethalin	Herbaceous	2–9	North America	Henderson et al., 2006
Trifolium spp.	African clover	2,4-D	Herbaceous	varies	varies	Anderson et al., 1993; ITRC PHYTO 3
Tripsacum dactyloides	Eastern Gamagrass	Chlorpyrifos, Chlorothalonil, Pendimethalin, Propiconazole	Herbaceous	4–9	Eastern USA	Smith et al., 2008
Triticum aestivum	Wheat	2,4-D, Diazinon, MCPA, Mecoprop	Herbaceous	3–8	Asia	Anderson, Guthrie and Walton, 1993; ITRC PHYTO 3
Typha spp.	Cattail	Atrazine	Wetland	3–10	North America, Europe, Asia	ITRC PHYTO 3; Kadlec and Knight, 1996
Vetiveria zizanioides, (syn. Chrysopogon zizanioides)	Vetiver	Endosulfan, Atrazine	Herbaceous	9–11	India	Abaga et al., 2012; Marcacci and Schwitzguébel, 2007
Zea mays	Corn	Alachlor, Atrazine, Diazinon, Temik	Herbaceous	4–11	USA	Anderson et al., 1993; ITRC PHYTO 3; Paterson and Schnoor, 1992

Project name: Farmer's Flying Service (Sand Creek, 2013)

Location: Bancroft, WI

Consultants/Scientists: Sand Creek Consultants, Wisconsin; University of Wisconsin, Stevens Point – Mark Dawson, William DeVita and Christopher Rog

Date installed: 2000

Plant species: 834 hybrid poplars (*Populus* NM-6, DN-34 and DN-17), Eastern Cottonwood (D-105) and hybrid willows (*Salix* SX-61 and SV-1)

Contaminant: Dinoseb ⬤ at 6,600 ppb (herbicide used against broadleaf weeds)

Farmer's Flying Service was an aerial agricultural spraying service based in Bancroft, WI. Years of pesticide storage and handling on the site caused extensive contamination of groundwater and soil. Following the installation of a poplar and willow phytoremediation system, this site experienced close to a 100% reduction in the concentration of Dinoseb in groundwater at the downgradient monitoring location, decreasing from a high of 1,549 ppb observed two years after planting in 2000, to no detection (<5 ppb) in 2004 and 2005 after the trees had become fully established. In comparison, source-area concentrations remained elevated.

Phytohydraulics ⬤ are believed to be the primary phytoremediation mechanism observed, with phytodegradation 🌱 being a small fraction of this response, due to the known challenges of degrading Dinoseb in an aerobic environment. Nevertheless, the planting at Farmer's Flying

116

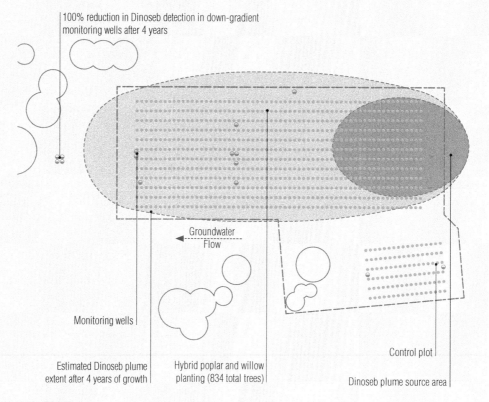

Figure 3.24 Case Study: Farmer's Flying Service, Bancroft, WI
Scale: 1 in = 100 ft

Barrels of pesticides (Dinoseb) excavated from the site. The pesticides had leached into the groundwater, polluting location drinking water supplies.

20 centimeter-long poplar cuttings were installed in rows.

Two weeks after planting.

Four weeks after planting.

One year after planting (5-6 ft height of growth)

Two years after planting. The tree in the center is dying because it has tapped into the polluted groundwater which was too toxic for successful plant growth. It is normal in a phytoremediation installation that plants in the most toxic areas do not survive.

117

Seven years after planting. The site has been conditionally closed and remediated, but the trees remain to maintain hydraulic control.

Seven years after planting. Taller trees indicate where the groundwater was originally less polluted and the shorter trees on the left demarcate areas that originally had greater pesticide concentrations. The shorter trees on the right are test hybrids of *Populus* and *Salix* that were less robust than the hybrids used in the remaining stand.

Figure 3.25 Case Study: Farmer's Flying Service, Bancroft, WI

Service is among the most successful pesticide phytoremediation efforts to date. Today the site is closed, but the phytotechnology planting continues its function to maintain hydraulic control. Additional research is needed to evaluate potential Dinoseb degradation rates in the hybrid poplars by searching for Dinoseb metabolite compounds within both the rhizosphere and the tree biomass.

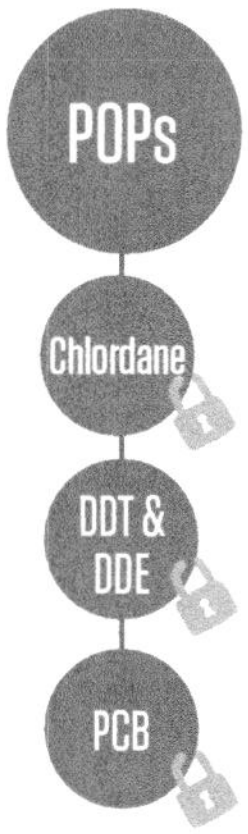

Figure 3.26 POPs

● Persistent Organic Pollutants (POPs)

Specific contaminants in this category: There are 24 toxic chemicals that originate from man-made sources and are categorized as POPs (US EPA, 2013c). The most commonly encountered POPs include a group of 12 'dirty dozen' toxic chemicals that can bioaccumulate in the soil, including:

- formerly used pesticides: DDT (DichloroDiphenylTrichloroethane), DDE (metabolite of DDT), Aldrin, Chlordane, Mirex, Toxaphene, Hexachlorobenzene, Dieldrin, Endrin and Heptachlor
- formerly used industrial chemicals: PCBs (Polychlorinated biphenyls) – mostly used as coolants
- byproducts from other chemical processes: Dioxins (polychlorinated dibenzo-p-dioxins) and furans (polychlorinated dibenzo-p-furans) (White and Newman, 2011)

Typical sources of POPs: Pesticides, industrial coolants in motors, fire retardants, transformers, air conditioners and in caulking around windows. Atmospheric deposition distributes POPs away from the original source or point of origin.

Typical land uses with POPs contamination: Industrial sites, transformers, railroad yards, buildings with old caulking especially around windows and doors. Locations where DDT, DDE and other pesticides were sprayed.

Why these contaminants are a danger: POPs bind tightly to soils and tend to bioaccumulate and biomagnify in food chains (White and Newman, 2011). PCBs, one category of POPs, were banned from use in 1979 in the US and worldwide in 2001, due to studies that linked them with cancer (Porta and Zumeta, 2002). The insecticide DDT (and its metabolite DDE) was banned in 1972 in the US and worldwide in 2001, due to its toxic effects on wildlife, especially birds (Moyers, 2007). Chlordane, an insecticide, was widely used not only on agricultural fields but on residential lawns and for termite control through the 1980s (Metcalf, 2002).

Mechanisms utilized: Phytostabilization ◯ ; some limited success with Phytoextraction and Rhizodegradation .

POPs adhere tightly to soil particles and enter groundwater only slowly, if at all. DDT and Chlordane and industrial uses of PCBs have contaminated soils, which remain polluted even after 100 years. Currently, there are no widely accepted field-scale phytotechnology systems that can be realistically applied to degrade or extract these contaminants (White and Newman, 2011). Some studies have shown that breakdown of DDT, PCBs and other POPs is possible with plants and associated microbes, but the degradation pathways and resulting metabolites are still unclear and findings are not consistent for field application. The structure of the molecules in this category makes them highly recalcitrant. Instead, plants can be used to help stabilize the contaminant onsite.

Some studies have found that certain species of plants can extract or degrade POPs at higher rates than other species. For example, recent research has identified one species of a zucchini/pumpkin cross that can accumulate dioxins (White and Newman, 2011). However, the extraction potential is likely only a small fraction of the contaminant concentration that can exist in polluted soils, therefore field application of this technology remains problematic (White, 2010). Even if the plants are harvested and removed from the site over repeated growing seasons, only the bioavailable fraction of the pollutant could be removed, and pollutants adhered to the soil may remain indefinitely. Mechanistic studies are being performed by scientists to understand why certain species take up some POPs more than others. In the future, potential species may be found and utilized for degradation or extraction and harvesting, but at this time these methods are not applicable.

In soil

 Stabilization:

- Planted Stabilization Mat: Chapter 4, see p. 202

One reason why these compounds are so persistent in the environment is their low solubility in water, so water vectors have not been considered here. Plant species for POPs extraction have shown some success at lab or pilot scale only, but should not be considered for field-scale implementation at this time. A list of plants tested in these studies is included here (Figure 3.27); however, for current field application, plants should be utilized only as stabilization covers. Any species that can be effectively grown to create a thick cover and prevent erosion or dust mobilization can be considered to help reduce the risk of contact with the soil.

Case Study
POPs

Project name: Etobicoke Field Site (Ficko et al., 2010; Ficko et al., 2011; Whitfield et al., 2007; Whitfield et al., 2008)

Location: Etobicoke, Canada (outside Toronto)

Institution/Scientists: Royal Military College of Canada, Queens University – Barbara A. Zeeb, Sarah A. Ficko, Allison Rutter

Date installed: Studies ongoing since 2004

Plant species installed: *Cucurbita pepo ssp. pepo* 'Howden' (Pumpkin), *Carex normalis* (Sedge) and *Festuca arundinacea* (Tall Fescue)

119

Figure 3.27 Persistent Organic Pollutants-Phytotechnology Plant List

Plants shown to have some extraction potential in scientific studies. These should *not* be considered for field-scale remediation

Latin	Common	Contaminant	Vegetation Type	USDA Hardiness Zone	Native to	Reference
Carex aquatica	Sedge	PCB	Herbaceous	3+	North America	Smith et al., 2007
Chrysanthemum leucanthemum	Oxe-Eye Daisy	PCB	Herbaceous	5–6	Europe	Ficko et al., 2010 Ficko et al., 2011
Cucumis sativus L. 'Dlikatess'	Cucumber	PCDD, PCDF	Herbaceous	Grown as annual	India	Hulster et al., 1994
Cucurbita pepo Cucurbita pepo L. 'Black Beauty' Cucurbita pepo L. convar. Giromontiina 'Diamant F1' Cucurbita pepo L. 'Raven' Cucurbita pepo L. 'Senator hybrid'	Zucchini	p,p'-DDE (Weathered DDT) DDT HCH 2,2-bis(p-chlorophenyl)-1,1-dichloroethylene(p,p'-DDE) Chlordane PCDD PCDF	Herbaceous	Grown as annual	North and Central America	Bogdevich and Cadocinicov, 2009 Hulster et al., 1994 Isleyen et al., 2013 Lunney et al., 2004 Mattina et al., 2004 Wang et al., 2004 White, 2001 Zeeb et al., 2006
Cucurbita pepo Cucurbita pepo L. 'Howden'	Pumpkin	Weathered DDT PCB	Herbaceous	Grown as annual	North and Central America	Kelsey et al., 2006 Lunney et al., 2004 Wange et al., 2004 Whitfield-Aslund et al., 2007 Whitfield-Aslund et al., 2008
Daucus carota	Queen Anne's Lace	PCB	Herbaceous	3–11	North America	Ficko et al., 2010
Festuca arundinacea	Tall Fescue	PCB (Delor 103 and 106)	Herbaceous	3–8	Europe	Pavlikova et al., 2007
Glycine max	Soybean	PCB	Herbaceous	Grown as annual	Asia	McCutcheon and Schnoor, 2003
Lagenaria siceraria	Calabash, Gourd	Heptachlor Heptachlor epoxide	Herbaceous	Grown as annual	Asia, Africa	Campbell et al., 2009
Lolium multiflorum	Ryegrass	p,p'-DDE (Weathered DDT)	Herbaceous	5+	Europe	White, 2000
Maclura panifera	Osage Orange	PCB	Tree	4–9	North America	Olson et al., 2003
Medicago sativa	Alfalfa	p,p'-DDE (Weathered DDT)	Herbaceous	3–11	Middle East	White, 2000
Mentha spicata	Spearmint	PCB	Herbaceous	3–11	Europe Middle East	Gilbert and Crowley, 1997
Morus rubra	Red Mulberry	PCB	Tree	4–9	North America	Olson et al., 2003
Phaseolaris vulgaris	Pole Bean	p,p'-DDE (Weathered DDT)	Herbaceous	Grown as annual	Central America	White, 2000

Latin	Common	Contaminant	Vegetation Type	USDA Hardiness Zone	Native to	Reference
Pinus nigra	Austrian Pine	PCB	Tree	5+	Europe	Leigh et al., 2006
Polygonum persicaria	Spotted Lady's Thumb	PCB	Herbaceous	4–8	Eurasia	Ficko et al., 2010
Rumex crispus	Curly Dock	PCB	Herbaceous	5+	Eurasia	Ficko et al., 2011
Salix caprea	Goat Willow	PCB	Shrub	4–8	Europe, Asia	Leigh et al., 2006
Sesamum indicum	Sesame	Lindane (γHCH)	Herbaceous	9–11	Africa	Abhilash and Singh, 2010
Solanum torvum L.	Turkey Berry	Lindane, HCH	Herbaceous	9+	North and Central America	Abhilash et al., 2008
Solidago canadensis	Canadian Goldenrod	PCB	Herbaceous	3+	Northeastern North America	Ficko et al., 2010 Ficko et al., 2011
Spartina pectinata	Prairie Cordgrass	PCB	Herbaceous	4–9	North America	Smith et al., 2007
Trifolium repans	White Clover	PCB	Herbaceous	3+	Europe	McCutcheon and Spoor, 2003
Vicia cracca	Cow Vetch	PCB	Herbaceous	5+	Eurasia	Ficko et al., 2010
Wilthania somnifera L. (Dunal)	Indian Ginseng	Lindane HCH	Herbaceous	9+	India	Abhilash et al., 2008

121

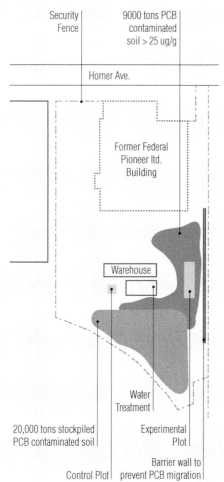

Security Fence

9000 tons PCB contaminated soil > 25 ug/g

Horner Ave.

Former Federal Pioneer ltd. Building

Warehouse

Water Treatment

20,000 tons stockpiled PCB contaminated soil

Experimental Plot

Barrier wall to prevent PCB migration

Control Plot

122

Figure 3.28 Case Study: Electric Field, Etobicoke, Canada
Scale: 1 in = 200 ft

Amendments: NPK fertilizer, rototilling

Contaminant: PCBs ● (Aroclors 1254/1260) @ 37ug/g

Target media and depth: Soils, top 60 centimeters (2 feet)

This field site is a former electric transformer manufacturing facility located in the town of Etobicoke just outside Toronto, Canada. Approximately 9,000 tons of PCB-contaminated soil are located at the site and have been secured with an asphalt cap. Groundwater flowing through the contaminated soil is treated with an on-site wastewater treatment facility before being discharged into the city sewer system. An area of asphalt was excavated and converted into a phyto research-study plot.

Studies on the site first started with field trials to see if various PCB-accumulating species that were identified in greenhouse studies might accumulate PCBs in field conditions, including *Cucurbita pepo ssp. pepo* "Howden" (Pumpkin), *Carex normalis* (Sedge) and *Festuca arundinacea* (Tall Fescue) (Whitfield et al., 2007). Further studies focused on different pumpkin planting methods, including the use of a trellis system and planting in potting soil (Whitfield et al., 2008). This study showed the importance of the pumpkin root structure (adventitious root system) in the uptake of PCBs and also revealed that increased planting density might be

Cucumbers, Tall Fescue and Sedge were planted on an old electric transformer site contaminated with PCBs to test for extraction potential. Spontaneous vegetation emerged on the edges of the plot, and those species were also tested for extraction. Of the 27 plant colonizers, 17 native invader "weeds" were found to extract greater amounts of PCBs than the original test species. These species are currently being tested on other sites. Phytoremediation potential for PCBs has not yet been verified, and plants are not viable means for extraction/remediation of contaminated PCB sites at this time.

Figure 3.29 Case Study: Electric Field, Etobicoke, Canada

123

counterproductive to overall PCB uptake. The study still showed, however, that although this particular subspecies of pumpkin can take up more PCBs than other plants, it likely does not take up enough to be useful for remediation at contaminated sites.

Further investigations have utilized a unique method to identify additional PCB-extracting species by analyzing weedy species that naturally colonized the bed edge of planting plots used in previous studies. A 2010 study by Sarah Ficko et al. revealed 27 natural plant colonizers of the PCB-contaminated soil. When analyzed for PCB concentrations within their tissues, it was found that 17 of the species had accumulated PCBs at levels similar to or greater than the previously identified PCB-accumulator pumpkin. Three of the species, *Solidago canadensis* (Canada Goldenrod), *Chyrsanthemum leucanthemum* (Ox-Eye Daisy) and *Rumex crispus* (Curly Dock), were further tested at the site (Ficko et al., 2011). They were found to have a greater extraction capacity than the pumpkin. However, further research is needed to identify the specific mechanism of PCB extraction within the identified weed species, as well as to optimize the planting density and methods of harvest. These studies reveal the importance of identifying natural colonizers of contaminated sites so as to advance knowledge of species capable of contaminant extraction, as well as of identifying commercially viable species for phytoextraction. Phytoextraction of PCBs is still being investigated in scientific field trials, and research over time will show if it may be viable for future field-scale remediation. Field-scale remediation of areas of low PCB soil concentration may be viable in the future if the time of remediation is flexible (i.e. it may take many, many years).

Other organic contaminants of concern

Specific contaminants in this category: This category is a catch-all for other organic contaminants of concern that have not been included in other categories. The following are included:

- Aircraft de-icing fluids and coolants: Ethylene glycol (EG) and propylene glycol (PG), two types of alcohol hydrocarbons
- Embalming fluids: Formaldehyde and methanol, two types of hydrocarbons that are Volatile Organic Compounds (VOCs)
- Pharmaceuticals and personal care products: A broad range of man-made medicines and lotions, including antibiotics, hormones, anti-depressants, cosmetics and many others.

Typical sources of other organic contaminants of concern

- The primary source of ethylene glycol in the environment is from run-off at airports, where it is used in de-icing agents for runways and airplanes. It is also used in brake fluids and antifreeze for cooling and heating systems. Propylene glycol, the additional additive in de-icing fluids, is generally recognized as safe and, additionally, is used in making polyester and in food processing (NHDES, 2006).
- Formaldehyde and methanol are used in embalming and preservation processes and are also prevalently used in industrial processes and manufacturing. They are also found in automobile exhausts.
- Pharmaceutical pollution primarily occurs from both human and animal wastewater. Most municipal water-treatment facilities do not treat for pharmaceuticals and they are often released when the treated water is discharged. The same is true for individual septic systems, where pharmaceutical treatment is not targeted. Animal-production wastewater can also contain these pollutants.

Typical land uses with contamination from other organic contaminants of concern: De-icing: airports, military uses; Embalming: funeral homes, graveyards and cemeteries; Pharmaceuticals: suburban homes, wastewater treatment facilities, animal producers and feed-lots.

Why these contaminants are a danger: Formaldehyde is a known carcinogen. However, formaldehyde typically breaks down quickly, within one day in both air and water (ATSDR, 2013), and ethylene glycol typically breaks down in the natural environment in about 10 days. These contaminants do not tend to be a danger to humans outdoors unless ingested in large quantities. However, they can easily dissolve in water and migrate. Pharmaceuticals also quickly dissolve in water and can contaminate drinking water supplies and surface water bodies. The effects of low-level repeat exposures for many pharmaceuticals are unknown, but increasing concern is emerging.

Mechanisms utilized: Rhizodegradation ⬤, Phytohydraulics ⬤, Phytovolatilization ⬤, Phytodegradation ⬤, Phytostabilization ⬤.

The organic contaminants covered in this 'other' category are primarily a concern because they can quickly dissolve in water and potentially contaminate drinking water and surface water bodies.

De-icing fluids can be easily degraded in constructed wetlands. Formaldehyde can also quickly degrade in the natural environment. Research in plant-based remediation of pharmaceuticals in wastewater is just beginning, but promising results have been seen in constructed wetlands for degradation of many pharmaceutical compounds with planted systems. It is anticipated that future findings and recommendations will emerge.

In water

Groundwater control typologies: The time frame is dependent on how contaminated the groundwater is, and the speed and volume of the plume.

- Groundwater Migration Tree Stand: Chapter 4, see p. 213

Stormwater or groundwater degradation: De-icing and embalming fluids have been shown to be successfully degraded quite quickly in water with constructed wetlands. In addition, promising studies on pharmaceutical degradation in wetlands have also been released. Contaminated groundwater or wastewater is cleansed by being pumped up naturally by the plants. Plant species and case studies for constructed wetlands have not been included here, since they have been widely documented in publications by others.

- Stormwater Filters: Chapter 4, see p. 235
- Multi-Mechanism Buffers: Chapter 4, see p. 234
- Surface-Flow Constructed Wetland: Chapter 4, see p. 238
- Subsurface Gravel Wetland: Chapter 4, see p. 241
- Floating Wetland: Chapter 4, see p. 242

125

II   Inorganic contaminant classifications

Inorganic contaminants cannot be degraded with plant-based systems. Because inorganic pollutants are elements found on the periodic table, they cannot be broken down into smaller parts. Instead, the form of the inorganic contaminant can be changed, stabilized on site, or moved into and stored in plant tissues (Pilon-Smits, 2005). Plants can help to change the form of the element, for example from a solid to a gas or from an oxidized state to a different state, to mitigate risk. In addition, in limited cases, some plants can extract inorganic contaminants. The plants can then be cut down and harvested to remove the pollutant off site. For further information on general principles of inorganics, see Chapter 2, p. 55.

Plant macronutrients: nitrogen, phosphorus and potassium

Specific contaminants in this category: Nitrogen (including various forms of nitrogen: ammonium (NH_4^+), nitrate (NO_3^-), nitrite (NO_2^-)), Phosphorus (including phosphate), Potassium (K)

Figure 3.30 Nutrients

Typical sources of nutrient pollution: Fertilizer application, manure, human wastes and septic systems, atmospheric deposition, leachate from landfills, air pollution from vehicle and industrial exhausts.

Typical land uses with nutrient contamination: Agricultural fields, animal producers and feed-lots, residential lawns, parks and open spaces (including golf courses), roads, landfills, and any use with a septic system.

Why these contaminants are a danger: Excessive macronutrients leach into waterways and feed algal blooms that consume life-supporting oxygen in waterways (eutrophication). Typically, nitrogen has the most detrimental effect on saltwater systems and phosphorus affects freshwater systems. The resulting low-oxygen 'dead zone' destroys aquatic life. In addition, excessive nitrogen in drinking water can cause 'Blue Baby Syndrome,' where newborns that drink the water become seriously ill with low levels of oxygen in the blood (US EPA, 2013a). Nitrogen can quickly contaminate drinking water supplies since agricultural uses and septic systems with excessive nutrients often leach water into drinking water aquifers. Excess potassium is not known to pose significant health or ecological risks at this time.

Summary

1 Nitrogen

Nitrogen can exist in many forms, from atmospheric nitrogen gases (N_2, N_2O and NO_x) to ions of nitrogen with various charges that can be found in soil and water (ammonium+, nitrite- and nitrate-) to solid organic forms that can be bound in soils and plant life. In the nitrogen cycle, nitrogen moves readily between these various forms, and bacteria and plants play a significant role in these transformations. Excess amounts of nitrogen in the environment are typically a problem when the ionized forms of nitrogen leach into surface waters or groundwater, rather than when the nitrogen is in the form of a gas, bound to soil or incorporated into organisms as organic nitrogen. Denitrifying bacteria associated with plant roots and soils can turn these polluting forms of nitrogen back into atmospheric gas, removing them as pollutants from soils and water. Returning excess nitrogen to the atmosphere is regarded as the best remediation solution, since nitrogen gas makes up almost 80% of the earth's atmosphere.

Planted systems can speed up the work of the denitrifying bacteria in soils that convert nitrogen into a gas. By supplying denitrifying bacteria with the sugars, oxygen and root exudates they need in order to thrive, plants can create soil zones where nitrogen is quickly transformed and returned to the atmosphere. In addition, plants can also use the polluting forms of nitrogen for plant growth, converting the nitrogen into plant biomass and other forms of organic nitrogen, thereby removing it from its mobile state in water that causes risks to human and environmental health.

Removal of nitrogen from soils, groundwater and wastewater is one of the best applications of phytotechnology, with decades of notable field-scale project successes. The three most typical nitrogen-remediation scenarios are remediation of polluted groundwater, wastewater or surface water. For groundwater remediation, high evapotranspiration-rate plants are used as solar pumps to pump up the water, while associated bacteria transform the nitrogen into a gas or the plant turns it into a form of organic nitrogen. With wastewater, the water is typically irrigated onto plants, where either the plants take up the nitrogen or it is converted to a gas by bacteria in the root zone of the plant. Wastewater can also be treated with constructed wetlands. Lastly, for surface water, constructed wetlands can be utilized to remove the nitrogen, or stormwater filters can address excessive nitrogen at the source.

Planting specifics

The following planting typologies can be considered for use in conjunction with the plant lists below.

In soil

Typologies for removal from soil: Potential removal time frame 0–5 years

- Degradation Cover*: Chapter 4, see p. 222
- Degradation Hedge*: Chapter 4, see p. 220
- Degradation Living Fence*: Chapter 4, see p. 220
- Degradation Bosque*: Chapter 4, see p. 218

* *Note*: Nitrogen is not actually degraded in this system, but instead volatilized into the air or metabolized and turned into organic nitrogen within the plant tissues. However, degradation typologies are noted because the nitrogen is 'used up', no byproducts are left behind, and no harvesting is necessary.

In water

Typologies to control contaminated groundwater, stormwater, or wastewater and transform nitrogen back into a gas: Timeframe for removal typically within 0–10 years.

- Groundwater Migration Tree Stand: Chapter 4, see p. 213
- Phytoirrigation: Chapter 4, see p. 207
- Stormwater Filters: Chapter 4, see p. 235
- Evapotranspiration Cover: Chapter 4, see p. 204
- Surface-Flow Constructed Wetland: Chapter 4, see p. 238
- Subsurface Gravel Wetland: Chapter 4, see p. 241

Since all plants use nitrogen and support denitrifying bacteria, any kind of plant can provide some form of nitrogen remediation from soils and water. However, the method that provides the quickest remediation tends to be a system that includes plants with very high growth rates and evapotranspiration rates. Nitrogen is used up quickly, or the plant acts like a large reactor, priming the soil bacteria for speedy conversion of the nitrogen into a gas. Plant species that produce a lot of biomass have been those most successfully used in studies to remove high levels of nitrogen in soils and groundwater.

Consider species that produce high biomass. For a list of such species see Figure 2.17 in Chapter 2, p. 46.

*Case Study
Nitrogen*

Project name: Poplar Tree Farm at the Woodburn Wastewater Treatment Facility (Stultz and CH2MHill, 2011; Smesrud, 2012; Woodburn, 2013)

Location: City of Woodburn, OR

Consultants/Scientists: Mark Madison, Jason Smesrud, Jim Jordahl, Henriette Emond and Quitterie Cotten: CH2MHill, Portland, OR; Oregon State University Department of Biological and Ecological Engineering; Ecolotree; GreenWood Resources; Hydrologic Engineering, Inc.

Date installed: 1995–1997, 2.8 hectares (7 acres) poplar plantation pilot project developed to refine design criteria for full-scale tree production, including study of poplar tree irrigation water requirements; 1999, full-scale 34 hectares (84 acres) poplar plantation developed to support WWTP compliance with summer-time nitrogen (ammonia) load limits on the Pudding River; 2008–2009, additional pilot testing projects implemented to test high-rate irrigation, coppice management and use of constructed wetlands for temperature treatment.

Species installed: Hybrid poplar (*Populus*) at 2 meters × 4 meters (6.5 feet × 13 feet) spacing.

Amendments: Micro-spray irrigation of trees with advanced secondary-treated wastewater and surface application of Class B biosolids during growing season.

Contaminant: Nutrient-rich, warm wastewater with high levels of nitrogen (ammonia)   is toxic to nearby rivers.

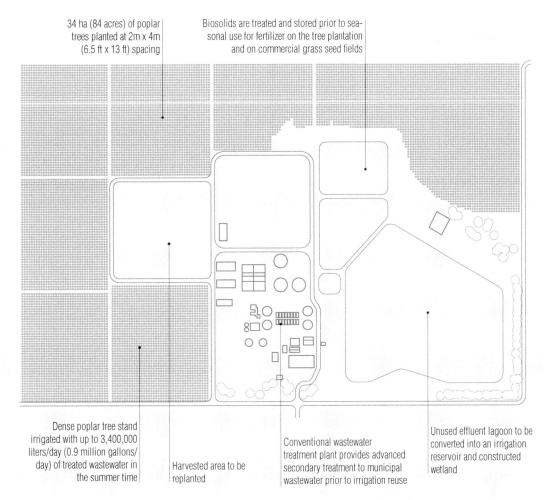

34 ha (84 acres) of poplar trees planted at 2m x 4m (6.5 ft x 13 ft) spacing

Biosolids are treated and stored prior to seasonal use for fertilizer on the tree plantation and on commercial grass seed fields

Dense poplar tree stand irrigated with up to 3,400,000 liters/day (0.9 million gallons/day) of treated wastewater in the summer time

Harvested area to be replanted

Conventional wastewater treatment plant provides advanced secondary treatment to municipal wastewater prior to irrigation reuse

Unused effluent lagoon to be converted into an irrigation reservoir and constructed wetland

Figure 3.31 Case Study: Poplar Tree Farm, Woodburn Wastewater Treatment Facility, Woodburn, OR
Scale: 1 in = 500 ft

The City of Woodburn, Oregon's Poplar Tree Farm is the first known phytoremediation planting completed in the United States designed to beneficially reuse treated municipal wastewater nitrogen while creating a commercial wood crop. Strict nitrogen (in the form of ammonia) discharge limits for the Pudding River encouraged designers to think 'outside the box' and develop a new natural treatment system for the wastewater generated by the city's approximately 23,000 residents.

A conventional wastewater treatment plant provides advanced secondary treatment for municipal wastewater. The partially treated wastewater is then used to irrigate a poplar tree farm on land surrounding the treatment facility, helping to reduce the amount of ammonia nitrogen discharged into the nearby Pudding River during the low-flow summer months. This beneficial reuse of nutrients and water encourages tree growth and creates a commodity crop for the city, with poplar trees harvested every 7–12 years. The harvested poplar trees are processed into solid-wood products and wood chips for creating paper or cardboard, creating an income stream that helps offset some of the management costs for the municipality.

130

Micro-spray irrigation at base of trees discharges nutrient-rich effluent during the summer time for irrigation and fertilization of poplar trees.

Canopy of 11-year-old poplars creates spectacular 75 ft tall outdoor room with dramatic, cathedral-like effect. Trees in this photo are ready for harvesting into wood chips for paper/ cardboard or as whole trees for milling into solid wood products.

Effluent storage lagoon in foreground will be converted into an irrigation regulation reservoir for temporary storage and constructed wetland for effluent cooling. Poplar tree stand, visible in background, blends well with the surrounding rural agrarian landscape.

Mobile tree harvest and processing operation in progress at the Woodburn Poplar Tree Farm. Poplar tree chips are used for cardboard manufacturing.

Figure 3.32 Case Study: Poplar Tree Farm, Woodburn Wastewater Treatment Facility, Woodburn, OR

In 1995, 2.8 hectares (7 acres) of poplar trees were planted as a pilot project to refine the design criteria needed to develop this system. This included a study of poplar tree irrigation water requirements. The pilot project proved successful, and in 1997 the full-scale system including irrigation and biosolids pumping with conveyance and monitoring facilities was constructed and an additional 31 hectares (77 acres) of trees were planted. The fast-growing poplars add about 2.4 meters (8 feet) in height every year. They reach their maximum water and nitrogen uptake rates after 4 years of growth and are harvested every 7–12 years as an agricultural crop. The treated wastewater and biosolids nitrogen application rates are kept within standard agricultural levels, to ensure the contaminated wastewater does not contact underlying groundwater. This poplar tree system has proven to be cost-effective in reducing the nutrient loading of surface water in comparison to other available conventional wastewater treatment technologies. The system also has lower energy demands than other conventional treatment methods and has developed broad support and acceptance from the public.

In the future, new temperature limits are anticipated to mitigate potential wastewater effluent discharge impacts on local cold-water fisheries (salmon, trout and steelhead). The existing poplar tree irrigation systems help to reduce the amount of thermal load discharged to surface water in the summer time. However, additional constructed wetlands utilized for passive effluent cooling prior to river discharge are also planned for the portion of effluent not consumed by the tree plantation.

Highlights of the project include:

- 84 acres of poplar trees are managed on a 7–12 year harvest rotation to beneficially reuse treated municipal wastewater nitrogen, while creating a commercial wood crop
- up to 3,400,000 liters/day (0.9 million gallons/day) of wastewater applied to the trees via microspray irrigation during the growing season
- up to 269 kg/hectare/year (240 lbs/acre/year) of nitrogen are applied as both irrigated effluent and biosolids to mature poplar trees.

2 Phosphorus

Unlike nitrogen, phosphorus cannot be removed from a terrestrial system and converted into a gas. As an inorganic mineral, it usually persists in the environment as phosphate, the oxidized form of phosphorus. Phosphorus contamination usually occurs in surface waters when small particles of soil phosphorus are picked up by wind or water and washed into water bodies. This often happens in stormwater, where run-off from roadways or agricultural fields moves into freshwater bodies and causes an explosion of the algae population that leads to depletion of oxygen, seriously affecting aquatic ecosystems.

The best method to remediate phosphorus is to capture and stabilize it on site. Since plants need phosphorus as an essential macronutrient, they can extract some phosphorus from the soil and metabolize it into the biomass of the plant. Phytoextraction of phosphorus-contaminated soils is demonstrated to be effective in extracting up to 30 lbs (13.6 kg) of phosphorus per acre per year (Muir, 2004). In temperate climates, if the leaves are allowed to fall and decay, the phosphorus will return to the soil, therefore plants must always be harvested and taken off site to remove the phosphorus. Generally, phytoextraction of phosphorus is not widely utilized because 30 lbs per acre removal rates are generally not large enough to make extraction and harvesting a useful option for remediation. Only in cases where high-biomass species are utilized can extraction of phosphorus from soils be considered.

Instead, most phytotechnology systems for phosphorus target phosphorus filtration from water, stabilizing it in surrounding soils. Phosphorus contamination in water is typically in two forms: (1) as sediment, meaning it is bound to soil particles as a sediment in water; and (2) dissolved, as soluble phosphorus dissolved in the water. As the contaminated water passes through remediation systems, phosphorus in the form of sediment can be physically removed with sedimentation basins and forebays via settling. The sediment must then be dug out and removed from the site. Dissolved phosphorus can be removed from the water when it comes into contact with the soil and is adsorbed. The phosphorus is bound to the soils and stabilized on site, letting the clean water pass through. When plants are added to soils, they can help create organic binding sites for both the sediment and the dissolved phosphorus particles to stick. Soil contact is the most important mechanism for phosphorus immobilization via adsorption to clays and organic matter and precipitation, when it forms phosphate compounds (for example with calcium, iron and/or aluminum). For each 1,000 cubic feet of soil, about of 40 lbs (18 kg) of phosphorus can be immobilized, significantly more than from plant uptake (Sand Creek, 2013). For this reason, stormwater filters and constructed wetlands created to remove phosphorus usually have carefully designed sedimentation areas and engineered soil media with infiltration to provide the maximum amount of binding sites and precipitation compounds for phosphorus removal, rather than extraction with plants. These soils may at some point reach a phosphorus 'carrying capacity'. Plants added to the system, however, help continuously renew the soils, creating new binding sites so that carrying capacity is not reached.

Planting specifics

The following planting typologies can be considered for use.

In soil

Extraction: at maximum rate of 30 lbs per acre per year

• Extraction Plots: Chapter 4, see p. 224

Stabilization: to prevent wind and water erosion of phosphorus in soils

• Planted Stabilization Mat: Chapter 4, see p. 202

In water

To control contaminated groundwater:

- Groundwater Migration Tree Stand: Chapter 4, see p. 213
- Phytoirrigation: Chapter 4, see p. 207

To remove from surface and groundwater: Primarily by physically trapping sediment and binding phosphorus to planting media

- Stormwater Filters: Chapter 4, see p. 235
- Multi-Mechanism Buffers: Chapter 4, see p. 234
- Surface-Flow Constructed Wetland: Chapter 4, see p. 238
- Subsurface Gravel Wetland: Chapter 4, see p. 241

Since all plants use phosphorus as a macronutrient, any plant species can provide some phosphorus extraction from soils and water. However, this is usually not enough to remediate polluted soils and water. The systems that provide the quickest remediation tend to be systems that create the most binding sites in the soil for phosphorus immobilization. Any living plant species needs phosphorus and can help create and maintain organic binding sites in soils. The best species to help maintain phosphorus-removal properties of soil will have dense, thick root systems, grow aggressively and entirely cover any open soils.

133

Case Study
Phosphorous

Project name: Willow Lake Water Pollution Control Facility (Eisner and CH2MHill, 2011; Salem, 2013)

Location: Salem, OR

Consultants/Scientists: Mark Madison, Henriette Emond, Dave Whitaker and Jason Smesrud, CH2MHill, Portland, OR; Bob Knight, GreenWorks; Stephanie Eisner, City of Salem, OR

Date installed: 2002, 4 hectares (10 acres) of constructed treatment wetlands

Plant species: 10 species originally planted in constructed treatment wetlands. Over time, diversity has decreased to five plant species. Cattail (*Typha spp.*), Soft-stem Bulrush (*Scirpus validus/Schoenoplectus tabernaemontani*), Rush (*Juncus effusus*), Duckweed (*Lemna minor*), and Hydrocotyle (*Hydrocotyle umbellata*) are currently the predominant species.

Contaminant: Advanced secondary-treated municipal wastewater effluent includes trace levels of nitrogen   , phosphorus   , heavy metals ▓, bacteria and pathogens and has an elevated temperature relative to receiving waters in certain instances.

The Willow Lake Water Pollution Control Facility serves the wastewater treatment needs of the 229,000 residents of Salem, Keizer and Turner, Oregon. In 2002, the facility constructed 4 hectares (10 acres) of constructed wetlands on former agricultural land to test natural treatment systems for potential use in providing additional advanced wastewater treatment. These wetlands included two approximately 1.6 hectare (4 acre) surface-flow wetlands, two

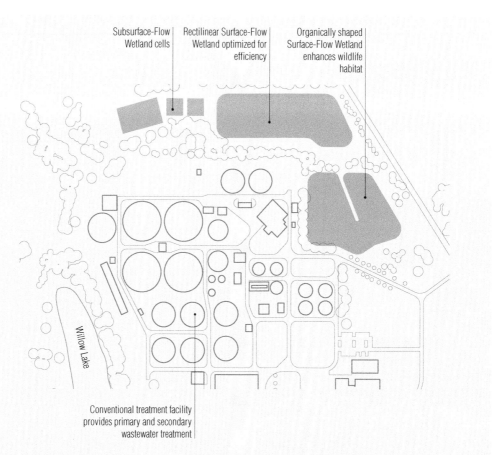

Subsurface-Flow Wetland cells

Rectilinear Surface-Flow Wetland optimized for efficiency

Organically shaped Surface-Flow Wetland enhances wildlife habitat

Willow Lake

Conventional treatment facility provides primary and secondary wastewater treatment

Figure 3.33 Case Study: Willow Lake Water Pollution Control Facility, Salem, OR
Scale: 1 in = 500 ft

134

0.4 hectares (1 acre) subsurface-flow wetlands and one 0.4 hectares (1 acre) surface overland discharge area.

As the system was constructed for the purposes of research and demonstration, each of the surface-flow wetlands was constructed with a different goal in mind. One was optimized for wildlife enhancement through its organic shape and deep open-water areas, while the other was optimized for maximum water shading and temperature-treatment efficiency by its rectilinear form and consistent shallow depth. Both surface-flow constructed wetlands and the subsurface gravel wetland system have provided valuable information on the ability of these systems to provide significant nitrogen and phosphorus removal and passive temperature reduction.

Ten wetland vegetation species were originally planted; however, today only five plant species dominate. Waterfowl grazing on wetland plants and vole damage to upland seedlings were a challenge to control during the original planting, while invasive species have also posed a problem.

The site is open to the public and is well used by many residents for recreation and wildlife viewing. The facility has also served as an educational tool to engage students and residents on the subjects of water quality control and wildlife habitat enhancement.

A surface-flow wildlife wetland both removes nitrogen and phosphorus from wastewater and provides deep pools and organic edges to maximize wildlife benefit.

Mixed shrub layers and tall cattails grow between recreation trails and wetland, minimizing direct recreation exposure to water treatment surfaces.

135

Recreation trails are provided between wetland areas and are open to the public.

A subsurface gravel wetland cell treats wastewater vertically. A pipe in the center brings the effluent to the surface which is treated as it slowly flows down through gravel to the bottom of the cell.

Installed bird houses not only benefit wildlife but provide attractive focal points among rolling fields of native grasses. Topography was crafted with cut soils from the wetland creation.

Pipes emit water in a grassy field where nutrients are removed in this experimental overland treatment system.

Figure 3.34 Case Study: Willow Lake Water Pollution Control Facility, Salem, OR

3 Potassium

Potassium ions are present in relatively large amounts in all living organisms and potassium is essential for life. Excessive potassium in soils and groundwater is currently not considered a human health risk and is not regulated by the US EPA, therefore phytotechnology options are not included here. Since potassium is a macronutrient and required by plants, all plants extract it to some extent.

Metals (and metalloids, categorized generally as metals henceforth)

Specific contaminants in this category

Easier to extract (tend to be more bioavailable to plants)

Arsenic (As), Cadmium (Cd), Nickel (Ni), Selenium (Se), Zinc (Zn)

Difficult to extract

Boron (B), Cobalt (Co), Copper (Cu), Iron (Fe), Manganese (Mn), Molybdenum (Mo)

Very difficult to extract

Chromium (Cr), Fluorine (F), Lead (Pb), Mercury (Hg), Aluminum (Al)

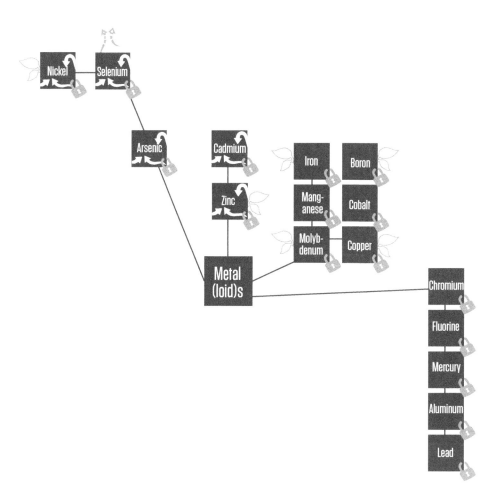

Figure 3.35 Metals

Typical sources of metals pollution: Mining tailings, smelting, pesticides application, manure and human wastes, road surfaces/stormwater, leachate from landfills, industrial applications, and atmospheric deposition/air pollution from vehicle and industrial exhausts.

Typical land uses with metals contamination: Mining sites and adjacent areas, industrial uses, agricultural fields, animal producers and feed-lots, landfills, roadways and areas where pesticides were historically sprayed, including rail corridors and utility and power lines.

Why these contaminants are a danger: Metals (and metalloids) naturally occur in the ecosystem and many are essential to plant growth (see Chapter 2. p. 55). However, high concentrations of metals can be toxic if ingested by humans and other animals. Metals can be found in soils and water and can contaminate drinking water and the air, and most high concentrations of these pollutants are created by human activities.

Metal toxicity can result in a host of impairments, but only when specific concentrations and forms are present in soil conditions. A summary of sources of metals contamination and risks to human health is provided in Figure 3.36. This summary however, is an overview and should not be directly applied to any site. Many soils are naturally high in metals and provide no risk to humans. An environmental expert should assess particular sites to determine if a true human health or environmental risk exists.

Figure 3.36 Metals Toxicity List

Contaminant	Notes
Aluminum (Al)	Aluminum is a widely used metal and contamination is most often centered on former metals production, smelting and mining areas. Overexposure to aluminum can cause bone and brain diseases. Inhalation of aluminum-contaminated dust has been known to cause lung problems (ASTDR, 2013).
Arsenic (As)	Arsenic contamination of soil and groundwater is widespread, especially in China, Bangladesh, South America and the western United States, where high levels of arsenic-containing bedrock are common. Additionally, it may be linked to contamination from older types of pesticides and pressure-treated lumber (http://water.epa.gov/lawsregs/rulesregs/sdwa/arsenic/Basic-Information.cfm). Due to its effects on many enzyme reactions in the body, toxicity affects multiple organs such as the heart, lungs and kidneys. Moreover, it may affect the neurological system and skin, causing headaches and confusion as well as skin lesions and hair loss (ATSDR, 2013).
Boron (B)	Boron contamination may be associated with glass manufacturing, pesticide use and leather tanning. In high doses, boron affects the cardiovascular system and may cause birth defects (ASTDR, 2013).
Cadmium (Cd)	Cadmium contamination is most often associated with metal smelting, battery and pigment production and former mining sites. It can also impact agricultural fields. Cadmium often takes years to accumulate in the body, but may seriously damage the lungs and kidneys, as well as soften bones when levels become toxic. Acute exposure can cause chills, aches and fever (ATSDR, 2013).
Chromium (Cr)	Chromium contamination is associated with the electroplating, automotive and tanning industries as well as the production of pressure-treated lumber. It can often be present in urban environments where byproducts of industry were used as urban fill. Chromium is a powerful carcinogen and can also cause respiratory and skin problems (ASTDR, 2013).
Cobalt (Co)	Cobalt has been used as a colorant in glass and ceramic production, as well as an alloy in aircraft manufacturing. Cobalt also has several useful radioactive isotopes used in the medical and commercial industries (http://www.epa.gov/ttnatw01/hlthef/cobalt.html). Effects on human health include damage to the heart, lungs and kidneys, as well as birth defects (ASTDR, 2013).

Figure 3.36 (*continued*)

Contaminant	Notes
Copper (Cu)	Copper contamination is often associated with metals, pipe and wire production, and pesticide and fungicide use. Some natural sources may contaminate water and soil. Copper poisoning results in kidney and liver damage, vomiting and coma (ASTDR, 2013).
Fluorine (F)	Small amounts of fluorine are an important component of healthy teeth and bones and are added to most municipal drinking water and toothpastes. Fluorine contamination is associated with phosphate fertilizer production as well as smelting, coal-fired power plants and mining. Some contamination may occur from naturally fluorine-rich groundwater. Fluorine poisoning may result in nausea, vomiting and tingling of the extremities (ASTDR, 2013).
Iron (Fe)	Iron and its compounds are widely occurring and essential to plant and animal life. It is a ubiquitous structural mineral in soil, and is typically considered a contaminant only when found in water in large quantities. Iron is key to metals and alloy production, and some iron compounds have been used in older forms of pesticides. High levels of iron in soil and groundwater are common. Many homes that get drinking water from a well may install a water softener to help remove it. While iron generally has few effects on human health, it can be toxic in extremely high doses, particularly to young children (http://www.nlm.nih.gov/medlineplus/ency/article/002659.htm).
Lead (Pb)	Lead was once a common additive to house paint and gasoline and has been banned from these uses in the United States since the 1970s. As such, lead contamination may be associated with older residential areas, gas stations and roadsides. Lead is a persistent element that is not very mobile, mostly contaminating soil. Overexposure to lead can cause severe organ damage, brain damage and developmental delays. Children are at risk of neurological damage, due to exposure to old paint chips and lead-contaminated soil (ASTDR, 2013).
Manganese (Mn)	Manganese is widely found in soils and is typically considered a contaminant only when mobilized in water. It is used in steel and battery production as well as in gasoline additives. Most contamination is related to former industrial uses; however, some naturally occurring manganese-rich bedrock may contaminate groundwater. In small amounts it is essential to human health and is found in many foods. However, in extremely high doses it may cause damaging neurological effects and birth defects (ASTDR, 2013).
Mercury (Hg)	Mercury was used as a component in fungicides and some paints. As well, it is strongly associated with coal-burning power plants, where it is deposited from atmospheric sources into soil and waterways. Mercury and its compounds are highly toxic. Poisoning most often occurs due to eating contaminated food, as mercury easily bioaccumulates in animal tissues. Chronic overexposure carries a greater risk of cancer, birth defects and neurological impairment (ASTDR, 2013).
Molybdenum (Mo)	Molybdenum is used in the metals, pigment and medical industries. Contamination is most often associated with former mining areas. In small amounts molybdenum is essential to human health. There is some evidence that high doses of molybdenum can cause cancer. Chronic long-term exposure can harm the liver and kidneys. Poisoning is often due to exposure to contaminated soils (ASTDR, 2013).
Nickel (Ni)	Nickel is a widely occurring element with many uses in metal alloys and battery production, where it may contaminate soil and groundwater. Nickel may also enter the atmosphere through oil, coal or waste burning, when it is later deposited onto the ground and waterways. At extremely high levels, nickel is a carcinogen. However, most overexposure results in dermatitis and respiratory and gastrointestinal problems (ATSDR, 2013).
Selenium (Se)	Selenium contamination is generally associated with waste products produced by petrochemical, agricultural and electronics industries. Additionally, activities like agriculture and mining may disturb naturally occurring selenium minerals, causing them to enter groundwater (http://www.epa.gov/ttnatw01/hlthef/selenium.html). Like many other trace elements, selenium is required for the body to function, and deficiencies may cause health problems. However, in large amounts selenium is toxic, causing hair and fingernail loss, respiratory problems or neurological symptoms such as numbness (ASTDR, 2013).
Zinc (Zn)	Zinc contamination is concentrated around mining and smelting areas and in urban areas from tire debris. Zinc may accumulate in soils and its compounds are readily dissolvable in water. Overexposure is often caused by breathing contaminated dust. The toxic effects of zinc can cause fever, chills and neurological impairment (http://www.nlm.nih.gov/medlineplus/ency/article/002570.htm).

138

Summary

Metals are inorganic pollutants; some may be extracted by plants, harvested and removed from sites, but they cannot be degraded. (See Chapter 2, p. 55 for fundamentals and an explanation of why inorganic contaminants cannot be degraded.) In addition, the structural form of an element may be changed, for example from a liquid to a gas or to an oxidized state, potentially removing some health and environmental risks. To remove metals from a polluted site, plants that accumulate metals in high concentrations (called hyperaccumulators, see Chapter 2, p. 55) or fast-growing plants that produce a lot of biomass (such as willow and poplar) may be able to be grown to take up the contaminant. They are then harvested, removed and disposed of as landfill or incinerated to remove the metals from a site. However, extraction of metals with plants is currently quite difficult, showing potential viability for only a few metals. Phytoextraction is likely only useful as a 'polishing strategy' for lands marginally contaminated with the few more highly bioavailable metals that have been identified (Dickinson, 2009). Practical field-applied phytoextraction projects are difficult to find, but the technology may be effectively applied in some specific situations. The few metals where phytoextraction may be considered are summarized in the following section.

Most often, metals are tightly bound to soils and not bioavailable enough to be taken up into plants. When low bioavailability is present, the only applicable technology is to stabilize metals on site (with or without plants) or remove them with conventional dig-and-haul remediation. The most applicable use of phytotechnologies for lowering risk of exposure to metals is by capping and stabilizing the metals on site and adding plants to assist in this. Almost all plant-based metals-stabilization projects utilize amendments to further bind the metals in place so as to prevent migration. Many of the species most commonly used for stabilization are able to tolerate and grow in metals-rich soil. Some research on metals stabilization has identified plants that release root exudates that help immobilize pollutants, essentially repairing the environment by changing phytotoxic forms of metals to create a safe growing environment for other plants (Gawronski et al., 2011). Some plant species will not let certain metals into their roots, and therefore will not move the metals into above-ground parts. These 'excluder' plants are often preferred for replanting sites, since the risk of animals eating the plants or being exposed to contaminated above-ground portions is minimized. Plant species that are considered excluders of certain metals are listed in Figure 3.37. This is not meant to be a complete list, but is a compilation of example plants.

Detailed information where extraction vs stabilization should be considered on sites is included in the following pages, organized by type of metal from most applicable for extraction to least applicable. However, before the details of each individual metal are addressed, a few general notes about metals and phytotechnology applicability are included below.

- **Greenhouse and short-term studies:** Many scientific studies can be found that claim effective plant extraction of metals from soils, extrapolating this assumption from greenhouse studies or studies using spiked soil (Dickinson et al., 2009). However, many of these plant species have not been tested in the field, or have not been proven viable at field scale under aged soil conditions, therefore the studies and lists of extraction species can be misleading. Readers are cautioned on the use of greenhouse studies to assess if field application will work. At this time, recommendations here are based on peer-reviewed, published field-scale trials, not on greenhouse studies.

139

Figure 3.37 Metals Excluders (not a complete list, but rather a list of representative species)

Latin	Common	Contaminant Excluded	Vegetation Type	USDA Hardiness Zone	Native to	Reference
Acacia mangium	Mangium, Black Wattle	Pb	Tree	9+	Australia	Meeinkuirt et al., 2012
Agrostis tenuis	Bentgrass	As, Cu, Pb, Zn	Herbaceous	3–10	Asia	Alvarenga et al., 2013 Dahmani-Muller et al., 2000
Carduus pycnocephalus L. subsp. pycnocephalus	Italian Thistle	Cd, Cr, Cu, Ni, Pb, Zn	Herbaceous	7+	Southern Europe	Perrino et al., 2012 Perrino et al., 2013
Chelidonium majus var. asiaticum	Celandine	As	Herbaceous	4–8	Europe, Western Asia	Zhang et al., 2013
Erica andevalensis	Heather	Pb, As, Cu, Fe	Herbaceous	7+	Europe	Monaci et al., 2012 Mingorance et al., 2011
Eschscholzia californica	California Poppy	Cu	Herbaceous	9–11	Western USA	Ulrikson et al., 2012
Festuca rubra 'Merlin'	Red Fescue	Zn, Cu, Ni	Herbaceous	3–8	Northern USA	Lasat, 2000
Ficus goldmanii	Ficus	Cu, Zn, Pb	Tree	10+	Central America	Cortés-Jiménez et al., 2012
Fuchsia excorticate	Tree Fuschia	As	Tree	8–10	New Zealand	Craw et al., 2007
Gentiana pennelliana	Wiregrass Gentian	Pb, Cu, Zn	Herbaceous	8+	Florida	Yoon et al., 2006
Griselinia littoralis	Broadleaf	As	Shrub	8–9	New Zealand	Craw et al., 2007
Guardiola tulocarpus	Guardiola	Cu, Zn, Pb	Herbaceous	not available	Mexico, Southwestern USA	Cortés-Jiménez et al., 2012
Hibiscus cannabinus L.	Kenaf	As, Fe	Herbaceous	10+	Southern Asia	Meera and Agamuthu, 2011
Jatropha curcas L.	Barbados Nut	Al, Cu, Pb, Zn, Cd	Tree/Shrub	11	Central America	Wu et al., 2011
Juniperus flaccid	Mexican Juniper	Cu, Zn, Pb	Tree	8+	Mexico, Southwestern USA	Cortés-Jiménez et al., 2012
Leptospermum scoparium	Manuka	As	Shrub	9–10	New Zealand	Craw et al., 2007
Lolium spp.	Ryegrass	Cu	Herbaceous	3+	Europe	Ulrikson et al., 2012
Oenothera glazioviana	Evening Primrose	Cu	Herbaceous	3–8	North America	Guo et al., 2013
Quercus ilex subsp. Ballota	Holm Oak	Cd	Tree	7–11	Mediterranean	Dominguez et al., 2009
Silene paradoxa	Silene	As, Cd, Co, Cr, Cu, Fe, Mn, Ni, Pb, Zn	Herbaceous	6+	Southern Europe	Pignattelli et al., 2012
Silybum marianum	Milk Thistle	Cd, Cr, Cu, Ni, Pb, Zn	Herbaceous	5–9	Southern Europe	Perrino et al., 2012 Perrino et al., 2013
Sinapis arvensis L.	Wild Mustard	Zn, Cd, Pb, Cu	Herbaceous	6+	Mediterranean	Perrino et al., 2012
Stipa austroitalica Martinovský subsp. Austroitalica	Stipa	Zn, Cd, Pb, Cu	Herbaceous	not available	Italy	Perrino et al., 2012
Triticum aestivum L.	Wheat	Ni	Herbaceous	2–7	Asia	Massoura et al., 2005
Ulex europaeus	Gorse	As	Shrub	7–10	Europe	Craw et al., 2007

- **Cations and anions:** The pH (charge) of a soil can greatly affect whether a metal is bioavailable for plant extraction. Many metals (particularly those that exist as cations) are less likely to form complexes with the soil particles and are therefore more mobile in acidic soils. Alternatively, positively charged metals tend to be less bioavailable when the soils contain higher amounts of free, negatively charged surfaces, so the metal will tend not to be available for plant uptake. Charge in soils can be determined with a pH test. pH is a measure of the amount of H^+ and OH^- ions available. When soil is acidic (with a low pH and more H^+ ions), mineral surfaces tend to be positively charged, and when it is alkaline (with a high pH and hydroxyl ions (OH^-)), mineral surfaces are typically more negatively charged (Hettiarachchi, 2011). Metals that are pollutants tend to exist in soils as ions with positive or negative charges. The charge of both the soil and the pollutant will affect how much of the metal is 'bioavailable' for ultimate plant uptake. In summary, the bioavailability of a contaminant can be manipulated by changing the soil pH, as well as several other factors such as the amount of oxygen, water or organic matter in the soil. In many cases with metals contamination, several different types of pollutants are found on sites, and complications from reactions between pollutants can arise as well. Bioavailability to the plant is a tricky subject. For this reason, potential metals phytoextraction projects should not be attempted without the help of a trained phytoextraction specialist.

- **Chelants:** In addition to manipulating pH with minerals to make metals more bioavailable for plant uptake, chemical additives called chelants can be added to the soil to change the soil chemistry. Common chelants for metals include EDTA (Ethylene diamine tetra-acetic acid) and other organic acids. In field conditions, chelants can mobilize contaminants and increase the risks of leaching the metal beyond the root zone, so the addition of chelants to obtain results is not recommended. Research studies must be carefully reviewed when looking for new plants and opportunities for phytoextraction, because chelants may have been used, artificially increasing the concentrations in the harvested plant parts, or metals may have been mobilized off site (Chaney et al., 2003). Plant lists provided in this section do not include studies where metal chelants were used as an amendment to obtain results.

- **Hyperaccumulators vs high-biomass 'accumulator' species:** When the chemical properties of the soil favor potential plant uptake of metals, there are two approaches to phytoextraction.

    1 **Hyperaccumulators:** Known hyperaccumulator plant species can be specified to take up a significant amount of a metal. The plants are then harvested and removed from site. The advantage of this approach is that hyperaccumulator species may be interbred to create hybrids that thrive in particular site conditions and can take up significant amounts of inorganic pollutants. However, the disadvantage of using hyperaccumulators is that they may not be native to a region in some instances and can prove to be weedy or invasive, not hardy, or difficult to cultivate. In addition, they may not produce enough biomass to be useful for harvesting and extraction. There may even be great variation in extraction potential between different varieties and ecotypes within the plant species. Lastly, the list of known hyperaccumulators is small, and the known plants may not be well suited to growing in the soil and climate conditions of a particular contaminated site. "Hyperaccumulators have been confirmed for nickel, zinc, cadmium, manganese, arsenic and selenium. However, to date, hyperaccumulation of lead, copper, cobalt, chromium and thallium remain largely unconfirmed" (Van der Ent et al., 2013).

141

2 **High-biomass species:** The alternative approach is to use plants that are considered 'accumulators' rather than hyperaccumulators. These plants are typically species that can take up a contaminant, but not to the levels of hyperaccumulator species. They instead have fast growth rates and produce a lot of biomass. This allows for a large total uptake, even if concentrations in the tissues of the plant are lower than in hyperaccumulators. With this approach, plant species that thrive and produce the most biomass in the specific contaminated soil are chosen, and are cut down and harvested over time to extract and remove the contaminant from the site. When these high-biomass accumulator species are used they are often paired with other amendments to change the soil chemistry and improve metals uptake. In some cases, these high-biomass species may be more effective to use in field conditions than hyperaccumulator species. They may be easier to grow, readily available as seed and better adapted to soil conditions and climate. This approach has been used with crop species such as sunflowers, mustard, soybeans and corn, sometimes in combination with chemical chelants like EDTA to remove metals. However, as mentioned previously, the use of chelants is not recommended. Confusion over 'accumulator' species is common, with designers assuming that a plant can accumulate something, when this is not true unless the soil chemistry is drastically altered by chelants. For this reason, extraction of metals is discouraged without the engagement of expert phytoscientists (i.e. do *not* attempt the remediation of lead, arsenic, nickel, etc. without a trained professional!).

- **Other soil additives:** The installation of densely rooted plant systems can significantly increase soil organic matter. Increasing soil organic matter is an established means of immobilizing metals in the subsurface, as metals tightly bind to organic material, reducing bioavailability and mobility. Therefore, where metals phytoextraction is not an option, the installation of plants can help to bind the contaminant on a site and help to reduce the toxicity and mobility of metals.

- **Sampling, testing and monitoring:** One additional challenge with metal contamination is the sampling and testing protocol to determine how much contamination is on the site. Testing protocols vary widely, but typically test for only the 'total' concentration of a metal, as recovered in an acid digestion process. If the metal has oxidized or changed form in a way that affects its bioavailability or solubility, it may not be picked up in the testing. In addition, contaminant distribution is typically very heterogeneous in soil and concentrations may greatly vary even several inches away, and so when only a tiny fraction of the soil is tested the results may not be accurate. This proves to be a challenge when working with contaminants occurring as particulates (such as lead- or cadmium-based paint residues) or when monitoring to determine if a treatment is actually working. Where metals extraction with plants is considered, composite testing, where several samples are lumped together, homogenized and averages are taken may prove to be useful. Composite testing can help to obtain more accurate readings both of the existing contamination and in long-term monitoring (Blaylock, 2013). Regulatory protocols should be researched and followed where applicable.

- **Food crop contamination:** For agricultural land uses, metals contamination is a concern in food crops where the plants take up excessive amounts of metals and translocate them into the edible portions of the plant, making them a risk for consumption. For example, arsenic contamination in rice and cadmium contamination in corn and grain crops are an increasing concern. In the case of

142

cadmium, human and animal consumption of these products may lead to bioaccumulation in fatty tissues over time, making repeated intake a significant concern. For arsenic, even one initial toxic exposure may create a significant health risk. One new area of phytotechnology that is increasingly in demand is the identification of alternative agricultural commodity crops that are 'phytoexcluders,' species that do not take up the excessive metals in soils. These phytoexcluders can be considered when metals concentrations in agricultural soils are above stated limits and the risk of exposure to pollutants in the above-ground portions of a plant is a concern (see plant list in Figure 3.37 above).

- **Phytotoxicity:** Many metals are toxic to plant growth in high concentrations. Where phytotechnologies are proposed for these sites, careful agronomic soil analysis and recommendations must be completed by an expert to ensure the plants can survive.

- **Volatilization:** Phytovolatilization of metals is limited to the elements selenium (Se) and mercury (Hg). Since there is a deficiency of selenium in the atmosphere, its release can be considered as a positive process; however, volatilization of mercury has only been developed with genetically modified plants and will merely transport the pollutants elsewhere in a different form (Gawronski et al., 2011).

- **Phytomining:** In some cases where inorganics are phytoextracted into plants, the plants may be able to be harvested and turned into an ore for mining purposes. For example, nickel has been successfully mined with plants, extracting these elements from soil. The plants are harvested, dried and turned into ash and smelted. This technique is an attractive proposition, because whereas phytoextraction for remediation costs money, phytomining is aimed at making money by removing inorganic contaminants from the soil. Researchers are investigating how additional metals such as gold might be mined from the soil with plants, but to date phytomining practices are likely only applicable for nickel and selenium extraction.

- **Halophytes:** Halophytic plants (salt-tolerant plants that thrive in saline environments) are of special interest in the identification of potential species for metals accumulation since these plants "commonly live in environments with an excess of toxic ions, mainly sodium and chloride. Several studies have revealed that these plants may also tolerate other stresses including heavy metals … [since the processes] rely on common physiological mechanisms" (Manousaki and Kalogerakis, 2011). Some halophytic species use storage or excretion mechanisms in their leaves for salt uptake, and in the future this mechanism may also be found viable for removal of metals from soils (Lefevre et al., 2009). At this time, the use of halophytes for metals extraction is still in experimental research phases, but in the future may lead to discoveries of new plants for metals extraction.

The most common metal contaminants are individually detailed below, to provide an overview of the applicability of phytotechnology for metals.

4 Metals with high bioavailability

Arsenic (As), Cadmium (Cd), Nickel (Ni), Selenium (Se), Zinc (Zn) (ITRC, 2009, p. 16; Van der Ent et al., 2013)

143

a Arsenic

Background: Arsenic is a widespread pollutant in both urban and rural areas. In some localities it occurs naturally in high concentrations in soils and groundwater. However, most anthropogenic arsenic pollution is a result of its use as a component in pesticides, pressure-treated wood or byproducts of industry, mining or military operations. Manufacturers of treated wood used arsenic for the same reason that it is undesirable in food: arsenic kills organisms such as fungi, insects and bacteria that might otherwise eat the wood. Low-level arsenic pollution is commonly found under residential decks in the US that were built before 2004 using wood pressure treated with copper-chromated arsenate (CCA), or on sites previously used as orchards, where arsenic-based pesticides were historically applied.

Applicability: On some sites, low levels of arsenic contamination have been effectively phytoextracted with plants that were then harvested and used as landfill off site. Although several hyperaccumulator species have been identified by various studies, only a few species of sub-tropical ferns have actually been shown to extract enough arsenic to be useful in the field. Many greenhouse-based studies have shown that species of *Brassica*, beans, beets and lettuce, among others, take up more arsenic than do other crop plants. However, researchers currently do not recommend utilizing these alternative species in the field, as they have been found not to extract satisfactory amounts to make them viable remediation alternatives.

In the US, the company Edenspace has historically been the primary producer and installer of projects that utilize the Chinese Brake Fern, *Pteris vittata*, for arsenic extraction. Edenspace has partnered with the US EPA and Army Corps of Engineers on several projects to phytoremediate soils with elevated arsenic concentrations and demonstrate the effectiveness and economic viability of this species. While the fern is perennial in climates with a hardiness of US Zone 8 or warmer, it has been used as an annual in remediation projects in cooler climates. Field-application results have been mixed. In several instances where hyperaccumulating ferns have been planted, sufficient arsenic was not removed to make it a useful strategy, either because of poor biomass yields or low concentrations in the plant. A cold-tolerant hyperaccumulator plant species for arsenic has not yet been identified for effective field-scale remediation in northern and temperate climates. Several researchers are working on genetically modifying ferns and other species to take up greater quantities of arsenic, but these are not yet proven or available at the time of this publication. The applicability of phytoextraction for arsenic removal should be assessed on a site-by-site basis, with an experienced professional.

Planting specifics

Currently, ferns in the *Pteris* genus and a few other tropical ferns (such as *Pityogramma calemelanos*) are the only hyperaccumulators that can be recommended for successful field application for arsenic remediation (Van der Ent et al., 2013). *Pteris vittata* has been shown to effectively remove up to 20 ppm of arsenic per year into above-ground biomass (Blaylock, 2013). Phytoremediation of arsenic with ferns is usually most effective for moderately elevated sites (with respect to the

144

remediation target concentration), where remediation targets can be expected to be met within one to three years. The ferns can be grown in a wide range of arsenic concentrations and plant uptake is typically correlated with the soil concentration, such that greater uptake and removal occurs at higher soil concentrations, instead of at very low soil concentrations where available arsenic may be limited. When using *Pteris* ferns as annuals, plant spacing is typically one plant per square foot and the ferns are typically installed as plugs provided in 4-inch pots. In climates where the fern can be grown as a perennial, planting density can be decreased, allowing plants to fill in over time. Edenspace also recommends that plants be fertilized to provide enough macronutrients for sufficient plant growth (Blaylock, 2013). It is also suggested that liming sites to raise the pH can assist in arsenic extraction by the plants (Blaylock, 2013) and generate higher biomass yields. Arsenic tends to exist as an anion such as arsenate or arsenite (with a negative charge) in the field (Hettiarachchi, 2012). Therefore, the surrounding soils and additives (like lime) which are more alkaline may make arsenic more bioavailable to plants for extraction (Blaylock, 2013).

Even low, widespread concentrations of arsenic which may seem ideal for fern extraction can prove problematic. Arsenic can be bound to soils in forms that are not bioavailable to the plant for uptake. In addition, concentrations can be quite heterogeneous in soils and arsenic may prevent the uptake of other essential nutrients. Lastly, as with any remediation approach, the way in which samples are taken and tested can affect observed concentration measurements. Maintaining consistent, repeatable sampling and analytical techniques is critical for reliable long-term monitoring and results. Arsenic can change ionic form quickly, making testing and monitoring additionally challenging.

The following planting typologies can be considered for arsenic remediation use in conjunction with the plant list in Figure 3.38.

In soil

Potential extraction rates: 10–20 ppm average removal from soil per year (Blaylock, 2013)

- Extraction Plots: Chapter 4, see p. 224

Stabilization: To prevent wind and water erosion of arsenic in soils

- Planted Stabilization Mat: Chapter 4, see p. 202

In water

To control contaminated groundwater:

- Groundwater Migration Tree Stand: Chapter 4, see p. 213

To remove from surface and wastewater: The systems described below have been used for arsenic removal from water primarily by physically trapping arsenic and binding or precipitating it into planting media. Since constructed wetlands have been widely documented, species for these purposes have not been included in this publication.

145

Figure 3.38 Arsenic-Hyperaccumulating Ferns

Latin	Common	Vegetation Type	USDA Hardiness Zone	Native to	Reference
Pityrogramma calemelanos	Dixie Silverback Fern Gold Dust Fern	Perennial	11	Central and South America	Francesconi et al., 2002 Niazi et al., 2011
Pteris creticam (var. nervosa)	Cretan Brake Fern	Perennial	8–10	Europe, Asia, Africa	Wang et al., 2006 Zhao et al., 2002
Pteris longifolia	Longleaf Brake Fern	Perennial	10	Southeastern USA, Central America	Zhao et al., 2002 Molla et al., 2010
Pteris multifida	Spider Fern Huguenot Fern	Perennial	7–10	Asia	Wang et al., 2006
Pteris oshimensis		Perennial	8–10	Asia	Wang et al., 2006
Pteris umbrosa	Jungle Brake Fern	Perennial	10–12	Australia	Zhao et al., 2002
Pteris vittata	Chinese Brake Fern	Perennial	8–10	Asia	Blaylock, 2008 Ciurli et al., 2013 Danh et al., 2014 Hue, 2013 Kertulis-Tartar et al., 2006 Ma et al., 2001 Mandal et al., 2012 Niazi et al., 2011 Ouyang, 2005 Salido et al., 2003

146

- Stormwater Filters: Chapter 4, see p. 235
- Multi-Mechanism Buffers: Chapter 4, see p. 227
- Surface-Flow Constructed Wetland: Chapter 4, see p. 238
- Subsurface Gravel Wetland: Chapter 4, see p. 241

The hyperaccumulator plant species in Figure 3.38 have been shown to provide effective extraction of arsenic without the use of chelator additives, and can be considered for potential field-trial use in certain situations with low-level contamination.

The plant species in Figure 3.39 have been demonstrated in other studies to potentially take up arsenic as accumulators, but are currently not considered hyperaccumulators as verified in field trials (Dickinson, 2009). These species should not be considered for field application unless additional scientific research is completed to verify extraction capabilities, in collaboration with experts in this field.

Figure 3.39 Potential Arsenic 'Accumulators' (not Hyperaccumulators) Plant List

Not recommended for field extraction projects

Latin	Common	Vegetation Type	USDA Hardiness Zone	Native to	Reference
Agrostis delicatula	Bentgrass	Herbaceous	4–9	Southwestern Europe, North Africa	Gomes et al., 2014
Bassia scoparia	Summer Cypress Mexican Firebush	Herbaceous	Grown as annual	Spain	Gisbert et al., 2008
Beta vulgaris	Beet	Herbaceous	Grown as annual	Mediterranean	ITRC PHYTO 3 Speir et al., 1992
Brassica juncea	Indian Mustard	Herbaceous	Grown as annual	Central Asia	Anjum et al., 2014
Chelidonium majus var. asiaticum	Nipplewort	Herbaceous	5–9	Europe, Asia	Zhang et al., 2013
Colocasia esculanta	Taro	Herbaceous	8–10	Malaysia	Molla et al., 2010
Cynodon dactylon	Bermuda Grass	Herbaceous	7–10	Middle East	Molla et al., 2010
Cyperus rotundus	Purple Nut Sedge	Herbaceous	3–10 (invasive)	Africa	Molla et al., 2010
Dryopteris filix-mas	Male Fern	Herbaceous	4–8	America, Europe, Asia	Mahmud et al., 2008
Echinochloa crus-galli	Barnyard Grass Japanese Millet	Herbaceous	Grown as annual	Southeast Asia	Molla et al., 2010
Helianthus annus	Sunflower	Herbaceous	Grown as annual	North and South America	McCutcheon and Schnoor, 2003
Hirschfeldia incana	Shortpod Mustard Hoary Mustard	Herbaceous	6–9	Spain, Mediterranean	Gisbert et al., 2008
Inula viscosa (Dittrichia viscosa)	Sticky Fleabane Yellow Fleabane	Herbaceous	9–11	Spain, Mediterranean	Gisbert et al., 2008
Isatis capadocica	Isatis capadocica	Herbaceous	not available	Iran	Karimi et al., 2009
Jussiaea repens	Floating Primrose Willow	Herbaceous	8–11	North America	Molla et al., 2010
Lactuca sativa cv. Cos	Lettuce	Herbaceous	Grown as annual	Europe, Asia	ITRC PHYTO 3 Speir et al., 1992
Leersia oryzoides	Rice Cutgrass	Herbaceous	3–9	North America	Ampiah-Bonney and Tyson, 2007
Melastoma malabathricum	Malabar Melastome Straits Rhododendron	Herbaceous	10–13	Indonesia, Asia	Selamat et al., 2013

Figure 3.39 (continued)

Latin	Common	Vegetation Type	USDA Hardiness Zone	Native to	Reference
Oryza sativa	Asian Rice	Herbaceous	2–11	Indonesia, Southeast Asia	Molla et al., 2010
Phaseolus vulgaris cv. Buenos Aires	Bean	Herbaceous	Grown as annual	Central America	Carbonell-Barrachina et al., 1997 ITRC PHYTO 3
Populus alba Populus spp.	Poplar species and hybrids	Tree	2–9	varies	Ciurli et al., 2013 Madejón et al., 2012
Pteris cretica	Spider Brake Cretan Brake Fern	Herbaceous	7–10	Europe, Asia, Africa	Ebbs et al., 2009
Solanum nigrum	Black Nightshade	Herbaceous		Europe, Asia	Gisbert et al., 2008
Tagetes spp.	Hybrid triploid nugget Marigold	Herbaceous	9–11	Central America	Chintakovid et al., 2008
Thelypteris palustris	Marsh Fern	Herbaceous	3–7	Europe, Eastern North America	Anderson et al., 2010
Trifolium repens cv. Huia	Clover	Herbaceous	3+	Europe	ITRC PHYTO 3 Speir et al., 1992
Viola allcharensis G. Beck Viola arsenica G. Beck Viola macedonica Boiss. & Heldr. (Balkan)	Viola	Herbaceous	7–8	Macedonia and the Balkans	Bačeva et al., 2013
Xanthum italicum	Italian Nappola	Herbaceous	Not available	Italy, Mediterranean	Molla et al., 2010

Project name: Spring Valley Formerly Used Defense Site (SVFUDS) (Blaylock, 2008)

Location: Northwest Washington, DC

Institution/Scientists: US Army Corps of Engineers (USACE); US Environmental Protection Agency (US EPA); District Department of the Environment; Edenspace Systems Corporation; Dr. Michael Blaylock, 210 N 21st Street, Suite B Purcellville, Virginia 20132

Date installed: 2006

Plants installed: *Pteris vittata* "Victory", *Pteris cretica mayii* "Moonlight", *Pteris cretica parkerii*, *Pteris cretica nervosa, Pteris multifida.*

Amendments: NPK fertilizer, equivalent to 25 lbs/acre, and lime were added where soil pH < 6.5. All plantings were irrigated with a sprinkler.

Contaminant: Arsenic ■, in concentrations up to 150 ppm

Target media and depth: Surface soils, up to 0.75 meters (30 inches) depth.

Spring Valley (a former US Military defense site) is located in northwest Washington, DC. It is approximately 668 acres and currently includes 1,600 private residences, foreign embassies, an American university, a seminary and various commercial properties. The site was used during World War I to test chemical warfare materials. In 1993, explosives were unearthed at the site and triggered a series of environmental investigations. Neighborhood-wide soil sampling revealed that about 10% of the 1,600 properties in Spring Valley contained elevated arsenic levels.

With large existing trees within an established historic neighborhood, conventional soil removal and backfilling would have been invasive and extremely costly. Plantings of the arsenic hyperaccumulator fern *Pteris vittata* "Victory" were chosen to minimize destruction of existing trees and to reduce restoration costs within the residential area. Phytoremediation plantings were implemented, beginning in 2004 with a field verification study at three properties using about 3,000 plants. The effort was expanded in 2005 to include 33 sampling grids at 11 properties and approximately 10,000 plants. The effort reached its peak in 2006 when 13 properties containing 48 sampling grids were planted with 11,000 plants. Sampling grids were removed when soil concentrations had reached the project goal of 20 mg/kg (or 43 mg/kg where certain other criteria were met). Nineteen sampling grids at six properties were planted in 2007 and the project concluded in 2008 with plantings on sixteen sampling grids at three properties. Soil was amended with fertilizer and lime if necessary prior to planting and ferns were planted at 30 centimeters (12 inches) on center, except in 2008, when 20 centimeter (8 inch) spacing was used to increase root density and arsenic uptake on the remaining properties. In 2004–2006 multiple *Pteris* fern species (*P. cretica mayii, P. cretica parkerii, P. multifida* and *P. nervosa*) were evaluated for performance and it was determined that *P. vittata* was best suited for the properties in this location. Planted ferns were irrigated with sprinkler systems during establishment and dry periods.

149

Grids of *Pteris vittata* 'Victory' fern installed on 22 separate properties in Spring Valley district. This particular property, Lot 15, has 9 separate gridded areas where 60% of the arsenic was removed over a 2-year time frame.

Residential neighborhood made conventional excavation treatment difficult and expensive

Figure 3.40 Case Study: Spring Valley, Washington, DC
Scale: 1 in = 100 ft

150

Extraction ferns planted under a residential deck. Plants were harvested and tested after one growing season. Any ferns with arsenic concentrations over regulatory limits were sent to a hazardous waste landfill.

Extraction ferns planted along fence on Lot 15, as shown on the site plan.

The attractive ferns prefer a shaded environment, and are only cold tolerant to US Zone 8, making them an annual plant and not typically cost-effective for phytoextraction in cooler climates.

Extraction ferns integrated into existing residential landscapes.

Figure 3.41 Case Study: Spring Valley, Washington, DC

Over the 5 years of implementation of the fern-based phytoextraction ☐ technology for the removal of arsenic from the SVFUDS area in Washington, DC, 71 sampling grids at 22 different properties were examined. Although arsenic concentration in fronds and biomass yields varied in each year of the study, soil arsenic concentrations tended to continually decrease following fern-based phytoextraction treatment. Average measured removal rates ranged from approximately 9 mg/kg in 2004 to just over 1 mg/kg in 2007. During the 5-year project, 61 of the 71 sampling grids achieved concentrations below project goals, many within 1 or 2 years. However, six of the sampling grids (located at a single property) did not achieve a measureable reduction during 5 years of plantings. Four other sampling grids did not meet project goals in the 2–3 year time frame anticipated by the property owner and were subsequently assigned for other remedial activities.

b Cadmium and zinc ▩

Background: Cadmium and zinc contamination are usually found together, often with 100–200 times more zinc contamination than cadmium. Almost all naturally cadmium-contaminated soils also have elevated amounts of zinc, so plants that can extract one of these elements can often extract or tolerate high levels of the other (Van der Ent et al., 2013). Cadmium has assumed importance as a significant environmental contaminant over the last 50 years and is listed among the top 20 toxins (Yang et al., 2004).

Contamination risk to food chains within edible crop production is the most frequently encountered problem with cadmium. As field crops are grown, the plants extract cadmium. This occurs because plants take up zinc, and since cadmium is chemically similar, the cadmium is taken up with the zinc-transporter mechanism in roots. When cadmium in soil is too high, crops can accumulate excessive levels in their tissues. Prolonged consumption of high cadmium-content food crops can lead to cadmium toxicity in livestock and humans. Because of the widespread risk of food crop contamination in the European Union, Japan and New Zealand, many field studies have investigated the use of phytoextraction with plant harvesting for cadmium removal from soils. Agricultural fields can become contaminated with cadmium due to smelter or mine-waste dispersal, and there is concern about accumulation from continuously repeated fertilizer or manure application.

Zinc contamination can be found in mining and industrial sites where metals were processed, and urban soils often have high levels from smoke-stack and vehicle emissions, tree debris paint residues and application of phosphate fertilizer. Zinc toxicity in edible food-crop production is not often a priority concern.

Applicability: In certain soil conditions, both cadmium and zinc can have high bioavailability and can be slowly extracted from soils. However, even for minimally impacted sites, cadmium and zinc extraction via plants can take decades or even centuries, and can prove to be quite difficult for aged soils. In addition, zinc and cadmium in high concentrations can also inhibit

151

plant growth, making extraction a difficult proposition (Van der Ent et al., 2013). Because of these challenges, instead of utilizing extraction and harvesting, cadmium and zinc are often instead left on sites and stabilized in soils. Phytoexcluders that do not take up these elements are planted, rather than any edible food crops with extraction capabilities (see plant excluder list, Figure 3.37, p. 140).

However, recent research has considered how plants with an ability to accumulate cadmium slowly, over time, may offer a technology for removing cadmium that is dangerous to food chains. After the cadmium is removed to the required extent, liming the soil to prevent zinc phytotoxicity returns the site to productive status. Cadmium and zinc phytoextraction does not have adequate value for phytomining, so there is no financial benefit to processing the biomass and recycling the metals after harvesting. Because cadmium removal is often the focus of remediation, larger 'accumulator' species that are focused on biomass production, such as willow, poplar and corn, are being studied.

Planting specifics

Around 12 hyperaccumulator species for zinc and two for cadmium have been verified at the date of this publication (Van der Ent et al., 2013). The hyperaccumulator species *T. caerulescens* has been proposed for potential field-scale phytoextraction (Chaney et al., 2010). However, zinc extraction has not been found to be viable in the field. In addition to these hyperaccumulators, several 'accumulator' species that produce high biomass have been proposed for the extraction of cadmium. These high-biomass plants are not considered hyperaccumulators, but extract these elements faster than most plants, due to their fast growth rate (Dickinson et al., 2009).

Cadmium and zinc tend to exist in field conditions as cations with a positive charge, therefore soils and additives that are more acidic (i.e. more positive) tend to make them more bioavailable to plants for extraction. In contrast, adding more alkaline, higher-pH substances tends to bind cadmium and zinc more strongly to soils (Hettiarachchi, 2012; Wang et al., 2006; Yanai et al., 2006; and Kothe and Varma, 2012).

The following planting typologies can be considered for use in conjunction with the plant list in Figure 3.42.

In soil

Extraction: Extraction of cadmium and/or zinc with plants should only very tentatively be considered if long-term growing time frames are possible and agronomic conditions to ensure bioavailability are carefully studied and controlled. Plants must be harvested to remove the contaminants from the site.

• Extraction Plots: Chapter 4, see p. 224

Stabilization: This is the preferred phytotechnology treatment for cadmium- and zinc-contaminated soils. Stabilization plantings can prevent wind and water erosion of these elements in soils, minimizing the risk of exposure. Stabilization, however, is not applicable to food-producing soils or soils contaminated with cadmium without the typical 100-fold higher zinc concentrations. Plants that exclude zinc and cadmium can be used to ensure that the elements do not translocate into above-ground tissues (see plant excluder list, Figure 3.37, p. 140).

• Planted Stabilization Mat: Chapter 4, see p. 202

In water

To control contaminated groundwater:

• Groundwater Migration Tree Stand: Chapter 4, see p. 213

To remove from surface and wastewater: The primary mechanism for removal from water is to physically trap cadmium and zinc by binding or precipitation into the planting media. The systems described below have been used for cadmium and zinc removal from water. Since plant species for constructed wetlands have been widely documented, species for these purposes have not been included in this publication.

• Stormwater Filters: Chapter 4, see p. 235
• Multi-Mechanism Buffers: Chapter 4, see p. 234
• Surface-Flow Constructed Wetland: Chapter 4, see p. 238
• Subsurface Gravel Wetland: Chapter 4, see p. 241

153

Zinc extraction has not been found to be viable with plants. Phytoextraction of cadmium is not recommended for field application without detailed assistance from a scientist who specializes in metals phytoextraction. Cadmium extraction may not be viable for field application, since it is often not bioavailable to plants. If extraction is applicable, it can take decades or even centuries for these metals to be removed to below regulatory levels. The plant species listed in Figure 3.42 have been demonstrated in studies to be hyperaccumulators of cadmium and zinc.

The higher-biomass 'accumulator' plant species listed in Figure 3.43 have shown some extraction in studies for cadmium and zinc in phytoextraction research, with extraction over a much longer time frame.

Figure 3.42 Cadmium and Zinc Hyperaccumulators

Latin	Common	Contaminant	Vegetation Type	USDA Hardiness Zone	Native to	Reference
Arabidopsis halleri (*Cardaminopsis halleri*)	Rockcress	Cd, Zn	Herbaceous	6+	Europe	Baker, 2000 Baker and Brooks, 1989 Banasova and Horak, 2008 Reeves, 2006 Zhao et al., 2006
Dichapetalum gelonoides	Gelonium Poison-Leaf	Zn	Herbaceous	Not available	Philippines	Reeves, 2006
Minuartia verna	Spring Sandwort	Zn	Herbaceous	6–11	Europe	Reeves, 2006
Polycarpaea synandra	Polycarpaea	Zn	Herbaceous	Not available	Western Australia	Reeves, 2006
Rumex acetosa	Rumex	Zn	Herbaceous	3–7	Europe	Reeves, 2006
Thlaspi brachypetalum	Pennycress	Zn	Herbaceous	Not available	Europe	Baker and Brooks, 1989 Reeves, 2006 Reeves and Brooks, 1983
Thlaspi caerulescens (syn. Noccaea caerulescens and Thlaspi tatrense)	Alpine Pennycress	Cd, Zn	Herbaceous	6	Europe	Baker et al., 2000 Broadhurst et al., 2013 Chaney et al., 2005, 2010 ITRC PHYTO 3 Lasat et al., 2001 McGrath et al., 2000 Reeves, 2006 Rouhi, 1997 Saison et al., 2004 Salt et al., 1995 Schwartz et al., 2006 Simmons et al., 2013, 2014
Thlaspi capaeifolium ssp. Rotundifolium	Pennycress	Zn	Herbaceous	6–9	Central Europe	Baker and Brooks, 1989 Rascio, 1977 Reeves, 2006
Thlaspi praecox	Pennycress	Zn	Herbaceous	6	Central Europe	Baker and Brooks, 1989 Reeves, 2006 Reeves and Brooks, 1983
Thlaspi stenopterum	Pennycress	Zn	Herbaceous	Not available	Central Europe	Baker and Brooks, 1989, Reeves, 2006
Thlaspi tatrense	Pennycress	Zn	Herbaceous	Not available	Europe	Baker and Brooks, 1989 Reeves, 2006 Reeves and Brooks, 1983
Viola caliminaria	Viola	Zn, (2) Cd, Pb	Herbaceous	Not available	Central Europe	Baker and Brooks, 1989 Reeves, 2006

Note: Other hyperaccumulators of Zn may include (Baker and Brooks, 1989). *Haumaniastrum katangense* (Africa), *Noccaea eburneosa* (Europe), *Thlaspi alpestre* (Europe), *Thlaspi bulbosum* (Greece), *Thlaspi calaminare* (Europe), *Thlaspi limoselifolium* (Europe)

Figure 3.43 Cadmium, Zinc, Nickel and Other Metal 'Accumulators' (not Hyperaccumulators)

Latin	Common	Contaminant	Vegetation Type	USDA Hardiness Zone	Native to	Reference
Agrostis delicatula	Bentgrass	Zn, As, Cu, Mn	Herbaceous	4–10	Southwestern Europe, North Africa	Gomes et al., 2014
Amaranthus hypochondriacus	Amaranth Prince-of-Wales Feather	Cd	Herbaceous	Not available	Mexico	Li et al., 2013
Arabis flagellosa	Rock Cress	Cd	Herbaceous	Not available	Asia	Chen et al., 2009
Arabis gemmifera	Rockcress	Cd	Herbaceous	Not available	Japan	Kubota et al., 2003
Arrhenatherum elatius	False Oat Grass	Ni, Cu, Cd, Co, Mn, Cr, Zn	Herbaceous	4–9	Europe	Lu et al., 2013
Athyrium yokoscense	Hebino-negoza Fern	Cd, Cu	Herbaceous	7	Japan	Chen et al., 2009
Atriplex hortensis var. purpurea	Golden Orache	Zn	Herbaceous	11	Europe, Asia	Kachout et al., 2012
Averrhoa carambola	Star Fruit	Cd	Tree	9–11	Southeast Asia	Li et al., 2007
Bidens pilosa	Beggar Ticks Spanish Needle	Cd	Herbaceous	Not available	North America, South America	Wei and Zhou, 2008
Brassica carinata	Ethiopian Mustard Abyssinian Cabbage	Ni	Herbaceous	Not available	Africa	Purakayastha et al., 2008
Brassica juncea *Brassica juncea cv., 182921* *Brassica juncea cv. Pusa Jia Kisan* *Brassica juncea cv., 426308*	Indian Mustard	Cu, Cd, Cr(VI), Ni, Zn	Herbaceous	2–11	Eurasia	Bauddh and Singh, 2012 Blaylock et al., 1997 Bluskov et al., 2005 ITRC PHYTO 3 Kumar et al., 1995 Lai et al., 2008
Brassica napus	Rapeseed	Cd, Cu, Zn	Herbaceous	2–11	Eurasia	Thewys et al., 2010 Van Slycken et al., 2013 Witters et al., 2012
Chicorium intybus var. foliosum	Chicory	Ni, Cd	Herbaceous	4–11	Mediterranean	ITRC PHYTO 3 Martin et al., 1996
Chromolaena odoratum	Cd	Shrub	9–11	Thailand	Phaenark et al., 2009, 2011
Conyza canadensis	Canadian Horseweed	Cd, Ni, Zn	Herbaceous	2–11	North America	Wei et al., 2004

155

Figure 3.43 (continued)

Latin	Common	Contaminant	Vegetation Type	USDA Hardiness Zone	Native to	Reference
Erigeron canadensis	Canada Fleabane	Cd, Ni	Herbaceous	4–8	USA	ITRC PHYTO 3 Martin et al., 1996
Eupatorium capilifolium	Dogfennel	Cd, Ni	Herbaceous	4–9	Southern USA	ITRC PHYTO 3 Martin et al., 1996
Festuca arundinacea	Tall Fescue	Zn	Herbaceous	3–8	Europe	Batty and Anslow, 2008
Gynura pseudochina	Purple Velvet Plant	Zn, Cd	Herbaceous	10–11	Asia	Phaenark et al., 2009
Helianthus annuus L. cv. Ikarus *Helianthus annuus*	Sunflower	Cd, Zn, As, Ni	Herbaceous	Grown as annual	USA	Adesodun et al., 2010 Cutright et al., 2010 ITRC PHYTO 3 Kumar et al., 1995 Nehnevajova et al., 2005 Nehnevajova et al., 2007 Padmavathiamma and Li, 2009 Salt et al., 1995 Stritsis et al., 2014
Helianthus tuberosus	Jerusalem Artichoke	Cd	Herbaceous	3–9	Eastern USA	Chen et al., 2011
Impatiens violaeflora *Impatiens walleriana Hook. f.*	Impatiens	Cd	Herbaceous	Not available	Not available	Lin et al., 2010 Phaenark et al., 2009
Justicia procumbens	Water Willow	Cd, Zn	Herbaceous	Not available	Thailand, India	Phaenark et al., 2009, 2011
Kalimeris integrifolia	Japanese Aster	Cd	Herbaceous	5–9	Asia	Wei and Zhou, 2008
Limonastrium monopetalum	Limonastrium	Cd	Herbaceous	10–11	Greece	Manousaki et al., 2014
Linum usitatissimum *L. ssp. usutatissimum cv. Gold* *Merchant*	Flax	Cd	Herbaceous	4–10	Mediterranean Middle East	Stritsis et al., 2014
Medicago sativa	Alfalfa	Zn, Cd, Ni	Herbaceous	3–11	Asia	ITRC PHYTO 3 Tiemann et al., 1998 Videa-Peralta and Ramon, 2002
Nicotiana tabacum	Tobacco	Cd, Zn	Herbaceous	Grown as annual	North America	ITRC PHYTO 3 Kumar et al., 1995 Vasiliadou and Dordas, 2010 Yancey et al., 1998

156

Latin	Common	Contaminant	Vegetation Type	USDA Hardiness Zone	Native to	Reference
Oryza sativa	Rice (extraction highly specific by cultivar)	Cd	Herbaceous	7+	Asia	Chaney et al., 2010 Murakami et al., 2007
Pelargonium roseum	Scented Geranium	Cd, Ni	Herbaceous	10–11	South Africa	Mahdieh et al., 2013
Populus spp. Populus alba L. var. pyramidalis	Hybrid poplar	Zn, Cd	Tree	3–9	varies	Hu et al., 2013 Ruttens et al., 2011 Van Slycken et al., 2013 Thewys et al., 2010 Witters et al., 2012 Hinchman et al., 1997 ITRC PHYTO 3
Potentilla griffithii	Potentilla	Zn, Cd	Herbaceous	Not available	China	Qiu, 2006
Pseudotsuga menziesii	Douglas Fir	Cd	Tree	5–7	North America	Astier et al., 2014
Raphanus sativus cv. Zhedachang	Radish	Cd	Herbaceous	Grown as annual	Europe	Ding et al., 2013
Ricinus communis	Castor Oil Plant	Cd	Herbaceous	10–11	Mediterranean East Africa	Bauddh and Singh, 2012
Rorippa globosa	Globe Yellowcress	Cd	Herbaceous	6	Europe	Wei and Zhou, 2006
Rumex crispus	Curly Dock	Cd, Zn	Herbaceous	1–11	Europe, Asia	Zhuang et al., 2007
Salix spp. 'Belders' (S. alba L. var. alba), 'Belgisch Rood' (S. × rubens var. basfordiana) (Zwaenepoel et al., 2005), 'Christina' (S. viminalis), 'Inger' (S. triandra × S. viminalis), 'Jorr' (S. viminalis), 'Loden' (S. dasyclados), 'Tora' (S. schwerinii × S. viminalis) and 'Zwarte Driebast' (S. triandra). Salix viminalis L.	Willow	Cd, Zn	Shrub	varies	varies	Algreen et al., 2013 Evangelou et al., 2012 Ruttens et al., 2011 Thewys et al., 2010 Van Slycken et al., 2012 Van Slycken et al., 2013 Witters et al., 2012
Sedum alfredii	Sedum	Cd, Zn	Herbaceous	Not available	Asia	Li et al., 2011 Lu et al., 2013 Wang et al., 2012 Xiaomei et al., 2005 Xing et al., 2013 Yang et al., 2013 Zhuang et al., 2007

157

Figure 3.43 *(continued)*

Latin	Common	Contaminant	Vegetation Type	USDA Hardiness Zone	Native to	Reference
Sedum jinianum	Sedum	Cd, Zn	Herbaceous	Not available	China	Xu et al., 2009
Sedum plumbizincicola	Sedum	Cd, Zn	Herbaceous	Not available		Liu et al., 2011
Sesbania drummondi	Rattlebush, Poison Bean	Cd	Herbaceous	8–11	Southeastern USA	Israr et al., 2006
Solanum elaegnofolium	Purple Nightshade	Cd	Herbaceous	6+	Western USA, South America	Gardea-Torresdey et al., 1998 ITRC PHYTO 3
Solanum nigrum	Black Nightshade	Cd, Ni, Zn	Herbaceous	4–7	Eurasia	Ji et al., 2011 Wei et al., 2004 Wei et al., 2012
Solanum tuberosum cv. Luyin No.1	Potato	Cd	Herbaceous	3–12		Ding et al., 2013
Sonchus transcaspicus	Sowthistle	Ni, Cu, Cd, Co, Mn, Cr, Zn	Herbaceous	4–9	Europe Asia	Lu et al., 2013
Spinacia oleracea L. cv. Monnopa	Spinach	Cd	Herbaceous	Grown as annual	Asia	Stritsis et al., 2014
Tagetes patula	French Marigold	Cd	Herbaceous	Grown as annual	North America, South America	Lin et al., 2010
Thlaspi ochroleucum	Pennycress	Zn	Herbaceous	Not available	Greece	Kelepertsis and Bibou, 1991 Reeves, 2006
Tithonia diversifolia	Mexican Sunflower Tree Marigold	Zn	Herbaceous	9–11	Eastern Mexico	Adesodun et al., 2010
Tripsacum dactyloides	Eastern Gamagrass	Zn	Herbaceous	4–9	Eastern USA	Hinchman et al., 1997 ITRC PHYTO 3
Vetiveria zizanioides	Vetiver Grass	Zn, Cd, Cu	Herbaceous	8–10	India	Danh et al., 2009
Viola baoshanensis	Viola	Cd	Herbaceous	Not available	China	Wu et al., 2010 Zhuang et al., 2007
Zea mays *Zea mays L. cv. Cascadas*	Corn (specific hybrids)	Cd	Herbaceous	3–11	North America	Broadhurst et al., 2014 Stritsis et al., 2014 Thewys et al., 2010 Van Slycken et al., 2013 Witters et al., 2012

Project name: Lommel Agricultural Fields "Der Kempen" (Ruttens et al., 2011; Van Slycken et al., 2013; Thewys et al., 2010; Witters et al., 2012)

Location: Flanders region, Belgium

Institution/Scientists: Centre for Environmental Sciences (CMK), Hasselt University, Agoralaan, Building D, 3590 Diepenbeek, Belgium, led by Dr. Jaco Vangronsveld. The project is one of the 17 sites coordinated by the GREENLAND project supported by the European Commission (FP7-KBBE-266124, GREENLAND), website: http://www.greenland-project.eu/.

Date installed: Initially installed in 2004, research ongoing

Species installed: Corn (*Zea mays*), rapeseed (*Brassica napus*), willow (*Salix spp.*) and poplar (*Populus spp.*)

Contaminant: Cadmium, zinc (and lead – cannot be phytoextracted) ▦

In Belgium and the Netherlands, an area of over 700 km^2 (270 square miles) is contaminated with heavy metals (cadmium, zinc and lead) as a result of historic zinc-smelting activities. Metals deposition declined significantly in the 1970s as the industry shifted toward different production processes, but soil contamination remains an issue. Adding to this problem is the fact that soils in the area are sandy and acidic, which renders the cadmium and zinc metals more mobile. Additionally, local land use is largely agricultural. The Belgian Federal Agency for Food Safety (FAVV) has seized several agricultural fields with vegetable crops because cadmium levels exceeded the legal threshold for human consumption.

159

Research at the Lommel site in this region has focused on repurposing these contaminated agricultural lands to produce biomass and energy crops instead of food. In transitioning to

Hybrid poplar trees planted at the Lommel site are slowly extracting cadmium over time. The trees are being tested to see if the species would additionally serve as a good bioenergy crop in Belgium. It is estimated that phytoremediation biomass crops in this area would need to be grown and harvested for 50-100 years to extract soil cadmium before regulatory soil limits are achieved.

Figure 3.44 Case Study: Lommel Agricultural Fields, Flanders Region, Belgium

biomass and energy species, agricultural lands remain profitable to individual farmers, despite contamination. Additionally, metals can be phytoextracted ⬚ from soils over time through continual harvest of biomass, eventually remediating the land. This research signals a shift in focus away from the use of metal hyperaccumulator species, which usually do not produce enough biomass for quick remediation times, and instead use high-biomass species. The overall goal of research at the Lommel sites is to obtain remediated soils that can safely be used for food production.

Energy crops being evaluated at Lommel are primarily corn (*Zea mays*), rapeseed (*Brassica napus*), as well as willow (*Salix spp.*) and poplar (*Populus spp.*) varieties. The results have shown that corn offers the best option for energy production, through biodigestion and burning in a combined heat and power system. However, corn's ability to extract metal contamination from the soil is far lower than that of willow. At the site, willow and poplar species have been grown using short-rotation coppice systems in which they are harvested every few years as a bioenergy crop. So far, results have shown that harvestable willow biomass far exceeds that of the poplar varieties tested, and would be best for remediation goals. The researchers have calculated that it would take a minimum of 55 years of willow harvest to reduce the cadmium levels from 5 mg Cd/kg in soil to safe levels of 2 mg Cd/kg in soil. However, if all willow leaves were collected in the autumn every year, instead of being allowed to fall, the time could be reduced to 36 years.

c Nickel ◼

Background: Nickel is used for the production of stainless steel and other metal alloys. Nickel contamination is often spread through mining and industrial activities (Cempel and Nikel, 2006). It can easily migrate into air and water and be dispersed, causing large areas of low-level contamination. In addition, naturally high nickel content in ultramafic soils is common globally. Nickel can be easily accumulated in plant and animals tissues and can bioaccumulate (Cempel and Nikel, 2006). High nickel concentrations can be phytotoxic to sensitive agricultural crops and cause lower agricultural yields (Chaney et al., 2003).

Applicability: Nickel is one of the more promising metals to be considered for phytoextraction and harvesting (Van der Ent, 2013). Phytoextraction ⬚ of nickel has proved in field studies to be efficient in some cases (Chaney et al., 2007). Because of nickel scarcity and high market prices, phytomining of nickel for industrial use may have potential in the future. Several plants have been shown to accumulate 1–3% nickel in dry matter, providing biomass ash for refineries that is richer than conventional ore materials (Chaney et al., 2007). In addition, biomass from nickel phytomining can be used as organic nickel fertilizer where nickel deficiency in soil has previously killed pecan trees and other agricultural crops (Wood et al., 2006).

However, nickel-contaminated soils can also be phytotoxic to plants, mostly when soils are acidic. Nickel phytotoxicity often does not affect agricultural food chains because, unlike cadmium and zinc, a substantial reduction in crop yield occurs when nickel contamination is high, therefore food cannot be grown on these sites (Chaney et al., 2003). Phytoextraction is best utilized for nickel as a 'polishing strategy' for sites with lower levels of contamination that are not phytotoxic to plants, unless a detailed soil risk assessment and amendment strategy is completed. Alternatively, higher nickel concentrations can be stabilized in soil with plants, using amendments to prevent nickel toxicity. Between phytoextraction and inactivation of soil nickel through addition of amendments and stabilization, phytotechnologies can potentially be used to remediate nickel-affected soils at a reasonable cost (Chaney et al., 2003).

According to Dr. Rufus Chaney, in 2003 soil removal and replacement cost about $2 million/ hectare-30 cm deep for nickel-contaminated sites. However, a phytoextraction and amendment remediation may only cost $3,000–$10,000/hectare. This does not include the potential for the nickel to be collected from the harvested plants and sold, which may add an additional revenue source where soil-nickel concentrations are high (Chaney et al., 2003).

Planting specifics

At this time, more than 450 plant species have been documented as hyperaccumulators of nickel (Van der Ent et al., 2013), but only a small fraction of these accumulate over 1% Ni, which would be required for effective phytoextraction or phytomining. High nickel-accumulating species such as *Alyssum spp.* are well documented and grow well in most climates (Chaney, 2013). The large number of nickel-accumulating plant species is likely because of the great extent of nickel-rich soils worldwide (Van der Ent et al., 2013).

Phytoextraction of nickel should only be attempted after careful study by an agronomist for potential profit from extraction or phytotoxic effects, and after consideration of any risks associated with translocating nickel to the above-ground portions of the plant. Phytomining of nickel is likely one of the most useful opportunities for phytoextraction. The following planting typologies can be considered for use in conjunction with the plant list in Figure 3.45.

In soil

Extraction and harvesting:

- Extraction Plots: Chapter 4, see p. 224

Stabilization: Amendments are added to restore soil fertility and prevent nickel uptake, and plants are installed to prevent erosion of nickel in soils.

- Planted Stabilization Mat: Chapter 4, see p. 202

In water

To control contaminated groundwater:

- Groundwater Migration Tree Stand: Chapter 4, see p. 213

To remove from surface and wastewater: The systems described below have been used for nickel removal from water. Since constructed wetlands have been widely documented, plant species for these purposes have not been included in this publication.

Figure 3.45 Nickel Hyperaccumulators (not a complete list, rather a representative list of species)

Latin Name	Common	Vegetation Type	USDA Hardiness Zone	Native to	Original Reference
Alyssum bertolonii	Alyssum	Not available	Not available	Italy	Robinson et al., 1997b
Alyssum bracteatum	Alyssum	Not available	Not available	Iran	Ghaderian et al., 2007
Alyssum lesbiacum	Alyssum	Not available	Not available	Not available	Kupper et al., 2001
Alyssum murale	Yellowtuft	Herbaceous	2–5	Balkans	Bani et al., 2007 Chaney et al., 2003, 2007, 2010 Prasad, 2005
Arenaria humifusa Wahlenb.	Low Sandwort	Not available	Not available	Eastern North America, Northern Canada, Europe	Phytorem Database Rune and Westerbergh, 1992
Arenaria rubella	Sandwort	Herbaceous	4	Western USA	Kruckeberg et al., 1993
Berkheya coddii	South African Aster	Not available	Not available	South Africa	Keeling et al., 2003 Morrey et al., 1989 Robinson et al., 1997a
Bornmuellera tymphaea	Bornmuellera	Herbaceous	Not available	Greece	Chardot et al., 2005
Brassica juncea	Indian Mustard	Herbaceous	Grown as annual	Asia Europe, Africa	Saraswat and Rai, 2009
Leptoplax emarginata	Leptoplax	Herbaceous	Not available	Greece	Chardot et al., 2005
Pearsonia metallifera	Pearsonia	Herbaceous	Not available	Zimbabwe	Wild, 1970 Brooks and Yang, 1984
Phyllanthus serpentinus	Phyllanthus	Not available	Not available	New Caledonia	Kersten, 1979
Phyllomeli coronata	Phyllanthus	Not available	Not available	Caribbean	Reeves et al., 2006
Ruellia geminiflora	Ipecacuanha	Herbaceous	Not available	South America	Jaffré and Schmid, 1974 Brooks et al., 1992
Sebertia acuminata	Latex Rubber Tree	Tree/Shrub	Not available	New Caledonia	Cunningham and Berti, 1993 ITRC PHYTO 3 Van der Ent et al., 2013
Senecio pauperculus	Balsam Groundsel	Herbaceous	1–10	North America	Baker and Reeves, 2000 Roberts, 1992
Solidago hispida	Hairy Goldenrod	Herbaceous	3–8	Eastern North America	Baker and Reeves, 2000
Streptanthus polygaloides	Milkwort	Herbaceous	9	Western USA	Baker and Reeves, 2000
Thlaspi caerulescens	Alpine Pennycress	Herbaceous	6	Western USA, Europe	ITRC PHYTO 3 Rouhi, 1997 Salt et al., 1995
Thlaspi montanum L. var. montanum	Fendler's Pennycress	Herbaceous	6–10	Western USA	Boyd et al., 1994 Phytorem Database Prasad, 2005

Alyssum murale plots planted for phytoextraction and phytomining of nickel at Port Colborne, Ontario (Chaney et al., 2003).

Above-ground plant parts were harvested and incinerated into an ash, ready to be used in the nickel smelting process (Dr. Scott Angle pictured).

The plant ash is smelted to extract nickel for industry. The plant-derived ash is richer in nickel than conventional mined ore materials (Chaney et al., 2007).

Figure 3.46 Case Study: Port Colborne, Ontario, Canada

- Stormwater Filters: Chapter 4, see p. 235
- Multi-Mechanism Buffers: Chapter 4, see p. 234
- Surface-Flow Constructed Wetland: Chapter 4, see p. 238
- Subsurface Gravel Wetland: Chapter 4, see p. 241

The selected nickel hyperaccumulator plant species listed in Figure 3.45 have been cited frequently as potential species for nickel uptake. For nickel, it has been found that the use of fast-growing hyperaccumulators for extraction is likely more effective than high biomass-producing crop plants (Chaney et al., 2010).

163

Case Study | Nickel

Project name: Port Colborne Nickel Refinery (Chaney et al., 2003; Chaney et al., 2007; Kukier and Chaney, 2004)

Location: Port Colborne, Ontario, Canada

Institution/Scientist: The USDA-Agricultural Research Service (R. L. Chaney and Y.-M. Li), the University of Maryland (J. S. Angle and E. P. Brewer), the Environmental Consultancy of the University of Sheffield (A. J. M. Baker and R. D. Reeves), Oregon State University (R. J. Roseberg)

Date installed: 1990s

Species installed: Oats, radish, corn, soybeans (to assess phytostabilization ⬚ of Ni), *Alyssum murale* and *A. corsicum* (for phytoextraction ⬚).

Amendments: Dolomitic limestone, to raise pH, manganese fertilizers and NPK fertilizers.

Contaminant: Nickel

Target media and depth: Top 50 centimeters (20 inches) of soil.

Twenty-nine square kilometers of surface soils in the Port Colborne, Ontario area became contaminated with nickel from a nearby refinery operating over 60 years. Copper (Cu) and cobalt (Co) concentrations in soil were also elevated near the refinery in these surrounding lands used for agriculture. Because nickel levels were so high, the agricultural crop plants were affected and reduced crop yields occurred, due to phytotoxicity. Removal, hauling and landfilling of contaminated soils was considered, but because the area of contamination was so large and spread by air emissions, and because a hugely unsustainable amount of topsoil would have been required for revegetation, phytostabilization ⬚ on site by the addition of amendments and phytoextraction ⬚ and harvesting to remove the nickel were both tested as remediation options (Chaney et al., 2003).

To test the opportunity to stabilize the nickel on site and prevent it from contaminating and harming the agricultural crops, soil pH was raised with amendments as an effective method of remediating the phytotoxicity of nickel-contaminated soils. High rates of limestone were applied to Port Colborne soils to reduce the amount of nickel bioavailable to the plants. Liming soil to a pH of about 7.5 (calcareous, with excess limestone remaining to keep pH high for centuries) substantially reduced nickel concentrations in the shoots of all species and enabled survival and normal plant growth. In addition, other adjustments to soil fertility, such as adding manganese, were made to ensure optimal yields. Nickel phytotoxicity in an industrially contaminated soil was highly dependent on soil pH and on plant species, where some agricultural species were more sensitive to nickel than others. Grass species (*Poacea spp.*) were found to be more tolerant of nickel than the other species tested (Kukier and Chaney, 2004). Phytostabilization ⬚ was found to be successful using soil amendments and tolerant plant species.

In addition, in a separate set of tests on the same site, *Alyssum murale* and *A. corsicum* were planted to phytoextract ⬚, harvest and remove nickel from soils. Dr. Chaney and his collaborators developed a nickel phytoextraction technology to mine nickel (phytomining) with plants using these hyperaccumulating, high biomass-production species. Over 200 ecotypes (genetically distinct geographic varieties within the species) of nickel hyperaccumulator *A. murale* were collected in southern Europe, and new cultivars were developed by plant breeding to optimize the favorable extraction characteristics of this perennial, which regrows after cutting to harvest biomass. *A. murale* was then tested at two sites with high levels of nickel in the soil: (1) in soils contaminated by a nickel refinery in Port Colborne, Ontario and (2) on historic serpentine soils in Oregon with high levels of naturally occurring nickel. The crops were harvested in early flowering stage, allowed to air-dry for several days, then baled, handled and stored away from the production field. The plants attained significant amounts of nickel in above-ground parts and the shoots were dried and burned to make ash for nickel recovery in smelting. The dried hay can be burned in a biomass generator, which supplements the cost of growing the crop, and the resulting ash is a high-grade nickel ore. With high biomass-yield potential, *A. murale* and *A. corsicum* can phytoextract 200–400 kg/hectare/year. It was shown that the value of the plant ash as a metal ore could offset the costs of soil remediation and provide more profit than conventional crops on these soils. Commercial phytoextraction from mineralized soils in Oregon resulted, though the application of this phytomining technology is limited, due to specific soil requirements.

As with all crop plants, the phytomining plants must be fertilized and managed for optimum economic production of nickel. Challenges may exist for growing plants in certain soils and climatic conditions. In addition, the potentially invasive characteristics of plants should be considered before any plants are introduced into a new, non-native environment. *A. murale* was found to have invasive characteristics when used for phytomining in North America.

d Selenium

Background: Selenium contamination has become a challenge in parts of the western United States due to irrigation of naturally enriched seleniferous soils that leach selenium into other, low-lying wetland areas (Chaney et al., 2007, Bañuelos et al., 2005). Typically, selenium contamination of drinking water is of greatest concern. Contaminated water typically results from naturally occurring selenium being mobilized by irrigation run-off or industrial water pumping through oil shales, including natural-gas fracking and oil drilling, or from agricultural run-off.

Applicability: Selenium is efficiently extracted and also volatilized by some hyperaccumulating plants and high-biomass species. The volatile form of selenium is 2–3 orders of magnitude less toxic than the inorganic selenium forms found in the soils (Terry et al., 2000). One of the more promising phytomanagement uses is for recycling selenium using accumulating forage-crop species (Dickinson et al., 2009). Selenium-enriched seed meals have been effectively grown on selenium-rich sites, harvested and used as a nutritional supplement to feed livestock (Chaney et al., 2007; Bañuelos et al., 2010). Selenium is an essential mineral for animals, including cattle, yet most soils are selenium deficient, and the feed grown in such areas lacks the selenium concentrations necessary for animal nutrition (Bañuelos et al., 2010). Approximately 20 species of plants have been verified as selenium hyperaccumulators, with some such as *Stanleya pinnata* performing better than others in field conditions (Van der Ent et al., 2013; Freeman and Bañuelos, 2011),

Planting specifics

Species for extraction of selenium have been identified and successfully used for remediating field sites with low levels of selenium contamination. Some of the selenium may be accumulated into shoots, though selenium changes form and can also be volatilized into the air.

In soil

Extraction and volatilization: Selenium is extracted into the above-ground portions of the plant and the plants are harvested to remove the contaminant from the site. In addition, selenium is also volatilized, and in some situations the above-ground parts do not necessarily need to be removed.

- Extraction Plots: Chapter 4, see p. 224

Stabilization: Plants are used to keep the contaminant on site, preventing it from moving and creating exposure risk. Soil amendments should be added to decrease mobility.

- Planted Stabilization Mat: Chapter 4, see p. 202

In water

To control contaminated groundwater: High evapotranspiration-rate species can be planted to control the groundwater plume, slowing it or redirecting it to prevent the spread of the contaminant in groundwater. Selenium can be held in the root zone or extracted into the above-ground portions of the plant.

- Groundwater Migration Tree Stand: Chapter 4, see p. 213

To remove from surface and groundwater: Surface or groundwater is run through a natural treatment system that primarily physically traps the selenium and binds or precipitates it to planting media. Since constructed wetlands have been widely documented, plant species for these purposes have not been included in this publication.

- Stormwater Filters: Chapter 4, see p. 235
- Multi-Mechanism Buffers: Chapter 4, see p. 234
- Surface-Flow Constructed Wetland: Chapter 4, see p. 238
- Subsurface Gravel Wetland: Chapter 4, see p. 241

The upland hyperaccumulator plant species listed in Figure 3.47 is a representative list of plants shown to extract selenium.

166

Case Study | **Project name:** Ridge Natural Area above Spring Creek Park (Freeman and Bañuelos, 2011;
Selenium | Freeman, 2014)

Location: City of Fort Collins, CO

Consultant/Scientist: John L. Freeman, Ph.D., Phytoremediation and Phytomining Consultants United (http://www.phytoconsultants.com)

Date installed: Spring 2007

Species installed: *Stanleya pinnata* (Prince's Plume)

Amendments: None

Contaminant: Selenium naturally present in cretaceous shale.

Selenium toxicity is a significant problem in the western United States, where it impacts drinking water supplies near oil shale-drilling operations, agriculture and other uses where soil is exposed to rain and erosion. Researchers evaluated several genotypes of the native Colorado species *Stanleya pinnata* for its ability to remove the highly toxic form of selenium from soil and volatilize it into the air as a less toxic organic form. It was found that the plant can remove 30% of selenium soil concentrations in one growing season, although the removal rates do decrease each year. Salt- and boron-tolerant genotypes were selected, since these elements are often present at selenium-contaminated run-off sites (Bañuelos and Freeman, 2011).

Stanleya pinnata was planted to remediate mobile selenium-contaminated soils after pipelines were installed in an area 150 ft wide x 1/2 mile long.

The attractive phytoremediation plantings blend with the adjacent park aesthetic.

Four years after installation, the mobile selenium was remediated and native Colorado grasses replaced the previous phytoremediation planting.

Figure 3.48 Case Study: Spring Creek Park, Colorado

167

Once select genotypes were identified, the chosen native Colorado *Stanleya pinnata* genotypes were planted at a selenium-rich site in Fort Collins, Colorado. At the Pine Ridge Natural Area above Spring Creek Park, new drainage improvements had to be made by the City of Fort Collins in areas of naturally occurring high-selenium soils. These soils would typically release selenium into local water supplies when disturbed. To prevent this, *Stanleya pinnata* was planted where selenium-rich soils were exposed in order to help stabilize, extract and volatilize the selenium. This perennial plant prefers dry and nutritionally poor soils, so it is ideal for large-scale plantings. The highly mobile selenium was controlled after four years of growing and the Colorado native grasses naturally succeeded the previous monoculture planting of *Stanleya pinnata*.

5 *Metals: moderately difficult to extract*

Boron (B), Cobalt (Co), Copper (Cu), Iron (Fe), Manganese (Mn), Molybdenum (Mo)

Background: All of the metals in this category, considered moderately difficult to extract, are essential plant micronutrients except cobalt, which is an essential element only for legumes.

Applicability: Phytoextraction ⬚ of all the moderately difficult-to-extract metals in this category is considered to be unfeasible at this time. Although hyperaccumulator species of these elements can be found in the literature, these metals are usually difficult to extract at field scale. Large amounts of these metals in soils can be phytotoxic to plant growth, or the metals may be

Figure 3.47 Selenium Hyperaccumulators (not a complete list, rather a representative list of species)

Latin	Common	Vegetation Type	USDA Hardiness Zone	Native to	Original Reference
Acacia cana	Boree	Tree	9–10	Australia	Baker and Reeves, 2000 McCray and Hurwood, 1963
Astragalus bisulcatus	Two-Grooved Milkvetch	Herbaceous	2–7	Western North America	Baker and Reeves, 2000 Byers, 1935 Byers, 1936 Lakin and Byers, 1948 Rosenfeld and Beath, 1964 Van der Ent et al., 2013
Astragalus grayi	Gray's Milkvetch	Herbaceous	Not available	Western North America	Baker and Reeves, 2000 Byers, 1935
Astragalus osterhouti	Osterhout Milkvetch	Herbaceous	Not available	Western North America	Baker and Reeves, 2000 Rosenfeld and Beath, 1964
Astragalus pattersonii	Patterson's Milkvetch	Herbaceous	Not available	Western North America	Baker and Reeves, 2000
Astragalus pectinatus	Narrowleaf Milkvetch	Herbaceous	2–6	Western North America	Baker and Reeves, 2000 Rosenfeld and Beath, 1964
Astragalus racemosus	Cream Milkvetch	Herbaceous	3–10	Western North America	Baker and Reeves, 2000 Byers, 1936 Chaney et al., 2010 Knight and Beath, 1937 Moxon et al., 1950 Rosenfeld and Beath, 1964 White et al., 2007
Atriplex confertifolia	Shadscale Saltbush	Herbaceous	6–10	Western North America	Baker and Reeves, 2000 Rosenfeld and Beath, 1964
Castilleja chromosa	Indian Paintbrush	Herbaceous	4–9	Western North America	Baker and Reeves, 2000 Rosenfeld and Beath, 1964
Haplopappus (sect. Oonopsis) condensate	Haplopappus	Herbaceous	Not available	Western North America	Baker and Reeves, 2000 Byers, 1935
Haplopappus (sect. Oonopsis) fremontii	Wards False Goldenweed	Herbaceous	Not available	Western North America	Baker and Reeves, 2000 Byers, 1935 Rosenfeld and Beath, 1964
Lecythis ollaria	Coco de Mono	Tree	Not available	Venezuela, Brazil	Baker and Reeves, 2000 Aronow and Kerdel-Vegas, 1965

Figure 3.47 (continued)

Latin	Common	Vegetation Type	USDA Hardiness Zone	Native to	Original Reference
Machaeranthera (Xylorhiza) glabriuscula	Smooth Woodyaster	Herbaceous	4–5	Western North America	Baker and Reeves, 2000 Rosenfeld and Beath, 1964
Machaeranthera (Xylorhiza) venusta	Cisco Woodyaster	Herbaceous	7	Utah, Colorado	Baker and Reeves, 2000
Machaeranthera parryi	Machaeranthera	Herbaceous	Not available	Western North America	Baker and Reeves, 2000 Byers, 1935
Machaeranthera ramosa	Machaeranthera	Herbaceous	Not available	Western North America	Baker and Reeves, 2000 Rosenfeld and Beath, 1964
Machaeranthera venusta	Machaeranthera	Herbaceous	Not available	Utah, Colorodo	Rosenfeld and Beath, 1964
Morinda reticulate	Mapoon	Herbaceous	Not available	Australia	Baker and Reeves, 2000 Knott et al., 1958
Neptunia amplexicaulis	Selenium Weed Water Mimosa	Herbaceous	Not available	Australia	Baker and Reeves, 2000 McCray and Hurwood, 1963
Stanleya bipinnata	Bipinnate Prince's Plume	Herbaceous	Not available	Western North America	Baker and Reeves, 2000 Byers et al., 1938 Moxon et al., 1950 Rosenfeld and Beath, 1964
Stanleya pinnata	Prince's Plume	Herbaceous	3–7	Western North America	Baker and Reeves, 2000 Byers et al., 1938 Rosenfeld and Beath, 1964 Van der Ent et al., 2013 White et al., 2007

Selenium Agricultural Accumulators (not Hyperaccumulators, but Used in Agriculture as Accumulators)

Brassica juncea	Indian Mustard	Herbaceous	9–11	Russia to Central Asia	Banuelos, 2000 Zayal et al., 2000
Festuca arundinacae	Tall Fescue	Herbaceous	2–10	Europe, North America	Banuelos, 2000
Hibiscus cannibinus	Kenaf	Herbaceous	9–11	Unknown	Banuelos, 2000
Lotus corniculatus	Birdsfoot Trefoil	Herbaceous	3–8	Africa	Banuelos, 2000

169

tightly bound to soils. Engineering soil amendments and vegetation for phytostabilization is the recommended phytotechnology for soil contamination with these metals at this time. In addition, conventional dig-and-haul or capping remediation strategies can be utilized for contaminated soils.

However, when these metals are mobilized in water, such as in groundwater plumes or stormwater, they can be filtered out of the water with the assistance of plants and held locally in the soils. With the exception of manganese, groundwater migration tree stands, stormwater filters and constructed wetlands can be used to filter the metals out of water and to trap the metals in the soil media.

In addition, when these metals are mobilized in groundwater, high-evapotranspiration plant species can slow down or delay the groundwater plume, potentially controlling the contaminant against spreading, as long as the concentration of the metal is not phytotoxic to plants. The target goal is not to extract these metals into a plant, but instead to control their spread in groundwater.

Planting specifics

In soil

Stabilization: Plants are used to keep the contaminant on site, preventing it from moving and creating exposure risk.

• Planted Stabilization Mat: Chapter 4, see p. 202

In water

To control contaminated groundwater: High evapotranspiration-rate species can be planted to control the groundwater plume, slowing it or redirecting it to prevent the spread of the contaminant in groundwater. As the water is drawn into the plant, the metals are typically held in the root zone and soils around the plant, filtering them from the water. This hydrological control prevents migration of the metals in the groundwater.

• Groundwater Migration Tree Stand: Chapter 4, see p. 213

To remove from surface and groundwater: The systems described below have been used for metals removal from water. Since constructed wetlands have been widely documented, plant species for these purposes have not been included in this publication. Metals that may enter the site via stormwater inflows will generally be of very low concentrations and will not be readily available for plant-based extraction methods. However, the metals can be filtered out through the organic soil media.

• Stormwater Filters: Chapter 4, see p. 235
• Multi-Mechanism Buffers: Chapter 4, see p. 234
• Surface-Flow Constructed Wetland: Chapter 4, see p. 238
• Subsurface Gravel Wetland: Chapter 4, see p. 241

A partial list of plant species for stabilization of metals is found at the beginning of this section (see plant excluder list Figure 3.37, p. 140). Additional species can be researched. Soil amendments should also be considered to additionally decrease mobility and to encourage plant growth, due to the potential phytotoxicity of the metals.

Case Study
Copper

Project name: Biogeco Phytoremediation Platform (Bes and Mench, 2008; Bes et al., 2010; Bes et al., 2013; Kolbas et al., 2011; 2014; Marchand, 2011)

Location: Gironde, France

Institution/Scientists: UMR BIOGECO INRA 1202 – University of Bordeaux, France, led by Dr. Michel Mench. This platform was initially supported by ADEME (French Environment and Energy Agency), Department of Urban Landfills and Polluted Sites, Angers, France and then developed as one of the 17 sites coordinated by the GREENLAND project supported by the European Commission (FP7-KBBE-266124, GREENLAND), website: http://www.greenland-project.eu/.

Date installed: Risk assessment started in 2005; first field plots installed in 2006; research ongoing

Species installed: *Populus nigra* L., *Salix caprea* L., *Salix viminalis* L., and *Amorpha fruticosa* L. with soil amendments. Test plots of perennials: *Agrostis capillaris* L., *Agrostis castellana* Boiss. & Reuter, *Agrostis delicatula* Pourr. Ex Lapeyr., *Agrostis gigantea* Roth., *Dactylis glomerata* L., *Holcus lanatus* L., *Festuca pratensis* and *Cytisus striatus* Hill Rothm. Other perennials tested: for phytostabilization/biomasss production: *Vetiver, Miscanthus*, L. Annual crops tested for phytoextraction/biomass production: tobacco, sunflower, sorghum.

Amendments in field plots: Compost (made of pine-bark chips and chicken manure), compost and dolomitic limestone, alumino-silicates (Linz-Donawitz slags), zero-valent iron grit and compost. (See Bes and Mench, 2008, p. 1130 and Bes et al., 2013, p. 41 for full lists of soil amendments used in these studies.)

Contaminant: Heavy metals: copper (Cu) ■ in the form of copper sulfate ($CuSO_4$) and chromated copper arsenate used as wood preservative; PAHs ● and hydrocarbons from creosote ●

The Greenland initiative is a series of European Union-funded research projects that includes various gentle remediation options (GROs) and plant-based approaches to both remediate contaminated soils at low costs with limited harmful effects to the environment and produce biomass for plant-based feedstock (phytomanagement). Additionally, the project creates a network of long-term remediation case studies across Europe for comparison and optimization of GRO techniques. The Greenland Initiative recognizes that while GRO technologies may be innovative and efficient, they are still not widely used. The project seeks to produce sustainable as well as profitable management techniques for contaminated soils and promote their deployment at the field scale. As part of ongoing research for the Greenland project, Michel Mench and colleagues have examined aspects of a former wood-preservation site in France.

171

The site in Gironde, southwestern France has been the focus of ongoing research into phytoremediation techniques that could be applied to many sites with copper and copper/PAHs contamination. The area itself is 10 hectares (25 acres) in size and has been used for over a century to produce and store wood timbers, posts and utility poles. Contamination from these processes has resulted in various soil levels of copper (67–2,600 mg/kg) and PAH (see section on petroleum, p. 65), depending on the sub-site. The research is focusing on (1) how to stabilize the metal and get plants to grow on this contaminated site, and (2) how to combine bioavailable copper stripping, rhizodegradation of PAHs and production of biomass with profitable returns (phytomanagement).

Many combinations of soil amendments were applied to reduce the availability of copper and increase plant growth. The researchers found that soil amendments of activated carbon and zero-valent iron grit helped the most to reduce concentrations of copper in the soil solution. These amendments could be important first steps toward improving soil properties that would allow vegetation to flourish, as well as reducing the mobility of copper contamination.

Vegetation at the site was examined in depth to identify species with remediation potential as well as accumulator, excluder and tolerant properties towards copper. Based on its investigation, the team identified "aided phytostabilization" options for the site that utilized existing vegetation. One option was to use soil amendments to reduce available copper contamination and then add Common Bent Grass (*Agrostis capillaris*) and Redtop (*Agrostis gigantea* Roth) as excluder species to provide vegetative cover. Other species found at the site, such as Black Poplar (*Populus nigra*) and Basket Willow (*Salix viminalis*) show commercial promise as bioenergy species and can be used in short-rotation coppice systems for biomass generation while stabilizing copper in the soil. Various genotypes of annual secondary Copper accumulators (tobacco, sunflower) are being investigated for bioavailable Copper stripping (around 150 g Cu/hectare/year), oilseed production, biosourced chemistry and other productive products. Research into the effectiveness of additional species (*Vetiver*, *Miscanthus*, Switchgrass, etc.) is ongoing.

The Greenland Initiative studies show the reality of metal-contaminated sites and how remediation must be focused on (1) containing pollutants and limiting their mobility rather than removing them, given scarce resources and high cleanup costs and (2) options for phytomanagement (combining bioavailable metal stripping and biomass production). Moreover, it points to trends in phytoremediation that attempt to minimize pollutant linkages and risk from the contamination while creating vegetative cover and biomass that can provide a financial return.

6 Metals: difficult to extract (tend to be less bioavailable to plants)

Lead (Pb)

Background: Lead is one of the most common widespread contaminants in US urban areas, due to the historic use of leaded gasoline, lead paint, lead batteries, lead pipes and the continuing use of lead in industry. Lead in soils and in urban dust builds up cumulatively over time and is prevalent, and exposure to these sources is common. Lead poisoning is the leading environmentally induced illness in children. Children under the age of six are at greatest risk because of their fast neurological stage of development and problems arising from lead ingestion, leading to the decline of mental development and acquisition of motor skills (OSHA, 2013).

Applicability: The chemistry of lead in soil severely limits its availability for plant uptake. As a result, lead extraction with plants is not considered feasible for field-scale remediation. Lead has low bioavailability in aged soils, and often exists in forms of limited solubility; plant uptake and extraction is limited without the aid of chemical additives (Zia et al., 2011) and there are no known sucessful hyperaccumulators of lead for remediation that live in temperate climates. Where extraction of lead has been claimed in past studies, typically a chemical (chelating agent) has been added to make the lead more bioavailable and a high biomass-producing crop plant has been grown. The most common chelating agent that has been utilized to enhance lead extraction is EDTA: ethylene diamine tetra-acetic acid. This approach enables traditional crop plants like sunflower and mustard to take up lead, but generates the risk that lead could be mobilized into groundwater, creating uncontrollable pollution migration. For this reason, the use of EDTA for lead extraction has been banned in the US and EU for over 10 years (Chaney, 2014). The cost of the chelating agent amendments may also be prohibitive, at a cost of more than $30,000 per hectare/year (Chaney et al., 2007). At some limited field-scale sites where EDTA was used, leaching into groundwater was not detected and this technique was found to be significantly more cost-effective than excavation and soil removal (Blaylock, 2013; Weston, 2014); however, this is still not recommended because of potential risks.

Many designers and practitioners are misinformed and believe that sunflowers, mustard and other crop plants will naturally hyperaccumulate lead and can be harvested to remediate a site. One of the reasons for this misconception is that the use of chelating agents may not be understood, or publications promoting metals phytoremediation may have misunderstood or not mentioned them (Kuhl, 2010; Ulam, 2012). When referencing past studies and searching for plant species for phytoremediation of lead, practitioners must consider and investigate if EDTA or other chelating agents were used.

On a positive note, one study has suggested that the risk to humans of lead in urban soils may be lower than was previously assumed (Zia et al., 2011). Rather than total lead in soil, bioavailability of soil lead is the important measure for protection of public health. Recent findings have revealed that the bioavailable fraction of lead in urban soils is only 5–10% of total soil lead, far lower than the 30% as presumed by the US EPA (Zia et al., 2011). Lead bioavailability in soils can vary and is largely controlled by the amount of phosphate, iron, oxides, organic matter and pH levels (Zia et al., 2011). Only when soils are very high in lead, phosphate deficient and very acidic can a very few species naturally accumulate small amounts of lead in above-ground tissues (Chaney, 2013). Lead bioaccessibility tests have been developed, and these can be considered when testing sites to determine the true risk of lead on a site. However, regulatory guidelines in most cases still rely on a measure of total soil lead in determining legal reporting requirements and required remedial actions, and property values are affected accordingly. The most risky pathway for lead exposure is uncovered soil or dust that may migrate onto skin and be directly ingested, or move in the wind and accumulate onto edible plant parts, rather than lead taken up by plants and incorporated into the plants' biomass.

Planting specifics

Since lead does not migrate freely in soil and groundwater, nor is it easily taken up by plants, the recommended solution is usually to stabilize it on site and prevent dust and wind from exposing soil particles. For this reason, the US EPA has established guidelines for covering lead-contaminated soils to prevent physical exposure. Adding organic matter such as compost and raising the soil pH with lime and adding phosphorous will help further to bind lead to soils. Adding a thick vegetative layer on top can then prevent soil erosion and lead exposure.

Residential vegetable gardens and community garden sites are frequently contaminated with lead. To limit edible crop exposure, raised beds should be constructed with new, non-contaminated soils. In addition, thick layers of mulch or vegetation should cover any lead-contaminated soils to prevent wind from picking up soil dust particles and contaminating food crops with lead dust (see Chapter 5, p. 278 for community gardens).

In soil

Stabilization: Plants are used to keep the contaminant on site and decrease lead movement and exposure risk. Amendments and pH level are typically adjusted to increase stabilization. Lead exposure is most cost-effectively mitigated by binding it in place. Any plant species with an active root zone that prevents soil from eroding can be utilized.

- Planted Stabilization Mat: Chapter 4, see p. 202

In water

Lead typically does not easily leach into water, so techniques for removing lead from water have not been included.

Case Study Lead–1

Project name: Magic Marker Site (US EPA, 2014g; Clu-In, 2014; Blaylock, 2013; Blaylock et al., 1997; US EPA, 2002)

Location: Trenton, NJ

Institutions/Scientists: US EPA; Phytotech (company acquired by Edenspace Systems Corporation in 1999)

Date installed: 1996–1998

Species installed: *Helianthus annuus* (sunflower) and *Brassica juncea* (Indian Mustard)

Amendments: Chelators (EDTA)

Contaminants: Lead

Target media and depth: Soil to an 18-inch depth.

Prior to its use as a manufacturing facility for Magic Markers, this 7-acre site in Trenton, NJ housed various industrial activities including the manufacture of lead-acid batteries. When the factories shut down in the late 1980s, lead contamination (as well as other chemicals) remained. While most of the site was occupied by abandoned buildings and hardscape, approximately 1.5 acres of exposed soil with varying lead concentrations required remediation. Phytoextraction was proposed as a new technology that could potentially address the need for a cost-effective remediation strategy. Because there are no natural hyperaccumulators of lead, the approach of using existing crop plants, such as Indian Mustard and sunflower, combined with adding EDTA, a chemical chelating agent, to artificially enhance lead bioavailability and plant uptake was proposed. The project was initiated on approximately one-third of the site in 1996 using a spring planting of Indian Mustard, harvested in early June, followed by a planting of sunflowers, harvested in August, and then a third crop using Indian Mustard, harvested in late September, with EDTA added during all crop cycles. Soil samples were collected at a 5-foot grid spacing at three depths prior to planting and after harvesting of each crop to assess any change in the measured soil concentration. In the second year (1997) of the project, the planted area was expanded and the project was conducted under the oversight of the EPA SITE (Superfund Innovation Technology Evaluation) program, where an EPA contractor conducted soil sampling and analysis to evaluate technology performance. The phytoremediation project in both years produced both promising and conflicting results. While the soil testing was showing significant lead decreases, the lead quantity in the plants when measured could not account for the decrease in soil concentrations. As described in the EPA project report, "Possible explanations for the discrepancy include: (1) chelating agents that were applied to the soil mobilized and transported the lead out of the system," (2) the variability in crop productivity was incorrectly estimated, (3) there was a mistake made in the soil sampling processes or (4) enough amendments were added and tilled into the soil that spread out the hot spots and diluted the lead, bringing the overall soil concentration down. The supporting data was unable to confirm that the plants had actually extracted the lead, and this technology for lead extraction remains unproven. However, on several websites this project is characterized as successfully remediating lead with sunflower and mustard, and the use of EDTA and the inconsistencies with the data are not clearly addressed in non-technical terms. It is important to clarify that yes, the lead was no longer shown to be in the soils, but that plant extraction was not the reason for the lead removal. Sunflowers and Indian Mustard cannot extract and remediate lead.

It is laudable that the EPA and Phytotech were open to exploring and evaluating this innovative technology at field scale in the 1990s. However, practitioners must be careful to read the fine print and fully understand the intricacies of the data, so that future projects are modeled after scientifically proven precedents.

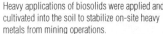

Heavy applications of biosolids were applied and cultivated into the soil to stabilize on-site heavy metals from mining operations.

Biosolids decreased the phytotoxicity of the heavy metals and the once-barren soils were vegetated.

Plants and the soil amendments work together to keep the heavy metals stablized and out of the Arkansas River.

Figure 3.49 Case Study: Alluvial Mine Tailling, Leadville, CO

Case Study
Lead–2

Project name: Alluvial Mine Tailings (Allen et al., 2007; Brown et al., 2005; Brown et al., 2007; Brown et al., 2009; National Research Council, 2003)

Location: Leadville, CO

Institutions/Scientists: US EPA, and URS Greiner, University of Washington

Date installed: 1997–2001

Species installed: Native grasses and shrub species for restoration.

Amendments: Municipal biosolids and limestone applied at 100 tons/acre.

Contaminants: Lead ■, zinc ■, cadmium ■ and acidity from historic mine tailings.

Target media and depth: Soil to a 12-inch depth.

Mining operations in Leadville, CO from the 1870s to 1980s created significant alluvial tailings contaminated with heavy metals including lead, zinc and cadmium. Contaminated tailings were found miles from the original source site, since the metals and sediments were moved and deposited in new locations by the Arkansas River during high-water events. The dispersed alluvial tailings were placed on the National Priorities List (list of Superfund sites) by the US EPA and were categorized for a removal action; however, the ecological impact and financial costs of excavating the tailings and sourcing replacement soil provided an incentive for scientists to instead use soil amending and phytostabilization ▢ to retain the metals on site.

Biosolids from Denver were mixed with limestone (to raise the pH) and tilled into the tailings deposits in an effort both to reduce the toxicity of the metals and to restore a plant cover. After amending, areas were seeded with native plant mixes for phytostabilization and prevention of erosion. Multiple tailings deposits along the Arkansas River were treated from 1997 to 2001. Close monitoring assured that the *in situ* treatment reduced metal availability and its associated risks to humans and animals. Plant cover, metal uptake by plants, species diversity, soil sampling and small-mammal trapping and analysis were carefully studied. The treated sites included private ranches as well as public lands. Many areas had very high levels of contamination. High Lonesome, one of the publicly owned treated lands is now open to the public as an access point to the Arkansas River for trout fishing.

176

■ *6 Metals: difficult to extract*

Chromium (Cr), Fluorine (F), Mercury (Hg), Uranium (U), Vanadium (V), Tungsten (W)

Background: The remaining metals in this category typically are not bioavailable and/or are phytotoxic to plants in soils. Some removal of chromium from contaminated water may be possible with aquatic plants. Mercury may be able to be taken up and volatilized into the air by some genetically modified plants but mercury in air is also problematic.

Applicability: Extraction of these metals in soils by plants is not feasible. However, plants can contribute by stabilizing metals on site and by physically filtering particulates containing these metals from water. In addition, high-evapotranspiration plant species can stop them from migrating by controlling contaminant plumes in groundwater, as long as the concentration of the metal is not phytotoxic to the plants.

Planting specifics

The most effective phytotechnology application for soils with low-bioavailable metals is to bind the contaminants in place utilizing phytostabilization. Any plant species with dense root mats that prevent erosion can be used. Species that can withstand the stress factors of these marginal sites should be considered.

177

In soil

Stabilization: Plants are used to keep the contaminant on site, preventing it from moving and mitigating an exposure risk. Amendments and pH level are typically adjusted to increase stabilization and enhance plant growth.

• Planted Stabilization Mat: Chapter 4, see p. 202

In water

To control contaminated groundwater: High evapotranspiration-rate species can be planted to control the groundwater plume, slowing it to prevent the spread of the contaminant in groundwater. As the water is drawn into the plant, the metals are typically held in the root zone and soils around the plant. This hydrological control prevents migration of the metals in the groundwater.

• Groundwater Migration Tree Stand: Chapter 4, see p. 213

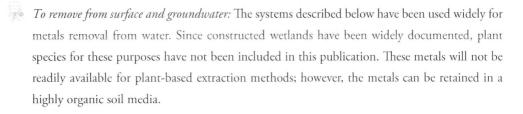 *To remove from surface and groundwater:* The systems described below have been used widely for metals removal from water. Since constructed wetlands have been widely documented, plant species for these purposes have not been included in this publication. These metals will not be readily available for plant-based extraction methods; however, the metals can be retained in a highly organic soil media.

- Stormwater Filters: Chapter 4, see p. 235
- Multi-Mechanism Buffers: Chapter 4, see p. 234
- Surface Flow Constructed Wetland: Chapter 4, see p. 238
- Subsurface Gravel Wetland: Chapter 4, see p. 241

Case Study
Chromium

Project name: BASF Rensselaer Landfill (Roux, 2014)

Location: Rensselaer, NY

Institution/Scientists: Roux Associates, Inc. (engineer); MKW Assoc. (landscape architect); BASF; New York State Department of Environmental Conservation (NYSDEC)

Date installed: 2008

Plant species: Mixed forest ecotype planting including many New York native higher-evapotranspiration rate species: *Alnus incana* (Grey Alder), *Acer rubrum* (Red Maple), *Aronia arbutifolia* (Red Chokeberry), *Betula nigra* (River Birch), *Castanaea dentata* (Chestnut), *Ceanothus americanus* (New Jersey Tea), *Cornus amormum* (Silky Dogwood), *Cornus racemosa* (Grey Dogwood), *Cornus sericea* (Red Osier Dogwood), *Clethra alnifolia* (Sweet Pepperbush), *Fraxinus americana* (White Ash), *Fraxinus pennsylvanica* (Green Ash), *Juniperus virginiana*

178

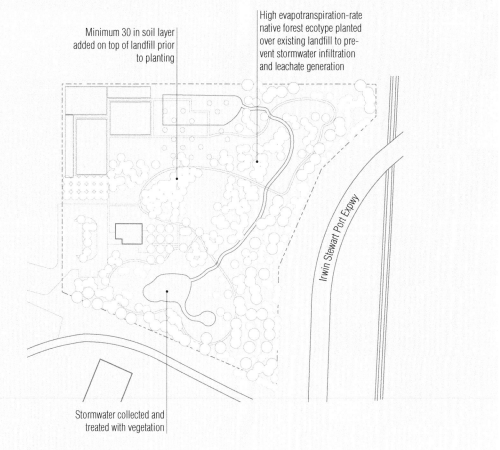

Minimum 30 in soil layer added on top of landfill prior to planting

High evapotranspiration-rate native forest ecotype planted over existing landfill to prevent stormwater infiltration and leachate generation

Irwin Stewart Port Expwy

Stormwater collected and treated with vegetation

Figure 3.50 Case Study: BASF Landfill Cover, Rensselaer, NY
Scale: 1 in = 250 ft

(Eastern Red Cedar), *Populus spp.* (poplar hybrids), *Salix discolor* (Pussy Willow), *Salix Nigra* (Black Willow), *Sassafras albidum* (Sassafras), *Sambucus nigra* (Elderberry).

Amendments: 0.7 meter (30 inch) soil cap over existing landfill.

Contaminant: VOCs (benzene ●, chlorobenzene ●, 1,2-dichlorobenzene ●, ethylbenzene ●, xylenes ●); heavy metals (arsenic ■, chromium ■, lead ■)

Target media: Soil and groundwater.

This 3.6 hectare (9 acre) former industrial landfill site is located in Rensselaer, New York. Wastes from a nearby chemical manufacturing plant were placed in the landfill until 1978, when the site was purchased by BASF. The site was listed by NYSDEC as a Class 2 Inactive Hazardous Waste Disposal Site. This triggered a series of environmental investigations which ultimately led to the installation of a soil cap in 1982 and a groundwater collection system in 1987.

In 2008, an alternative vegetated landfill cover (rather than a clay or plastic liner cap) was designed and installed to meet state landfill-closure regulations. The cover had to demonstrate its ability to prevent stormwater from entering the landfill so as to not generate polluted leachate. A densely planted scheme with a thick soil cap was developed that would evapotranspire the majority of rain-water, thereby minimizing infiltration through the former landfill. Additionally, the cover was designed as a phytoremediation planting that would provide phytodegradation and rhizodegradation of VOCs in the soil, while phytostabilizing heavy metals. The alternative landfill cover has been designed to include significant amenities, such as an environmental education center, walking trails and an amphitheater. Moreover, the plantings were designed to maximize ecological value, providing wildlife habitat.

179

Salt

Specific contaminants in this category: The term 'Salt' refers to any compound consisting of the cation from a base and the anion from an acid, which readily disassociates in water (Environment Canada, 2001). The following common salt compounds included in this category are sodium chloride (NaCl), calcium chloride (CaCl$_2$), magnesium chloride (MgCl$_2$) and potassium chloride (KCl). Aircraft de-icing fluids are typically ethylene glycol (EG) or propylene glycol (PG), which are hydrocarbons and are included in the section on other organic contaminants of concern earlier in this chapter.

Typical sources of salt pollution: De-icing maintenance activities, mining operations, fracking and oil drilling, fertilizer application, herbicide application.

Typical land uses with salt contamination: Roadsides, airports, and retail shops in cold-weather environments, natural gas and oil drilling sites, agricultural fields, industrial sites.

Figure 3.51 Salts

Why these contaminants are a danger: The spreading of salt on roads and highways prior to, during and after snow and ice events is the largest end-use of mined salt today, and an integral part of roadway maintenance. In 2008 alone, 22.6 million tons of road salt were used on US roads (Kostick, 2010). Salt dissolves easily in water and readily impacts groundwater and surface water. The most widely used road salt, sodium chloride, separates into the chloride anion and the corresponding cation, which move differently in the environment. Because sodium ions are positively charged, they exhibit a tendency to bond to negatively charged soil particles or be taken up in biological processes, while chloride ions, which are less reactive, can be quickly transported to surface waters through soil and groundwater, impacting water bodies hundreds of meters beyond roadways (Environment Canada, 2001).

In addition, globally, about 20% of agricultural land and 50% of cropland is considered salt stressed. (Manousaki and Kalogerakis, 2013). Accumulation of salt in the rooting zone of arable land is mostly due to irrigation, where salts are drawn up from deep layers of the soil profile by high rates of evaporation and transpiration in crop species (Rozema and Flowers, 2008). Coastal lowlands are also susceptible to increase in salinity as sea levels rise (Rozema and Flowers, 2008). Soil salinity inhibits plant growth, reduces yields and creates a significant threat to the global food supply (Chang et al., 2013). In addition, global oil and gas extraction (fracking) operations produce salt water as a byproduct, usually creating brine scars. Without significant soil amending and remediation practices, these salt-laden soils remain bare for decades. The salt wastewater produced from these operations is often ten times saltier than seawater and can also contain organic petroleum compounds. Soil salinity affects an estimated 95 million hectares of land worldwide (Szabolcs, 1994), trending with a 10% per year increase (Saboora et al., 2006).

About 1% of plant species are considered halophytes, salt-tolerant plants that can grow and reproduce in saline soils, with some species having extraction capabilities to remove soil salts into plant tissue. However, extraction of salt with plants is usually not practically achieved because enough biomass cannot be grown to eliminate the amount of salt polluting the system (Qadir et al., 2003). High amounts of plant biomass would be required to attain acceptable rates of remediation. Research into finding high biomass-producing halophytes that could be useful in remediation is currently being conducted (Ghnaya et al., 2005). If this is found to be feasible in the future, the plants would need to be harvested after the salt was extracted, to remove it from the contaminated site. Desalinization of soils utilizing halophytes may have more applicability in warmer climates, where species with high biomass production could potentially be grown and harvested. Since high-biomass plants for salt remediation have not been identified to date for temperate climates, salt-tolerant plants are instead typically grown for stabilization, to prevent soil from eroding, holding the salt on site.

Many soils heavily impacted with salt prevent plant growth. In this case, the remediation goal is the establishment of vegetation and phytostabilization. Soil amendments and bacteria can be used for soil improvement, and halophyte plant species that are salt tolerant are installed for restoration.

Unlike other uses of constructed wetland and stormwater filter systems that can effectively remove inorganics from water and store them in the soil, salt usually moves through the impacted stormwater or wastewater systems, and constructed wetlands and bio-swales usually do not have a beneficial effect.

In some limited instances, salt-impacted water may be irrigated in low concentrations onto salt-tolerant plants (Rozema and Flowers, 2008). This has been most successfully applied where saline leachate has been irrigated onto salt-tolerant, high evapotranspiration-rate species and the salt has been retained in the soil. As organic matter continues to build up from the roots and plant growth, new binding sites for salt removal become available in the soil.

Planting specifics

The most effective phytotechnology application for soils with high salt content is to reestablish vegetation and keep the salt on site utilizing phytostabilization. Salt-tolerant halophytes (called salt excluders or facultative halophytes) are typically used for these applications. Some obligate halophytes that take up salt and may be useful in the future for phytoextraction are in the genera *Atriplex*, *Brassica*, *Helianthus*, *Kochia*, *Pelargonium*, *Pinus*, *Salicornia*, and *Thlaspi* (Tsao, 2003).

In soil

Stabilization: Use salt-tolerant plants to establish a vegetative cover for ecological restoration purposes.

- Planted Stabilization Mat: Chapter 4, see p. 202

Extraction: This can only be considered utilizing high biomass-producing halophyte species. It currently has limited applicability. No applications are known that have utilized cold-hardy plants that produce enough biomass for effective removal of salt.

- Extraction Plots: Chapter 4, see p. 224

In water

To control contaminated groundwater: Salt-tolerant high evapotranspiration-rate species can be used to help control migrating groundwater plumes.

- Groundwater Migration Tree Stand: Chapter 4, see p. 213
- Phytoirrigation: Chapter 4, see p. 207

Some plants can be irrigated with salty water if they are salt tolerant. The salt is bound to new organic matter created by the plant roots when irrigated onto plants. The salt is tied up so it is no longer harmful. Plant lists for salt-tolerant species have not been provided, as this subject has been extensively covered by other publications.

Because the list of halophyte species that can be considered for stabilization on salt-impacted soils is quite extensive and varies by climate and region, plant lists have not been included in this publication, but can be easily found through literature research.

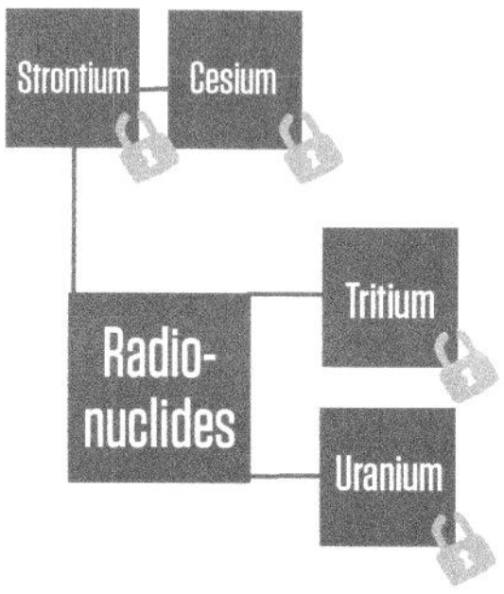

Figure 3.52 Radionuclides

■ Radioactive isotopes

Specific contaminants in this category: Energetic forms of Strontium (^{90}Sr), Cesium (^{137}Cs), Uranium (^{238}U), Tritium (T or ^{3}H – a radioactive isotope of Hydrogen)

Typical sources of pollution: Nuclear reactors, munitions, buried radioactive waste.

Typical land uses with energetics contamination: Munitions manufacturing and storage facilities, nuclear reactor sites, landfills with nuclear waste.

Why these contaminants are a danger: Extensive areas of contaminated land exist with low amounts of radionuclides, due to accidental release and activities of the nuclear industry (Dutton and Humphreys, 2005). Due to their relatively long half-lives and potential to transfer into the food chain (because of structural similarity to calcium and potassium, which are required by plants), cesium and strontium are of primary concern (Dutton and Humphreys, 2005). Additional types of radionuclides, such as tritium and uranium, are also of concern, since they can migrate into groundwater.

Summary

Cesium and strontium have been extracted in small amounts by plants under some conditions, however, at this time extraction and harvesting in the field are not viable. It has been demonstrated that these contaminants can be extracted into plant parts by fast-growing, high biomass-producing species and that strontium tends to be more bioavailable and less tied to the soil than cesium (Dutton and Humphreys, 2005). However, the half-life breakdown of these elements through natural attenuation can be faster than completing extraction and harvesting with plants (Dutton and Humphreys, 2005). The only potential future radionuclide extraction application for plants is when these contaminants cannot be treated by other conventional technologies and research demonstrates that plant extraction is calculated to be faster than the half-life breakdown.

Caution should be used when referencing past studies, because chelating agents (chemical additives) may have been used to make the radionuclides more bioavailable to the plants. Chelating agents can mobilize contaminants into adjacent soils and groundwater, and are also expensive. They likely are not useful for field application. Some common chelating agents that have been used to enhance uptake of uranium and cesium include citric acid and ammonium nitrate (Dodge and Francis, 1997; Riesen and Bruner, 1996). In addition, many agronomic factors further reduce the potential field application of extraction, since high organic matter, fine clay soil textures and high phosphorus levels have been shown to decrease the uptake of radionuclides (Negri and Hinchman, 2000). Plant species selection and the soil properties strongly influence the potential for extraction.

Short-rotation biomass crops, such as willow or poplar, can be grown to extract contaminated groundwater that is mobilizing the pollutant; however, this occurs very slowly over a long period of time. In addition, water contaminated with radionuclides can be mechanically pumped up and irrigated onto trees to potentially filter out or volatilize some radionuclides such as tritium.

In soil

 *Stabilization:* To immobilize radionuclides in soil.

- Planted Stabilization Mat: Chapter 4, see p. 202
- Evapotranspiration Cover: Chapter 4, see p. 204. Can be used to capture and transpire rainfall, preventing rain-water from mobilizing radionuclide-contaminated soils.

In water

To control contaminated groundwater: High evapotranspiration-rate species that produce high biomass may be used to help control migrating groundwater plumes.

- Groundwater Migration Tree Stand: Chapter 4, see p. 213
- Phytoirrigation: Chapter 4, see p. 207

183

Plant species from two taxonomic families and orders, *Asteraceae* from *Asterales* and *Betaceae* from *Caryophyllales*, are recognized in research as extractors of radionuclides, with the former showing better performance (Tang and Willey, 2003). In addition, a past study near the Chernobyl site showed some extraction potential with sunflowers. However, even though these plant species have been found to have some extraction capabilities in the research environment, they are still not recommended for field application. A list of plant species that have extracted radionuclides in research studies is provided in Figure 3.53. These plants may have future merit, but at this time the amount of radionuclides remediated is so small that field application is not viable without further scientific validation.

Figure 3.53 Radionuclides Phytotechnology Plant List

Plants shown to have some extraction potential in scientific studies. These should *not* be considered for field-scale remediation.

Latin	Common	Contaminant	Vegetation Type	USDA Hardiness Zone	Native to	Reference
Acer rubrum	Red Maple	226Ra	Tree	3–9	USA	ITRC PHYTO 3 Pinder et al., 1984
Alopecurus pratensis	Meadow Foxtail Grass	90Sr, 137Cs	Herbaceous	4–9	Europe, Asia	Coughtery et al., 1989 ITRC PHYTO 3 Vasudev et al., 1996
Amaranthus retroflexus	Redroot Pigweed	Cs	Herbaceous	3–10	North America	Negri and Hinchman, 2000 from Lasat et al., 1997
Beta vulgaris	Beet	Cs	Herbaceous	Grown as annual	Mediterranean	Broadley and Willey, 1997 Negri and Hinchman, 2000 Willey et al., 2001
Brassica juncea *Brassica juncea cv., 426308*	Indian Mustard	137Cs, 238U	Herbaceous	Grown as annual	Asia, Europe, Africa	Dushenkov et al., 1997b ITRC PHYTO 3 Vasudev et al., 1996
Brassica rapa	Turnip	99Tc, 137Cs	Herbaceous	Grown as annual	Europe	Bell et al., 1988 ITRC PHYTO 3
Cakile maritima	European Sea Rocket	Th, U	Herbaceous	6–10	Europe	Hegazy and Emam, 2011
Calluna vulgaris	Common Heather	137Cs	Herbaceous	4–10	Europe	Bunzl and Kracke, 1984 ITRC PHYTO 3
Caltropis gigantea	Giant Milky Weed	Sr, Cs	Herbaceous	10–11	Asia	Eapen et al., 2006
Carex nigra	Black Sedge	137Cs	Herbaceous	4–8	Europe, Eastern North America	ITRC PHYTO 3 Olsen, 1994
Cerastium fontanum	Chickweed	134Cs	Herbaceous	4+	Europe, Asia	ITRC PHYTO 3 Salt et al., 1992
Chenopodium quinoa	Quinoa	Cs	Herbaceous	8–10	South America	Negri and Hinchman, 2000 from Arthur, 1982
Chrysopogon zizanioides	Vetiver Grass	137Cs, 90Sr	Herbaceous	9–11	India	Singh et al., 2008

Latin	Common	Contaminant	Vegetation Type	USDA Hardiness Zone	Native to	Reference
Cucumis sativus	Cucumber	Co, Rb, Sr, Cs	Herbaceous	Grown as annual	India	Gouthu et al., 1997
Emilia baldwinii	Tassel Flower	224Ra	Herbaceous	Not available	India	Hewamanna et al., 1988 ITRC PHYTO 3
Eriophorum angustifolium	Tall Cottongrass	137Cs	Herbaceous	4+	North America	ITRC PHYTO 3 Olsen, 1994
Eucalyptus tereticornis	Eucalyptus	137Cs; 90Sr	Tree	9	Australia	Entry and Emmingham, 1995 ITRC PHYTO 3
Festuca arundinacea	Tall Fescue	137Cs	Herbaceous	4–8	Europe	Dahlman et al., 1969 ITRC PHYTO 3
Festuca rubra	Red Fescue	134Cs	Herbaceous	3–8	Northern USA	ITRC PHYTO 3 Salt et al., 1992
Helianthus annuus *Helianthus annuus 'Mammoth',* *'SF-187'*	Sunflower	I, U, 226Ra, 238U, 90Sr, 238U, 137Cs	Herbaceous	Grown as annual	North and South America	Dushenkov et al., 1997a, 1997b Soudek et al., 2004 Soudek et al., 2006a Soudek et al., 2006b Tome et al., 2008
Holcus mollis	Wild Millet	134Cs	Herbaceous	6–9	Northern Europe	ITRC PHYTO 3 Salt et al., 1992
Juniperus monosperma	Oneseed Juniper	U	Shrub/Tree	4+	Western USA	Ramaswami et al., 2001
Liquidamber stryaciflua	Sweet Gum	226Ra	Tree	5–10	Eastern USA	ITRC PHYTO 3 Pinder et al., 1984
Liriodendron tulipifera	Tulip Poplar	226Ra	Tree	5–10	Eastern USA	ITRC PHYTO 3 Pinder et al., 1984
Lolium perenne *Lolium perenne 'Premo'*	Perennial Ryegrass	134Cs, 58Co	Herbaceous	Grown as annual	Europe, Asia	ITRC PHYTO 3 Macklon and Sim, 1990 Salt et al., 1992
Lycopersicon esculentum	Tomato	Co, Rb, Sr, Cs	Herbaceous	Grown as annual	South America	Gouthu et al., 1997
Medicago truncatula L.	Barrel Clover	Th, U	Herbaceous	Not available	Mediterranean	Chen et al., 2005
Melampyrum sylvaticum	Cow Wheat	137Cs	Herbaceous	6	Britain, Ireland	ITRC PHYTO 3 Olsen, 1994

Figure 3.53 (continued)

Latin	Common	Contaminant	Vegetation Type	USDA Hardiness Zone	Native to	Reference
Melilotus officinalis	Sweet Clover	Cs	Herbaceous	4–8	Europe, Asia	Negri and Hinchman, 2000
Menyanthes trifoliate	Buckbean	137Cs	Herbaceous	Grown as annual	USA	ITRC PHYTO 3 Olsen, 1994
Miscanthus floridulus	Giant Miscanthus	Ba	Herbaceous	5–9	East Asia	ITRC PHYTO 3 Li et al., 2011
Panicum virgatum Panicum virgatum 'Alamo'	Switchgrass	90Sr, 137Cs	Herbaceous	2–9	USA	Entry et al., 1996 Entry and Watrud, 1998 ITRC PHYTO 3
Parthenocissus quinquefolia	Virginia Creeper	Sr	Herbaceous	3–10	Eastern USA	Li et al., 2011
Phaseolus coccineus cv. Half White Runner	Bean	238U	Herbaceous	Grown as annual	Southern USA	Dushenkov et al., 1997b ITRC PHYTO 3
Phleum pratense	Common Timothy Grass	90Sr, 137Cs	Herbaceous	5+	Europe, Asia	ITRC PHYTO 3 Vasudev et al., 1996
Phragmites australis	Phragmites	Th, U, 137Cs	Herbaceous	4–10	Europe, Asia	Li et al., 2011 Soudek et al., 2004
Picea mariana	Black Spruce	U	Tree	3–6	North America	Baumgartner et al., 1996
Pinus ponderosa Dougl. ex Laws	Ponderosa Pine	137Cs, 90Sr	Tree	3–7	USA	Entry et al., 1993 ITRC PHYTO 3
Pinus radiata D Don	Monterey Pine	137Cs, 90Sr	Tree	8+	California	Entry et al., 1993 ITRC PHYTO 3
Pisum sativum	Peas	137Cs, 106Ru, 99Tc, 144Ce	Herbaceous	Grown as annual	Europe, Asia	Bell et al., 1988 ITRC PHYTO 3 Vasudev et al., 1996
Poa spp.	Common meadow grasses	134Cs	Herbaceous	varies	varies	ITRC PHYTO 3 Salt et al., 1992
Populus grandidentata	Largetooth Aspen	226Ra	Tree	3–9	Northeastern USA	Clulow et al., 1992 ITRC PHYTO 3
Populus simonii	Poplar	137Cs	Tree	2–6	Northeast Asia	Soudek et al., 2004

Latin	Common	Contaminant	Vegetation Type	USDA Hardiness Zone	Native to	Reference
Populus tremuloides	Trembling Aspen	226Ra	Tree	2–8	Northern North America	Clulow et al., 1992 Dutton and Humphreys, 2005 ITRC PHYTO 3
Rumex acetosa	Common Sorrel	137Cs	Herbaceous	3–9	Europe, Asia	ITRC PHYTO 3 Olsen, 1994
Rumex pictus	Dock	Th, U	Herbaceous	Not available	Middle East	Hegazy and Emam, 2011
Salix caprea	Goat Willow	Sr, Cs	Shrub	4–8	Europe, Asia	Dutton and Humphreys, 2005
Salix spp.	Willow	137Cs, 90Sr	Shrub/Tree	varies	varies	Vandenhove et al., 2004
Salsola kali	Russian Thistle	Cs, Sr	Herbaceous	8–10	Europe, Asia	Negri and Hinchman, 2000 from Arthur, 1982 and Blanchfield and Hoffman, 1984
Senecio glaucus	Jaffa Groundsel	Th, U	Herbaceous	Not available	Europe, Asia, Africa	Hegazy and Emam, 2011
Solanum tuberosum	Potato	137Cs; 106Ru	Herbaceous	Grown as annual	South America	Bell et al., 1988 ITRC PHYTO 3
Sorghum sudanense	Sorghum	Cs	Herbaceous	8+	Africa	Negri and Hinchman, 2000
Trifolium repens	White Clover	134Cs	Herbaceous	4+	Europe, Asia	ITRC PHYTO 3 Salt et al., 1992
Triticum aestivum	Wheat	137Cs, 106Ru, 99Tc, 144Ce	Herbaceous	Grown as annual	Asia	Bell et al., 1988 ITRC PHYTO 3
Typha latifolia	Cattail	226Ra	Wetland	3–10	North America, Europe, Asia	ITRC PHYTO 3 Mirka et al., 1996
Vaccinium myrtillus	Bilberry	137Cs	Herbaceous	3+	Western USA	Bunzl and Kracke, 1984 ITRC PHYTO 3

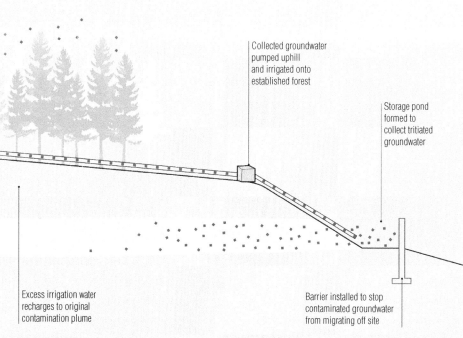

Forest on top of
contaminated plume
evapotranspires tritium
into atmosphere

Collected groundwater
pumped uphill
and irrigated onto
established forest

Storage pond
formed to
collect tritiated
groundwater

Excess irrigation water
recharges to original
contamination plume

Barrier installed to stop
contaminated groundwater
from migrating off site

Figure 3.54 Case Study: Savannah River Site, SC – Sectional Diagram

188

Project name: US Department of Energy Mixed Waste Management Facility, Southwest Plume Corrective Action Tritium Phytoremediation Project (Hitchcock et al., 2005)

Location: Savannah River Site, NC

Institutions: USDA Forest Service; University of Kentucky Department of Forestry; University of Georgia, Savannah River Ecology Laboratory; US Department of Energy; Westinghouse Savannah River Corporation; Cornell University, Department of Earth and Atmospheric Sciences.

Date installed: November 2000

Planting species: 22 acres (8.9 hectares) of existing mature native upland South Carolina forest consisting mainly of *Pinus taeda* (Loblolly Pine) and *Pinus elliottii* (Slash Pine), *Liquidambar styraciflua* (Sweetgum) and *Quercus hemisphaerica* (Laurel Oak).

Amendments: None

Contaminant(s) including initial concentrations: Tritium ■, 5,000 to 16,000 pCi mL^{-1}

Target media: Groundwater.

Tritium, a byproduct of nuclear materials production processes, was buried underground at an old radiological burial ground in South Carolina. The contaminant was being mobilized in groundwater to a nearby tributary of the Savannah River. A water containment and phytoirrigation strategy was implemented at the Savannah River Site to minimize the discharge of tritiated groundwater. First, a dam stops the groundwater seepage and creates a collection pond. The water from the pond is then mechanically pumped uphill and irrigated onto a 22 acre (8.9 hectare) mature upland forested area above the pond and above the contaminated plume. The trees then transpire the water and tritium. Any tritium not captured by the trees and soil migrates back to the groundwater and pond, where the water is recycled. The overall fate of the tritium is release into the atmosphere. The concentrations and risk factors of atmospheric release must be carefully studied before remediation systems are implemented, and it was found in this case that the danger of tritium in the migrating groundwater exceeded any risks associated with releasing it to the air.

As of March 2004, the system had irrigated approximately 133.2 million liters (35.2 million US gallons) and prevented approximately 1,880 Ci of tritium from entering the tributary. Prior to installation of the containment and disposition strategy, tritium activity in the tributary downstream of the seepage averaged approximately 500 pCi mL^{-1}. The system still functions and continues to be monitored.

III   Air pollution

Six contaminants are classified by the US EPA as air pollutants: ozone (O_3), carbon monoxide (CO), sulfur dioxide (SO_3), nitrogen dioxide (NO_2), fine particulate matter ($PM_{2.5}$) and all large respirable particulate matter that can be inhaled (PM_{10}), and also Volatile Organic Compounds (VOCs) in indoor environments.

Typical sources of outdoor air pollution: Car and industrial emissions, natural events such as volcanic eruptions, sandstorms, fires.

Typical sources of indoor air pollution: Indoor air pollution arises from the off-gassing from paints, finishes and other building materials; from additions to the space such as rugs, furniture, dry-cleaned fabrics, pets, household cleaning products, and from activities within the space such as cooking and use of electronic equipment. Combustion processes such as from wood stoves, gas heaters, gas appliances and tobacco smoke can additionally generate inorganic gaseous compounds such as carbon monoxide (CO), carbon dioxide (CO_2), nitrogen oxides (NO_x) and sulfur dioxide (SO_2) (Soreanu et al., 2013). Combustion gases and petroleum vapors arising from automobiles in attached or underground garages can migrate into living spaces to become a significant source of indoor air pollutants.

Typical land uses generating outdoor air pollution contamination: Roadways, industrial properties, lands adjacent to roadways and industrial properties.

Why these contaminants are a danger: Air pollution can compromise the human respiratory system, especially the smaller-size particulate matter ($PM_{2.5}$). The World Health Organization (WHO, 2002) estimates that more than 1 million premature deaths annually can be attributed to urban air pollution

189

Figure 3.55 Air Pollution Summary Chart

Contaminant	Notes
Ground-level Ozone (O_3)	Ground-level ozone is created by reactions between VOCs and nitrogen oxides as they are exposed to sunlight. Inhaling ozone can create a variety of health problems, mostly affecting the respiratory system. It can exacerbate diseases of the airway such as asthma and bronchitis and can impair lung function. Common symptoms of ozone overexposure include coughing, sore throat, pain or burning in the chest and shortness of breath (http://www.epa.gov/glo/).
Carbon Monoxide (CO)	Carbon monoxide is a toxic gas created by automobiles and the incomplete combustion of hydrocarbon fuels. It is a major contributor to smog creation. In the body, carbon monoxide attaches itself to hemoglobin in the red blood cells, which normally carries oxygen throughout the body. This inhibits the body's ability to take up oxygen. At low levels, it can cause fatigue and chest pain. At higher concentrations it can cause headaches, dizziness and confusion. High concentrations, particularly in enclosed indoor spaces, can cause death (http://www.epa.gov/iaq/co.html).
Nitrogen Oxides (NO_x - NO_2)	Nitrogen oxides are created by combustion of fossil fuels and automobile engines, in particular. They have similar health effects to ozone. Nitrogen oxides are major contributors to acid precipitation and smog. Overexposure can cause irritation to the airways. It can also irritate the mucosa of the eyes and nose. People with existing diseases of the airway are especially susceptible (http://www.epa.gov/air/nitrogenoxides/health.html).
Sulfur Oxides (SO_x - SO_2)	Sulfur oxides are formed by combustion of fossil fuels. Like nitrogen oxides, sulfur oxides are a major contributor to acid precipitation and smog. Health effects are also similar to those of nitrogen oxides and ozone, causing respiratory inflammation and impaired lung function (http://www.epa.gov/airquality/sulfurdioxide/).
Carbon Dioxide (CO_2)	Carbon dioxide is produced by fossil fuel combustion. At low levels, carbon dioxide has few health effects; however, at extremely high levels it can interfere with the ability of the body to take up oxygen. Carbon dioxide helps to create acid precipitation and is a strong greenhouse gas which actively contributes to global climate change (http://www.epa.gov/climatechange/ghgemissions/gases/co2.html).
Particulate Matter (PM_{10} and $PM_{2.5}$)	Particulate matter is generated from a variety of sources, including industrial practices and automobile emissions. The pollutant may comprise both liquid and solid particles found in the air. Due to their small size, particulates can travel deep into the lungs. Moreover, smaller particles pose a greater danger, due to their ability to travel greater distances in the air. Particulates most often cause irritation of the airways and reduced lung function. However, they have also been linked to cardiac diseases and some cancers (http://www.epa.gov/airscience/air-particulatematter.htm).
VOCs (Volatile Organic Compounds) Benzene, Toluene, Xylene etc.	VOCs are gases emitted from a variety of sources including paints, adhesives, cleaning products as well as fuels and automobile emissions. Many VOCs have been linked to increased risk of cancer. At low levels they irritate the airways, nose and eyes. VOCs also have powerful neurological effects, causing headaches, dizziness and memory impairment (http://www.epa.gov/iaq/voc.html).

190

in developing countries alone. In addition, ozone and carbon monoxide contribute to global warming. Concentrations of VOCs in indoor air can be up to 10 times greater than those found in outdoor environments and can greatly affect human health. Accumulation of indoor air pollutants appears to significantly contribute to 'sick building syndrome,' causing fatigue, allergies, poor productivity and headaches, among other symptoms (Soreanu et al., 2013).

Summary

The subject of air pollution remediation with plants is quite broad, with varying scientific opinions on its efficacy for different contaminants. Only a very brief introduction to the subject will be given here. Please refer to Chapter 6 for a guide to other references that cover this subject in greater depth.

A Outdoor air pollution

It is generally agreed that coarse particulate matter in air pollution (PM_{10}), nitrogen dioxide, sulfur dioxide, carbon dioxide and ozone can be removed by plants and that trees can contribute to the reduction of air pollution in cities (Yang et al., 2008; Nowak, 2002; Nowak et al., 2006; 2014; Rosenfeld et al., 1998; Scott et al., 1998). Nowak et al. (2006) estimated that urban trees remove a total of 711,000 metric tons of the US EPA's top five air pollutants annually in the US. These findings helped lead the US EPA to include tree planting as a recommended strategy for improving air quality in 2004 (US EPA, 2014e).

Plants remediate air pollution with several mechanisms that are different than those utilized for cleanup of soil and water. Trees serve to capture and filter out some components of air pollution, such as coarse particulate matter, as well as assimilating and mitigating other pollutants like nitrogen dioxide and ozone. These air pollution-filtration mechanisms are described below.

Phytoaccumulation (collects on leaf surfaces)

Deposition is the process by which aerosol particles collect or deposit themselves onto solid surfaces, decreasing the concentration of the particles in the air. Particulate matter can carry heavy metals, PAHs, and POPs attached to the particulates (Dzierzanowski and Gawronski, 2011), and the particulates can settle out through impaction and sedimentation (settling) onto leaf surfaces. Some particles can be absorbed into the tree, though most particles that are intercepted are retained on the plant surface. The intercepted particles are often resuspended to the atmosphere, washed off by rain or dropped to the ground with leaf and twig fall. Consequently, vegetation is only a temporary detention site (Nowak et al., 2006). Rain-water can also wash particulate matter from leaf surfaces into soils, so understory stormwater filters should be considered. It is also important to note that most of the phytoaccumulation studies describe the efficacy of all respirable particulate matter removal. However, it is the fine and ultrafine particles that present the greater concern for respiratory health.

Deciduous plants with 'sticky' leaves (waxy coatings and leaf hairs) and species with a greater leaf area index (see Figure 4.2a for a definition) have been shown to collect more particulate matter than other species (Dzierzanowski and Gawronski, 2011). In addition, some studies have also shown that conifers have the potential to collect ultrafine particles more effectively than do deciduous species due to the complex foliar structure of conifers (Beckett et al., 1998, 2000). Research in this field is nascent and it is assumed that individual species' effectiveness will be further validated over time.

Phytometabolism (becomes part of the plant)

Plants take up nitrogen dioxide from the atmosphere and assimilate it into organic nitrogen-containing compounds (Takashi et al., 2005, p. 634). This assimilation capability is dependent on the plant species. In examining 70 species, researchers found that four broadleaf deciduous species, *Robinia pseudoacacia*, *Sophora japonica*, *Populus nigra* and *Prunus lannesiana*, have high resistance to damage by nitrogen dioxide and assimilation, indicating that they would be good candidates for remediating urban air (Takashi et al., 2005).

191

VOC contribution and emissions

The world's vegetation generates about two-thirds of VOC emissions (US EPA, 2014e). Through the emission of VOCs, trees can contribute to the formation of ozone (O_3) (Chameides et al., 1988). The VOCs that are released through the leaves of trees combine with other elements in the air, such as nitrogen oxides. Tree species can be selected that release lower amounts of VOCs, and species selection may be important in industrial areas that already have high emission levels of nitrogen oxides, to prevent harmful reactions with these airborne chemicals. Some studies are revealing that planting urban trees, particularly low VOC-emitting species, can be a viable strategy to help reduce urban ozone levels (Nowak et al., 2006), particularly through tree functions that reduce air temperatures (transpiration), remove air pollutants (phytoaccumulation – dry deposition to plant surfaces) and reduce building energy and consequent power plant emissions (e.g., temperature reductions; tree shade). One study (Nowak et al., 2000) has concluded that for the US the positive physical effects of urban trees were more beneficial than the chemical release of VOCs in terms of affecting ozone concentrations (Nowak et al., 2006).

However, we must be careful not to over-quantify the beneficial effects of trees when it comes to air pollution. Though urban trees remove tons of air pollutants annually, the air-quality improvement in cities averaged less than 1% (Nowak, 2006). Percentage air-quality improvements were typically greatest for particulate matter, ozone, sulfur dioxide and nitrogen dioxide.

Planting specifics

192

For the air-quality improvements that have been reported, improvement increased with greater percentage of tree cover. The most beneficial air-quality enhancement from trees is likely their contribution to passive temperature cooling and the sequestration of carbon from the atmosphere, storing it in organic forms. It is estimated that urban trees in the United States currently store 700 million tons of carbon (Beattie and Seibel, 2007).

Results for air pollutant removal are greatly affected by air pollutant concentrations, weather conditions and the growth of plants. In temperate climates, typically the highest air-pollutant removal occurs during the in-leaf season, when the leaves of plants are fully expanded and the concentration of pollutants tends to be higher (Yang et al., 2008). One study, shown in Figure 3.56, illustrates the amount of air pollution reduction by several different kinds of vegetation types. In general, the larger the plant and the greater the leaf surface area, the greater the air pollution reduction.

Figure 3.56 Annual Removal Rate of Air Pollutants per Canopy Cover by Different Vegetation Types in Chicago between August 2006 and July 2007

Type of vegetation	SO_2 (g/m²/yr)	NO_2 (g/m²/yr)	PM_{10} (g/m²/yr)	O_3 (g/m²/yr)	Total (g/m²/yr)
Short grass	0.65	2.33	1.12	4.49	8.59
Tall herbaceous plants	0.83	2.94	1.52	5.81	11.1
Deciduous trees	1.01	3.57	2.16	7.17	13.91

Note: The non-vegetated surfaces were excluded from the calculation.
Source: Yang et al., 2008.

It is not only important which vegetation types are planted, but also what arrangement they are planted in. Air pollutants are dispersed primarily by wind, therefore the effects of emissions are not confined to the immediate vicinity but can also be observed at distances from the source. The concentration of pollutants decreases with distance from the source; however, the area impacted by significant concentrations can be quite large. The correlation between distance from roadways and concentrations of pollutants in the environment is described below. Forman describes the impacted distance along the edges of roadways as up to 50 meters for salt particles and nutrients generated by road dust (Figure 3.57) (Forman and Alexander, 1998; 2003). In addition, researchers in the European Union found greatly elevated levels of particulate matter up to 80 meters from freeways (Figure 3.58) (Zhua et al., 2002). Plant species installed in these zones can provide air-quality improvement by capturing particles and depositing them on leaves. Canada's Ministry of the Environment additionally recommends that the first 200 meters along roadways provide the most opportunities for plant-based sequestration of pollutants, since this is the distance impacted by increased levels of nitrogen oxides and particulate matter (Ministry of Environment, 2006).

Vegetation planted adjacent to road systems can be good for the phytoaccumulation and sequestration of airborne pollutants. However, other factors must also be considered. Urban trees can actually trap airborne contaminants at the street level when buildings form 'street canyons.' The vegetation canopy can prohibit the exchange of air between the atmosphere and the street environment, essentially forming a roof over the street. This reduced natural ventilation in urban streets can result in health impacts, so tree placement must be carefully considered (Vardoulakis et al., 2003). Where there are numerous pollutant sources below the canopy (e.g., automobiles),

193

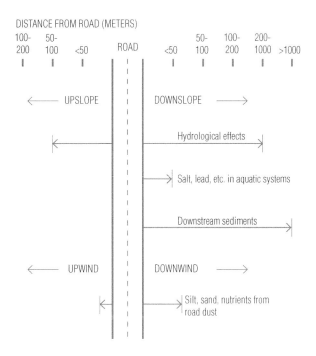

Figure 3.57 Road Effects
Source: From Forman, R. T. and Alexander, L. E. 1998. Roads and their major ecological effects. *Annual Review of Ecology and Systematics* 29, pp. 207–231.

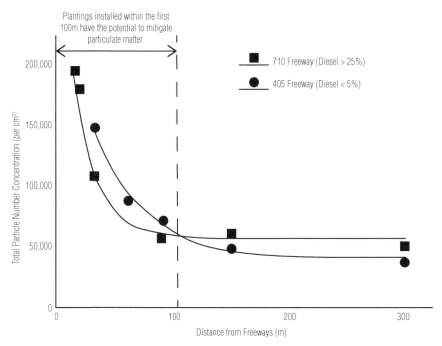

Particulate concentration is directly related to distance from roadways and fuel type (above), as well as wind direction.

(Redrawn from Zhua, Y., Hinds, W. C., Kim, S., Shen, S. and Sioutas, C. 2002. Study of ultrafine particles near a major highway with heavy-duty diesel traffic. *Atmospheric Environment*, 36, p. 4331)

194

Figure 3.58 Particulate Concentration: Distance from Freeways

the canopy could have a negative effect by minimizing the dispersion of the pollutants at ground level (Nowak, 2006). In the reverse rural scenario, forest canopies can limit the mixing of contaminated upper air with cleaner ground-level air, leading to significant below-canopy air quality improvements.

To remove particulate matter and nitrogen oxides from outdoor air: An Air-Flow Buffer can be considered adjacent to land uses that produce emissions with high particulate matter and nitrogen oxides. Particular species can be chosen that maximize the collection of particulate matter, or low VOC-emitting species can be selected where ozone formation is a concern (see plant lists in Figures 3.59 and 3.60). Air-Flow Buffers can be effective for distances as great as 200 meters from roadways.

- Air-Flow Buffer: see Chapter 4, p. 229

B Indoor air pollution

The US EPA ranks indoor air quality as one of the top five public health concerns (USEPA, 2013c). People living in urban environments spend about 85–90% of their time indoors, making the quality of the indoor air they breathe a big factor in their overall health. Indoor air pollutants include both large and small particulate matter, VOCs and inorganic gaseous compounds. Plant-based biological filtration systems appear to be a promising alternative to conventional methods (Llewellen and Dixon, 2011).

Rhizodegradation (soils in root zone remove air pollutants)

A widely referenced NASA study from the 1980s concluded that plants can remove VOCs from the indoor environment (Wolverton et al., 1989). This study was quickly debunked, showing that is was not the plant but the soil microbes in root zone of the plant that were degrading the VOCs (Godish and Guindon, 1989). For any VOC degradation to occur, the contaminated air must be in close contact with these soil microbes. Potted plants may act as passive air biofilters (Wolverton et al., 1989); however, the amount of passive interaction between the ambient air and the biologically active root zone is minimal, therefore only minimal air cleansing by these passive potted plants is likely (Godish and Guindon, 1989). To obtain greater biological remediation, active systems where water and air are pulled through the root zones of plants provide more substantial removal levels (Darlington, 2013).

To remove VOCs from indoor air: For VOCs to be substantially removed from indoor or outdoor air, the VOCs in the air must be brought into contact with the degrading microbes in the soil layer. Passive interaction of air flows and potted plants provides only minimal removal. Many interior plant system suppliers suggest that degradation of VOCs can occur by adding indoor plants or a Green Wall to a room. However, since only air that comes into direct contact with the soil has the potential to be remediated, and even in a well-ventilated space the indoor air is relatively still, these plant systems will not make a substantial impact on air quality. For this reason, a modified biofilter Green Wall, where air handling systems actively draw air through the root zone of plants, ensuring contact between soil

Figure 3.59 Top 20 Street Trees for New York City, Based on Ranked Environmental Criteria

195

Latin	Common	USDA Hardiness Zone	Native To
Liriodendron tulipifera	Tulip Poplar	5–10	Eastern USA
Magnolia grandiflora	Southern Magnolia	6+	Southern USA
Platanus occidentalis	American Sycamore	4+	Eastern North America
Platanus hybrida	London Plane Tree	4+	Europe
Ulmus glabra	Scotch Elm	5+	Europe, Asia
Ulmus americana	American Elm	2+	North America
Juglans nigra	Black Walnut	4+	Eastern USA
Cedrus atlantica	Atlas Cedar	6+	North Africa
Fagus grandifolia	American Beech	3–8	Eastern North America
Cedrus deodara	Deodar Cedar	7+	Asia
Cedrus libani	Cedar of Lebanon	5+	Mediterranean
Quercus nigra	Water Oak	6+	Southeastern USA
Quercus alba	White Oak	3+	Eastern North America
Quercus macrocarpa	Bur Oak	4+	Eastern North America
Quercus robur	English Oak	5+	Europe
Quercus rubra	Northern Red Oak	3+	Northeastern North America
Magnolia acuminata	Cucumber Tree Magnolia	3+	Eastern USA
Quercus shumardii	Shumard Oak	5+	Eastern USA
Pseudotsuga menziesii	Douglas Fir	5+	Western North America
Quercus prinus	Chestnut Oak	5+	Eastern USA

Note: Top species for NYC streets based on ranked criteria that included air quality, air temperature reduction, shading, energy conservation, carbon storage, low allergenicity and long life span.
Source: Nowack, 2006

Figure 3.60 Plant Species for Particulate Matter Removal

Latin	Common	Vegetation Type	USDA Hardiness Zone	Native to	Reference
Acer campestre	Field Maple	Tree	5–8	Europe	Dzierżanowski et al., 2011
Acer tataricum subsp. Ginnala	Amur Maple	Tree	3–8	Europe, Asia	Popek et al., 2013
Betula pendula Betula pendula 'Roth'	Silver Birch	Tree	2+	Europe, Asia	Dzierżanowski and Gawroński, 2011 Saebo et al., 2012
Corylus columa	Turkish Hazelnut	Tree	5–7	Europe, Asia	Popek et al., 2013
Forsythia × intermedia 'Zabel'	Forsythia	Shrub	5+	East Asia	Dzierżanowski et al., 2011
Fraxinus excelsior	European Ash	Tree	5–8	Europe	Dzierżanowski et al., 2011
Fraxinus pennsylvanica	Green Ash	Tree	2–9	Eastern North America	Popek et al., 2013
Ginkgo biloba	Ginkgo	Tree	3–8	China	Popek et al., 2013
Hedera helix	English Ivy	Vine	5–11	Europe, Asia	Dzierżanowski et al., 2011
Physocarpus opulifolius	Ninebark	Shrub	3–7	North America	Dzierżanowski et al., 2011
Pinus mugo	Mugo Pine	Shrub	2+	Europe	Saebo et al., 2012
Pinus nigra var. maritime	Corsican Pine	Tree	5–9	Corsica, Southern Italy	Beckett et al., 1998
Pinus sylvestris	Scots Pine	Shrub	3–7	Europe	Saebo et al., 2012
Platanus × hispanica Mill. ex Muenchh	London Plane Tree	Tree	4+	Europe	Popek et al., 2013
Populus simonii 'Corrière'	Simon's Poplar	Tree	2+	China	Dzierżanowski and Gawroński, 2011
Pyrus calleryana Decne. 'Chanticleer'	Callery Pear	Tree	5–8	China	Dzierżanowski and Gawroński, 2011
Quercus rubra	Northern Red Oak	Tree	3+	Northeastern North America	Dzierżanowski and Gawroński, 2011 Popek et al., 2013
Sambucus nigra	Common Elder	Shrub	3+	Europe, Asia	Popek et al., 2013
Sorbaria sorbifolia	False Spiraea	Shrub	2+	Northern Asia, Japan	Popek et al., 2013
Sorbus aria	Common Whitebeam	Tree	5–9	Europe, Asia	Beckett et al., 2000
Sorbus × intermedia	Swedish Whitebeam Hybrid Mountain Ash	Tree	4+	Northern Europe	Dzierżanowski and Gawroński, 2011
Spiraea japonica	Japanese Spiraea	Shrub	4–8	East Asia	Dzierżanowski et al., 2011 Popek et al., 2013
Stephanandra incise	Cut-Leaf Stephanandra	Shrub	4+	Japan, Korea	Saebo et al., 2012
Syringa meyeri 'Pallibin'	Meyer Lilac	Shrub	3–7	Europe, Asia	Popek et al., 2013
Taxus baccata	English Yew	Shrub	5+	Europe	Saebo et al., 2012
Taxus × media	Yew	Shrub	4–7	Japan, Korea	Saebo et al., 2012
Tilia cordata	Little-Leaf Linden	Tree	3–7	Europe	Dzierżanowski et al., 2011
Tilia tomentosa Moench 'Brabant'	Silver Linden	Tree	4–8	Europe, Asia	Popek et al., 2013
Viburnum lantana	Wayfaring Tree	Tree/Shrub	4+	Europe, Asia	Popek et al., 2013

microbes and the contaminated air stream, rather than a traditional Green Wall, is recommended for VOC removal in indoor environments (Darlington, 2013).

• Greenwall Air Filter: see Chapter 4, p. 231

Many existing publications list other applicable plant-based systems and plant species for remediating air pollution. Resources for additional research are included in Chapter 6. In addition, a short list of applicable plant species is provided in Figures 3.59 and 3.60.

Urban air pollution mitigation tree species: David J. Nowak of the USDA Forest Service has created a list of approximately 200 tree species for the city of New York ranked by functional attributes including: air pollution removal, air temperature reduction, tree shade, building energy conservation, carbon storage, pollen allergenicity and life span (Nowak, 2006, p. 93.) Not only does this list consider the removal of air pollutants, but it also considers other important functions for urban trees. The 20 highest-ranking trees are included in the list in Figure 3.59.

Case Study Indoor air pollution

***Project name**:* University of Guelph-Humber Living Wall Biofilter (Darlington, 2014)

***Location**:* Toronto, Ontario Canada

197

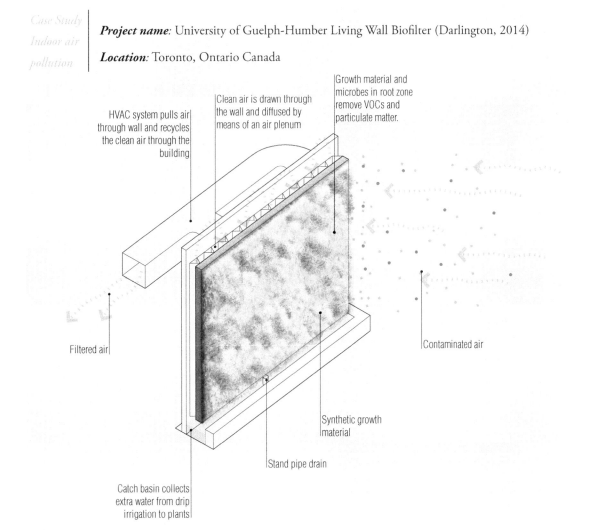

HVAC system pulls air through wall and recycles the clean air through the building

Clean air is drawn through the wall and diffused by means of an air plenum

Growth material and microbes in root zone remove VOCs and particulate matter.

Filtered air

Contaminated air

Synthetic growth material

Stand pipe drain

Catch basin collects extra water from drip irrigation to plants

Figure 3.61 Case Study: University of Guelph Green Wall Air Biofilter Diagram

A four-story green wall in the atrium cleanses and recycles the indoor air of this campus building.

Figure 3.62 Case Study: Green Wall, University of Guelph, Canada

198

Design team: Diamond and Schmitt Architects; Air Quality Solutions Ltd (now Nedlaw Living Walls Inc); Crossey Engineering (Mechanical)

Date installed: 2004

Plant species: Approximately 1,300 plants composed largely of *Schefflera arborcolia*, *Ficus spp.*, *Philodendron spp.* and *Dreceana spp.*

Growth media: Two layers, each ca. 2 cm thick, of open-spun synthetic fiber mat held together with epoxy resin.

Contaminant: VOCs, Particulate Matter and other indoor contaminants.

A four-story living wall was created in the atrium of University of Guelph-Humber as a piece of biological machinery for the building ventilation system. The plant wall is a large filter for the air circulating within the building and removes 90% of the VOCs brought through the wall in a single pass. Air is actively forced through the wall of plants, where naturally occurring microbes actively use the pollutants (such as VOCs) as a food source, degrading them into their benign constituents of water and carbon dioxide. The normal growth of the plants and their roots in the living-wall biofilter constantly adds new organics to the media and therefore constantly reinvigorates the microbial component. The clean air is then distributed throughout the space by the Heating Ventilation Air Conditioning (HVAC) system. This biofiltered air supplements or augments normal ventilation air coming in from outside.

This living wall is a significant energy saver for the building. Recirculating the air reduces the amount of cold or warm outside air that must be brought into the building and conditioned to room temperature, which can account for 30% of energy consumed by a building.

Behind the scenes, a pump constantly circulates water and nutrients from a reservoir at the base to the top of the wall. The water then flows down the wall through a porous synthetic root media in which the plants are grown.

Given the air flux through the biofilter, one square meter of the biofilter generates between 80 and 100 liters of virtual fresh air per second (16 to 20 cfm/ft^2), enough recycled air for up to 15 people. The system also removes significant amounts of inhalable dust and bacterial spores.

IV Summary

In summary, the phytotechnology applications that are most promising for integration into the field of landscape design include the following.

Degradation of petroleum wastes ●*:* Over half of the brownfield sites identified by the US EPA are considered petroleum brownfields, and many plants have been found to effectively degrade these contaminants in the root zone and leaves, eliminating any need for plant harvesting.

199

Groundwater plume control, including degradation: Chlorinated solvents ●, light fractions of petroleum ● and the explosive RDX ● can quickly move into groundwater and spread, contaminating large areas of drinking water supply. Plants can not only be used to control contaminated plumes from migrating, but the plants can potentially degrade the compounds in the process.

Evapotranspiration Covers, including landfill cap and closures and dewatering contaminated sludges/ sediments: High evapotranspiration-rate, high biomass-production plants can move huge amounts of water very quickly, preventing water from leaching through contaminated soils. This prevents the generation of leachate and contaminated groundwater. In the process, many organic contaminants may also be degraded.

Constructed Wetlands: These natural treatment systems can filter a host of contaminants, including metals, out of water. Generally the plants do not take up the contaminant, but rather, it is bound to the soil, which acts as a large filter as the water passes through. The plants' role in the system is to replenish oxygen and open binding sites in the soils.

Phytostabilization: Plants and amendments can be used to immobilize contaminants on site, especially heavy metals ■ that typically have limited bioavailability to be extracted by plants.

The next chapter will detail the specific suggested planting types indexed throughout this chapter.

Phytotypologies: phytotechnology planting types

This chapter illustrates 18 different types of phytotechnology plantings. These planting types, or 'phytotypologies,' can be used either individually or combined as a series of adaptive planting types to fit specific pollution prevention or remediation goals. In addition they can be combined and integrated with non-remediation planting methods. Many of the typologies are applicable for only a certain kind of contaminant (i.e. petroleum, nutrients, metals, etc.) within a certain target media (i.e. air, soil, groundwater, stormwater, wastewater, etc.) A system of icon shortcuts is provided at the beginning of each typology to quickly identify what contaminant and target media are being treated with the planting type and what mechanisms for removal are utilized. In addition, a diagram and description of each planting type is included, with notes about typical applications and plant selection.

When considering the application of planting types, the plants themselves must be selected based on the criteria described and the specific site conditions, including existing soils, groundwater, microclimate and contaminant(s) being addressed. The typologies described in this chapter should be utilized in conjunction with the contaminant-specific plant lists provided in Chapter 3 and any other published research on the contaminants of concern. This chapter is provided as a primer to introduce those unfamiliar with phytotechnologies to the spatial and functional requirements of various planting types. For ease of explanation, each separate typology highlights a particular phytotechnology mechanism. In actuality, many of the different typologies can be combined into a single planting scheme to achieve multiple remediation functions and design goals.

I   Introduction

For practitioners looking to implement remediation technologies, it is critical that an experienced phytotechnology specialist be engaged. Specific details such as bioavailability, phytotoxicity, hydrological factors and contaminant concentrations cannot be overlooked in the design and implementation of these remediation systems.

II Planting typologies

What follows is a sequence of 18 planting typologies with full descriptions and their applications.

Typology 4.1: Planted Stabilization Mat (holds contaminants on site)

Description: Introduced plants hold contaminants on the site to prevent them from migrating. No removal is provided; the objective is to minimize the risk of contaminant exposure to humans and natural systems.

Primary mechanism at work: Phytostabilization ◯, ▢

Target: Soil

Contaminants addressed: Most often used for Metals ▬, POPs ●, and Salts ▦ in soils. Can be used for all contaminant groups to some extent.

A Planted Stabilization Mat provides a similar function to a traditional clay cap often used in brownfield redevelopment, holding pollutants on site and minimizing human and environmental contact. The difference is that plants play a role in preventing the contaminants from moving, while water is still able

202

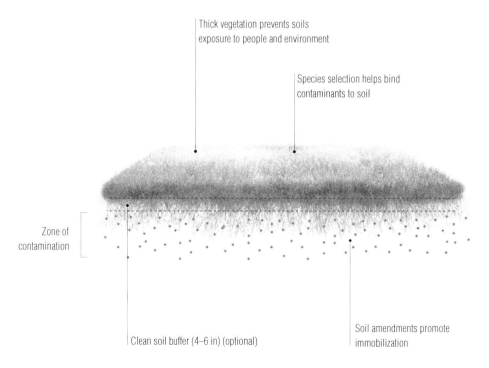

Thick vegetation prevents soils exposure to people and environment

Species selection helps bind contaminants to soil

Zone of contamination

Clean soil buffer (4–6 in) (optional)

Soil amendments promote immobilization

Figure 4.1 Stabilization Mat

to penetrate into the soils of the system. The roots of plants physically hold the contaminant in place and release root exudates that may further bind the contamination to soil particles to prevent future migration.

Planted Stabilization Mats are most often utilized at sites with widespread, non-bioavailable pollutants where site revegetation is a priority. In these situations, the soils are typically too toxic for many plants to be established. Species for the stabilization mat are carefully selected to withstand site contaminants, and amendments are often added to enhance plant growth and further bind toxins on site. 'Excluder' plant species are often used to prevent mobility of the pollutant into above-ground plant tissues, and minimize wind and soil erosion, while providing other ecosystem services such as habitat enhancement.

Typical applications

- **Former mining sites:** Large areas of unvegetated land laden with heavy metals, sulfur and salts are frequently the result of former mining operations. These landscapes are often very acidic, with pH levels beyond the normal ranges where sustained plant growth can occur. Often the primary goal for remediation on these sites is to stabilize the pollutants so they do not move into exposure pathways and become a hazard to humans and wildlife. Stabilization, however, is quite difficult because there is often a large land area to cover. Providing an impenetrable clay cap over such large sites, and importing 6–18 inches of clean soil on top for plant growth, is not sustainable or financially feasible. Planted Stabilization Mats are often the most cost-effective means of immobilizing polluted materials and creating an opportunity for ecological restoration. Soil amendments are typically added initially, to aid in plant establishment, and carefully selected plants are seeded to obtain stabilization goals. (For examples see case studies in Chapter 3, p. 176).

- **Former petroleum extraction sites:** When petroleum and natural gas are extracted from terrestrial sites, a large amount of salt is often brought to the surface. Brine scars, consisting of soils with salt content too concentrated for indigenous plant growth, are typically left on the surface at the location where the extraction was completed. These salts often cannot be remediated without large amounts of soil removal. Planted Stabilization Mats can be utilized to establish vegetation to prevent the salts from migrating and may initiate the return of more complex plant communities over time.

- **Painted residences/lead paint residue:** A significant amount of lead often persists in soils around homes that were painted with lead paints up through the 1970s, and in urban areas where leaded gasoline emissions were prevalent. Lead is not easily mobilized in water and can persist in soils indefinitely. The typical and most risky human exposure pathway is via minute soil particles being directly ingested by humans, especially children, in the form of airborne dust or dirt carried into the home on shoes, clothing or toys. The widespread occurrence of this pollutant makes it unfeasible to dig and haul away the impacted soils. Planted Stabilization Mats can be used to hold the soil in place, creating a protective barrier between the soil particles and the site's inhabitants.

Plant selection: To select species for a stabilization mat, consider:

1 *What species are tolerant to the contaminant?* Many species won't extract a particular metal, but can grow in high concentrations of it (see Figure 3.37, p. 140 for Metals Excluder Plant List).

2 *What species have been known to phytosequester the contaminant?* Some species have been tested for their ability to release root exudates that address the mobility of the contaminant itself.

3 *How can a thick site cover with no soil exposure be created?* Eliminating erosion by wind and water is the key to an effective Planted Stabilization Mat. Select species that will densely fill in voids and leave little soil exposure. Dense grasses are often utilized for this application because of their ability to form thick mats. Thick, densely planted, deep-rooted grasses, such as low-mow fescue lawn mixes, are some of the best plant species for this application, especially for residential applications with lead in soils.

Other design considerations

1 *Soil chemistry:* In Planted Stabilization Mats, it is important to remember that the soil chemistry is just as important as the plant species selected. Contaminants exist in many forms, and abiotic (non-living-related) mechanisms such as adsorption to soil particles, precipitation or sedimentation can play an even larger role in remediation than the role of plants, in some cases (ITRC, 2009). By changing the soil chemistry, including pH, availability of nutrients or other factors, contaminants can be mobilized or stabilized within the soil matrix. This can be done with the addition of carefully chosen soil amendments. As with any application of a phytotechnology, an experienced soil agronomist must be part of the advisory team on these issues.

2 *Amendments:* Fertilizers and organic products can be added to the soil to help stabilize the pollutant of concern via soil chemistry, and also aid in plant establishment.

3 *Soil buffer:* Where exposure risk is more pronounced, a thin, clean soil cap up to 6 inches in depth can be provided on the surface over the contaminated soils to further isolate the contamination from exposure risks before plants are installed.

4 *Topography:* If the landscape is to be altered through manipulation of the surface topography and new contours are to be added, grade slopes to promote run-off away from the contaminated area. During construction, adequate protection must be provided to ensure contaminated run-off does not leave the site and to protect workers from exposure.

5 *Contamination type and amount:* Not all instances of contamination can be treated by a Planted Stabilization Mat. Sometimes the contamination levels are too high and plants cannot be grown due to the toxicity of the pollutants. In other circumstances, the contamination is mobile in water. As Planted Stabilization Mats are permeable, allowing water to penetrate into the ground, the contaminant could still be mobilized, despite the installation of the mat. Regulatory requirements might also necessitate a different remediation approach. Consulting a soil chemist or environmental engineer to ensure that leaching will not occur is critical to the functioning of Planted Stabilization Mats.

Typology 4.2: Evapotranspiration Cover (minimizes water infiltration)

Description: Plants intercept rain-water and transpire it back to the atmosphere, preventing water from mobilizing contaminants. No contaminant removal is provided; the objective is to prevent contamination from migrating, thereby preventing exposure to humans and natural systems both on the site and downgradient in water supplies.

Primary mechanisms at work: Phytohydraulics ◌◌◌, Phytostabilization ◌◌◌

Target: The water vector – rain-water/stormwater

Contaminants addressed: All types of contaminants. Addresses the rain-water vector that could potentially mobilize any contaminant that could be leached into water.

Evapotranspiration Covers provide the function of an 'umbrella' over a contaminated site. When rain falls, it often moves through the soil medium, picking up contaminants along the way and leaching them into groundwater or nearby water bodies. The objective of an Evapotranspiration Cover is to intercept the rain-water and prevent it from infiltrating. There are two aspects to water protection: physical barriers and evapotranspiration.

1 *Physical barrier:* Many layers of leaves, created through a variety of plant species and canopy heights, can create a physical umbrella to soften the impact of heavy rains, slowing infiltration into soils.
2 *Evapotranspiration:* The water that does get through the canopy can be absorbed by the plant roots, transpired and released into the atmosphere preventing percolation beyond the top layer of soil.

The success of this typology relies on calculations to determine that the plants added will use more water than the amount of rain-water which falls on the site.

205

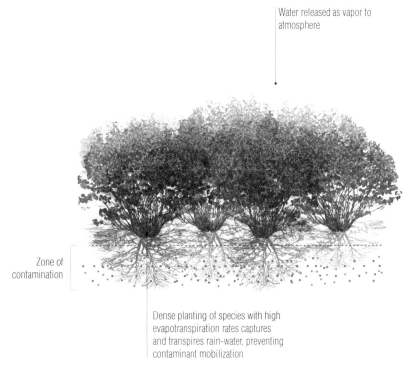

Water released as vapor to atmosphere

Zone of contamination

Dense planting of species with high evapotranspiration rates captures and transpires rain-water, preventing contaminant mobilization

Figure 4.2 Evapotranspiration Cover

Typical applications

- **Unlined landfills:** Current landfill practices in the US typically use Bithuthene™ or clay liners located both below and above the waste pile to keep water from penetrating into the waste material. The 'cap' is to prevent water from getting into the waste and subsequently picking up contaminants, leaching out through the bottom and polluting other adjacent waters. However, until the 1980s, landfills in the US were typically not lined. Furthermore, unlined landfills are still a common occurrence in other countries. In these unlined landfills, water can easily penetrate into the waste and leach out pollutants. The use of Evapotranspiration Covers to prevent water from entering unlined landfills is now common practice in the US and has gained regulatory acceptance. Typically, Evapotranspiration Covers utilize species with high evapotranspiration rates that are suited to the site location and the regional climate. Because the growth rate of these species is often quite high, the species can additionally be utilized as a biomass crop. Habitat creation and ecological restoration goals can be integrated into plantations of these evapotranspiration work-horse species, and successional strategies to provide multiple ecological functions can be created.

- **Contaminated groundwater plumes:** Unconfined contaminated groundwater plumes move faster when they are continually recharged by surface stormwater infiltration. Evapotranspiration Covers can be used upstream of contaminated groundwater plumes to slow down migration of the plume by minimizing the amount of water entering the system (ITRC, 2009).

Plant selection

1 *How can water removal by plants be maximized?* Plants with the highest evapotranspiration rates for a particular region are typically the best species to select for this typology. See Figure 2.18, p. 48 for a preliminary list of species with high evapotranspiration rates. It should be noted that the list provides only a small fraction of species that could be utilized. Regional species with high evapotranspiration rates should be considered. Willow and poplars are often chosen for this typology because they can additionally be used as a bioenergy or wind crop.

2 *How can water infiltration to soils be minimized?* Rain-water infiltration can also be controlled by choosing plants with a high Leaf Area Index (LAI). LAI is a measure of plant canopy thickness. A high LAI indicates that the plant generates significant leaf material between the ground and the top of the plant, so that a rain drop will hit a larger number of leaves or leaf area before it finally reaches the ground. Plants with a high LAI will typically minimize water infiltration under the canopy, as the leaves create a physical barrier between rainfall and the soil, slowing infiltrations and allowing the water to evaporate off the leaf surfaces before reaching the soil surface (Figure 4.2a).

Other design considerations

1 *Mass Water Balance:* In order for Evapotranspiration Covers to be successful, a detailed Mass Water Balance calculation must be completed by a hydrologist to ensure that the plants will be able to

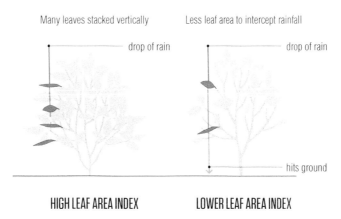

Figure 4.2a Leaf Area Index

transpire more water than the amount of water that falls onto the site. This is calculated with the following factors: typical annual rainfall, evapotranspiration rate of the plants selected (which is affected by seasonality and weather) and planting density. The varying microclimates of a site, including temperatures and wind exposures, are also considered.

2 *Dormancy:* Temperate climates with fall and winter seasons will typically have very low evapotranspiration rates during those seasons. Times of lower transpiration rates must be considered in the overall Mass Water Balance calculation. Often the ground is frozen during the winter months, so precipitation cannot penetrate through the soil medium. Therefore the seasonality of plant growth may not be a prohibiting factor for the use of Evapotranspiration Covers in temperate climates.

3 *Soil water capacity:* Water storage capacity in the soil medium should be maximized, so plants have a longer period of time to transpire the water after a rain fall (ITRC, 2009).

4 *Maximizing evaporation micro-climatic factors:* The amount of wind passing over the plants should be maximized. Increased air circulation will increase the amount of water transpired by the plants and the amount of water directly evaporated from the site. In addition, greater solar exposure and higher day-time temperatures will increase transpiration.

C Typology 4.3: Phytoirrigation (irrigating plants with contaminated water)

Description: Polluted water is irrigated onto plantings. Removal of the polluting compounds is provided by the plants; the objective is to completely degrade and remove the contaminants being irrigated onto the plants.

Primary mechanisms at work: Rhizodegradation   , Phytodegradation   , Phytovolatilization

Target: Wastewater or groundwater

Contaminants addressed: Nitrogen   : all forms including nitrate, nitrite and ammonia; Chlorinated Solvents , Petroleum , Selenium , Tritium

207

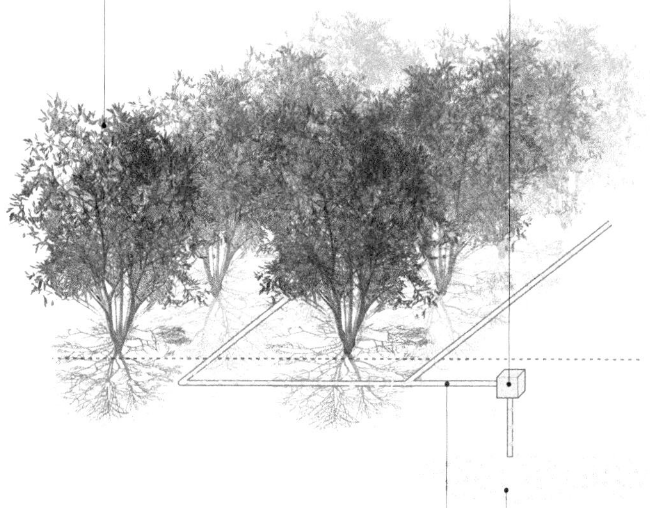

Fast growth and high
evapotranspiration-
rate plants

Irrigation pump (can be
solar powered)

Subsurface or spray
irrigation lines

Contaminated groundwater
or leachate

Figure 4.3 Phytoirrigation

Phytoirrigation is utilized to remove contaminants in water which can be volatilized or metabolized by the plants. With nutrient contaminants, the water is irrigated onto the plants and acts as a fertilizer to stimulate plant growth. This is often a win-win situation, where contaminated water can be cleaned while crop-plant production is maximized. For other types of contaminants listed, the water is irrigated into the plantings and the contaminant is either volatilized or degraded by the plant or microbes in the root zone, or trapped or held in the soil.

Subsurface drip irrigation, rather than spray heads, is often the preferred irrigation method, since the contaminant is thus released below the surface, minimizing exposure pathways. One challenge with this is that the water may have more than one contaminant and some pollutants, such as salts and particulate matter, can precipitate out and clog the drip-irrigation lines. This has been overcome in the past by flushing the system with clean water at intervals to prevent unwanted build-ups (CH2MHill, 2011).

Typical applications

- **Wastewater treatment facilities:** Both municipal and industrial wastewater treatment facilities have successfully utilized Phytoirrigation to remove nutrients. In this process, the wastewater is typically pretreated and then the final nutrient contaminants are removed when they are irrigated onto plantings. Successful Phytoirrigation applications are common in the central and western United States, but highly regulated to ensure that the wastewater does not oversaturate the system and move into the groundwater without being remediated. Plants selected for the Phytoirrigation planting are often chosen because they will use high amounts of water and will produce a commodity crop such as corn or alfalfa. Some of the best examples of Phytoirrigation are closed-loop water reuse systems, such as an industrial food-production facility that might water nearby agricultural fields with the nutrient-rich wastewater (Smesrud, 2012). Biomass energy production utilizing Phytoirrigation of hybrid poplars and willow is popular worldwide (see Case study p. 128, Chapter 3). Phytoirrigation of forests for hardwood or wood pellet production for heating is also gaining popularity.

- **Golf courses:** Golf courses are excellent receptors of Phytoirrigation systems because of their large water and nutrient requirements. The challenge in these landscapes is to ensure that the water used for irrigation has been pretreated to acceptable health standards. These systems are often difficult to permit, due to heavy human use at courses.

- **Contaminated groundwater plumes/fertigation wells:** Groundwater plumes contaminated with nitrogen can be controlled with pumping and Phytoirrigation systems. Wells can be drilled to intercept the groundwater plume and draw down the water with pumping, preventing the contamination from spreading into other water bodies. Instead of treating the pumped water through conventional methods, the water can be irrigated onto plantings for nitrogen removal. Wells that are utilized to tap nutrient-rich groundwater are often called fertigation wells. In addition, this same technique can be used for other groundwater contaminant plumes like selenium and tritium, where it can be irrigated onto existing stands of mature forest and volatilized into the air (see Case study, p. 188, Chapter 3).

Plant selection

209

1 *How can the amount of water that can be irrigated onto plants be maximized?* Plantings irrigated with contaminated water must be able to process water quickly, so that excess contaminated water does not run through the system into the groundwater below. Species selection must be carefully calibrated by a hydrologist to determine the water irrigation rates to be applied.

2 *How can species be selected to maximize nitrogen degradation?* All plants require nitrogen for the production of plant biomass. But also, nitrogen removal occurs because of increased biological activity in the soil as well, where denitrifying bacteria turn the nitrogen in the wastewater into a gas and release it back into the atmosphere. Typically, more biomass production means the more direct usage by the plant and the more denitrifying bacteria will be supported, hence greater nitrogen removal is achieved (for example high-biomass plant species list, see Figure 2.17, p. 46).

3 *How can species be selected for chlorinated solvents, petroleum, selenium or tritium volatilization or degradation?* Plants that grow fast with high evapotranspiration rates are the best species for Phytoirrigation systems. These plants can use a lot of water and irrigation rates can be higher (see high evapotranspiration-rate plant species list, see Figure 2.18, p. 48).

Other design considerations

1 *Mass Water Balance:* A Mass Water Balance calculation must be completed by a hydrologist to ensure that the plants will be able to utilize the amount of water irrigated onto the field area.

2 *Contaminants:* Detailed analysis is also required to determine the concentration of contamination that can be applied and remediated by the plants.

3 *Seasonality:* Phytoirrigation systems are only operable during active growing seasons. If wastewater or other contaminated water is generated year round, alternative storage accommodations must be made when the plants are not actively growing. Often, the water can be easily stored in lined retention ponds or constructed wetlands until plant growth resumes in the spring.

Typology 4.4: Green (and Blue) Roofs (minimize stormwater run-off)

Description: Evapotranspiration of water from roofs is maximized. Little or no contaminant removal is usually provided; the objective is to prevent water from entering onto contaminated areas (preventing contaminant mobilization).

Primary mechanism at work: Phytohydraulics ◯ ⊞

Target: The water vector – rain-water/stormwater

Contaminants addressed: All contaminants. This addresses the rain-water vector that could potentially mobilize any contaminant that could leach into water. Green and Blue Roofs are often used to prevent stormwater from washing over an impervious surface (especially roads, sidewalks, parking lots, etc.), thus preventing contamination by surface pollutants.

A Green Roof is a type of Evapotranspiration Cover specifically designed to exist on a building or infrastructure rooftop environment. The primary contamination mitigation benefit of a Green Roof is its ability to prevent stormwater from mobilizing downgradient contaminants.

At this time, it is suggested that Green Roofs be considered only for minimizing stormwater, and not for contaminant removal. There are mixed results from research documenting the contaminant-removal capabilities of Green Roofs. Several recent studies have shown that the Green Roof construction materials could actually contribute contamination to run-off flowing through the system, especially in the early years after construction (Harper, 2013; Hill, 2014). This can be due to mobilization of nutrients found within the plant growing media, or leaching of metals or toxins from insulation, filtration or structural system components. When selecting proprietary systems, the possibility of leachate generation should be strongly considered.

In addition, the amount of water that the plant component of the Green Roof will actually transpire is also debated. Much of the stormwater reduction may have nothing to do with the plants in the system; direct evaporation of water from hot roof-surface media may instead move water directly into the air, without any plant uptake or transpiration. Plants in the system might instead actually cool the roof surface, thereby preventing maximum evaporation potential, which is a more efficient water-removal mechanism than transpiration through the plants (Hill, 2014). For this reason, the concept of Blue Roofs has emerged as another best-management practice for stormwater reduction. The Blue Roof system is designed to provide short-term detention of rainfall and promote evaporation without the use of any plants. For Green Roofs, the increased aesthetic and environmental benefits of plants may at times outweigh the lower rate of evaporation.

Run-off minimized

Water vapor released
into atmosphere

Both transpiration and
evaporation occur with
plants and soil media

Drainage layer

Figure 4.4a Green Roof

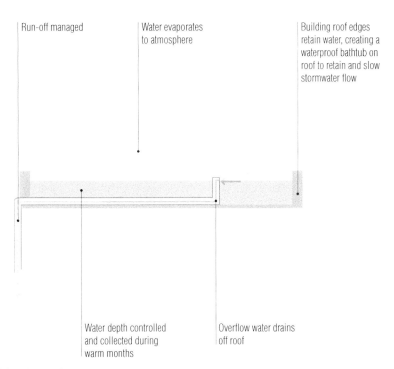

Run-off managed

Water evaporates
to atmosphere

Building roof edges
retain water, creating a
waterproof bathtub on
roof to retain and slow
stormwater flow

Water depth controlled
and collected during
warm months

Overflow water drains
off roof

Figure 4.4b Blue Roof

211

Typical applications

- **Convention centers, commercial and industrial buildings and infrastructure:** Buildings and infrastructure with large, flat or shallowly sloped roofs are ideal candidates for the introduction of Green and Blue Roof systems. Many industrial sites often have large building footprints and impervious areas for materials storage which can collect contaminants. Minimizing stormwater run-off from the buildings can reduce the potential for the water to pick up and transport contaminants during precipitation events.

Plant selection

- *What species will survive?* Green roofs are typically non-irrigated systems that survive in high-drought conditions, with elevated temperatures, wind speeds and other environmental stressors. Plant selection in these systems has historically been based on survivability of the plant species selected and ease of propagation within the nursery trade. There is increasing interest in planting Green Roofs to include native species and plants that provide other ecosystem services. These 'native' roofs have been completed at several installations in the US (Toland, 2013). Plants on Green Roofs are typically not selected to degrade or remove contaminants, since survival and water removal are the most important factors.

Other design considerations

1 *Soil depth and water-holding capacity:* There are two types of Green Roofs – intensive and extensive systems.
   - Intensive Green Roofs have a deeper soil profile, providing plants with greater than 15 cm (6 inches) of soil media for growth. These systems tend to be heavier and more expensive, but also tend to hold more water for detention and allow for a wider range of species growth.
   - Extensive systems utilize 5–15 cm (2–6 inches) of soil media for growth. Typical plant species include sedums, which are highly drought tolerant and have small root zones. The soil media may have less water-holding capacity, but also may have the potential to quickly evaporate water. If water retention is the goal, it has been found that the depth of planting media does not seem to make any difference in rooftop water-holding capacity (Hill, 2014). To detain the maximum amount of stormwater, a detailed cost-benefit analysis should be conducted by a hydrologist in conjunction with a weight analysis by a structural engineer to determine which system will provide the greatest, most cost-effective benefit.
2 *Leaching potential:* Specifications should be requested from the system provider on the potential for the selected Green Roof systems to discharge nutrients or other pollutants and review any available test data. If information is unavailable, consider planting a Stormwater Filter where the run-off from the Green Roof will discharge to assist in the removal of pollutants that may have been released (see 4.15 Stormwater Filter).
3 *Irrigation:* Consider pairing Green Roof and Blue Roof systems to increase the available irrigation water for the plants. Structural considerations must be carefully considered, due to the heavy weight of retained stormwater.

212

Typology 4.5: Groundwater Migration Tree Stand (trees pump and treat groundwater)

Description: Trees with deep tap roots and high evapotranspiration rates are planted to modify the groundwater hydrology and keep contaminants from migrating. The trees, through the pull of transpiration, can slow or stop a groundwater plume from migrating, or can change the plume's direction towards the trees. The objective is to control contaminated groundwater plumes from moving off site. As an additional benefit, many organic compounds as well as nitrogen can be degraded/removed in this process.

Primary mechanism at work: Phytohydraulics

Target: Groundwater (0–20 feet below the soil surface)

Contaminants that can be addressed: Most commonly used for Chlorinated Solvents , Petroleum and Nutrients in groundwater, since the contaminant can be degraded or volatilized as the

213

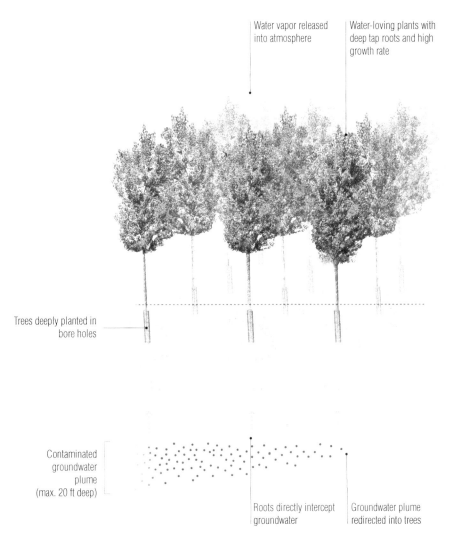

Figure 4.5 Groundwater Migration Tree Stand

groundwater plume is controlled. Also used to contain but not degrade Explosives ●, Radionuclides ■, POPs ●, Metals ■ in contaminated groundwater plumes.

Groundwater Migration Tree Stands act as large solar-powered pumps. Typically, when a plume of groundwater is contaminated, it is treated with a series of engineered wells downgradient of the plume. The wells intercept the migrating plume, which is pumped to the surface and treated with conventional filtration methods. With Groundwater Migration Tree Stands, instead of a mechanical pump, trees with high transpiration rates are planted in the ground. The tap-root system of each tree draws up the groundwater and the tree transpires the water. The evapotranspiration rate of the trees is significant enough to affect the groundwater movement, so the plume's movement is controlled. Ample space must be provided and the stands often must be quite large to intercept a dynamic plume. A Mass Water Balance calculation and soil analysis must be completed to determine the number of trees needed to stop a migrating plume. It is possible that soils may be too tight, so that the trees will be unable to access the water.

Groundwater Migration Tree Stands can also be planted to change the direction of the groundwater hydrology. Trees with high evapotranspiration rates act like a vacuum pulling the underground water towards the introduced planting. This can change the groundwater contours and direction of the plume below the surface. This technique can be utilized to divert groundwater away from an impacted area. Tree stands can be planted as security against future contaminant plumes. If the contaminant of concern within the groundwater is organic, with the log K_{ow} in the treatable range (see Chapter 2, p. 51), not only will the groundwater plume be controlled but the contaminant may also be degraded in the process.

Typical applications

- **Rail, military and industrial facilities with TCE/PCE plumes:** Chlorinated Solvents ○ such as TCE (Trichloroethylene) and PCE (Polychloroethylene) often move very quickly into a groundwater plume. These materials can disperse readily and are difficult to capture with conventional 'pump and treat' systems. Groundwater Migration Tree Stands can be an excellent way to treat a large dispersed area of contamination within groundwater. Not only may the trees be able to control the plume but, through the biological activity in the plants and associated root zone, the contaminant may also be degraded (see Case study, p. 96).

- **Dry cleaners:** TCE and PCE ○ are common dry-cleaning contaminants and chlorinated solvents. They move quickly in the groundwater and can be effectively pumped up and degraded with the introduction of tree stands (see Case study, p. 99).

- **Fuel tanks, gas stations and petroleum refinery operations:** Light fractions of petroleum products ●, such as BTEX (Benzene, Toluene, Ethylbenzene and Xylene) and MTBE (Methyl Tertiary-Butyl Ether) can easily migrate into groundwater. Releases from ruptured or leaking underground fuel tanks are common within the urban environment. Groundwater Migration Tree Stands may effectively stop the migration and degrade these petroleum contaminants (see Case studies, pp. 89 and 92).

Plant selection

1 *What species will tap the groundwater?* The plant species selected must be physically adapted and inclined to seek groundwater and be able to endure some root saturation in the capillary fringe between the groundwater and dry soil. For this reason, phreatophytes are most often utilized. These are species that seek out groundwater with long tap roots (for a list of species, see Figure 2.15, p. 45)

2 *How can the plant species maximize the pumping rate:* Plants with the highest evapotranspiration rates will move the most water. Hybrid poplars and willow are often used for this application, but more recently several field sites have tested native species to determine fast-transpiring plant options. When testing species, pilot-scale projects are first completed to compare potential species and the most effective are selected and planted for field-scale research (for a list of species with high evapotranspiration rates, see Figure 2.18, p. 48).

3 *How deep can the groundwater be?* In order to have an impact on the groundwater plume, the roots must be able to reach down and tap into the groundwater. Phreatophytes are known to have deep root systems that can reach as deep as 9 meters (30 feet); however, to ensure the system will work, Groundwater Migration Trees Stands are typically used where groundwater is within 6 meters (20 feet) of the surface. Shallower groundwater depths are more easily and quickly accessed by the trees. Since it may take several years for a tree's roots to grow down to the groundwater depth, trees are often 'deep root' planted. Holes are mechanically bored into the ground and cuttings are placed as far as 3 meters (10 feet) below the soil surface. This gives trees a 'head start' to reach the groundwater (see Figure 2.16, p. 46 for more information on this technique).

215

Other design considerations

1 *Mass Water Balance:* Groundwater Migration Tree Stands are effective only when a detailed Mass Water Balance has been calculated to ensure that a sufficient number of trees have been planted to intercept the plume. A hydrologist will calculate this using information regarding the typical amount of annual rainfall on a site, the evapotranspiration rate of the plants selected, the density of the planting and the speed, size and location of the groundwater plume. The microclimates of a site are also considered to calibrate the evapotranspiration rate, including temperature and wind conditions.

2 *Variability and dormancy:* Trees transpire water at different rates, depending on time of day, season and climate. Groundwater control may not be possible during fall and winter seasons in temperate climates, since the trees are dormant. Other groundwater plume-control measures may need to be used during this time, such as conventional pump-and-treat systems.

3 *Maximize evaporation climatic factors:* Greater solar exposure and higher temperatures will increase the uptake of water into the plant. The amount of wind passing over the plants should also be maximized. Increased air circulation will increase the amount of water transpired by the plants, and therefore the amount of groundwater that will be taken up.

Degradation typologies

The next five typologies (numbers 4.6–4.10) all target degradation and metabolism mechanisms to completely remove the contaminants from the site without the need for harvesting the plants. These degradation typologies typically cannot be used for inorganic contaminants. They can generally be used only for organic contaminants with good phytoremediation potential, as established in Chapter 3. Degradation typologies can also be used for a few essential plant nutrient inorganics such as nitrogen to metabolize and incorporate the contaminant into plant tissues or return it to the atmosphere as a gas. The main difference between the typologies listed in the following sections is that different sizes of plant material are used to achieve alternate aesthetic options. In addition to the plant-driven processes at work, biodegradation (microbial) oxidation/reduction (abiotic) and volatilization (abiotic) processes are also occurring (ITRC, 2009).

Typology 4.6: Interception Hedgerow

Description: Where groundwater is contaminated, a single row of trees tap into the water and assist in contaminant degradation. The objective is to remove some fraction of the contamination in the small amount of space planted as the groundwater moves below the surface. The groundwater plume is not contained and typically the contaminant is not fully remediated.

Primary mechanisms at work: Rhizodegradation   , Phytodegradation   , Phytovolatilization   , Phytohydraulics   , Phytometabolism

Target: Groundwater (up to 6 meters/20 feet deep)

Contaminants addressed: Organic contaminants: Petroleum ●, Chlorinated Solvents ◍, Pesticides ◍, Nutrients: Nitrogen ◌

Not applicable for: POPs ●, Explosives ●, Radionuclides ■, Metals ■, Salts ▨

Interception Hedgerows are used when there is not enough space on a site for a Groundwater Mitigation Tree Stand to completely remove an organic contaminant or nitrogen in groundwater, but when the goal is to at least remove some contamination in the modest amount of space that is available. Interception Hedgerows are typically planted around the edges of a site to aesthetically buffer a contaminated site from adjacent uses, tapping the groundwater and helping to degrade contaminants at the same time. Only partial degradation usually occurs because of the small amount of space available for planting.

Typical applications

- **Gas stations, auto-repair shops, dry cleaners, urban industrial site perimeters:** Interception Hedgerows can be placed around the edges of a property to degrade organic pollutants within the groundwater at the site perimeter. The buffer can serve aesthetic purposes to mitigate undesirable

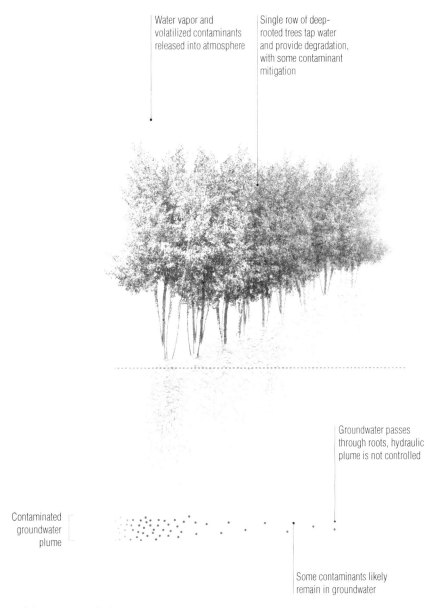

Water vapor and volatilized contaminants released into atmosphere

Single row of deep-rooted trees tap water and provide degradation, with some contaminant mitigation

217

Groundwater passes through roots, hydraulic plume is not controlled

Contaminated groundwater plume

Some contaminants likely remain in groundwater

Figure 4.6 Interception Hedgerow

views, define the site boundary or inhibit access to the site. Mixed species can be considered to provide other ecological functions such as habitat and wildlife corridors within the urban matrix.

- **Funeral homes and cemetery buffers:** Leaching of embalming fluids and nutrients into groundwater from ongoing operations, storage and landscape maintenance can occur at these sites. Interception Hedgerows can be used within the site and at the site perimeter to degrade these pollutants in groundwater.

- **Agricultural hedgerows:** Nutrients often mobilize and leach from crop-production fields, feed-lots and confined animal production operations, impacting adjacent streams and watersheds. Interception Hedgerows can be used to degrade contaminants before they migrate in groundwater.

Plant selection

1 *What species will access the groundwater?* To degrade contaminants within the groundwater plume, the plants must be able to access and transpire the water. Plants that seek groundwater (phreatophytes), with the ability to process water at a high evapotranspiration rate, are preferred (see Chapter 2, p. 45).

2 *How deep is the groundwater?* Depth to groundwater must be evaluated and plant species capable of reaching the groundwater should be selected.

3 *What species will best degrade the contaminant?* Once the plant species palette has been narrowed by the above criteria, plants should then be selected from the degradation plant lists that target for the contaminant of concern (see Chapter 3).

Other design considerations

1 *Mixed species:* Opportunities exist to provide habitat corridors for foraging and ecological connectivity. In addition, mixed species can promote a more diverse microbiology, which will typically enhance rhizodegradation.

2 *Layer with other phytotechnologies:* Interception Hedgerows can be paired with any of the other degradation, sequestration or metabolism typologies that target soil rather than groundwater.

Typology 4.7: Degradation Bosque

Description: Deep-rooted tree and shrub species degrade contamination areas within the soil profile. Contaminant removal is provided without harvesting the plant.

Primary mechanisms at work: Rhizodegradation, Phytodegradation, Phytovolatilization, Phytometabolism

Target: Deep soils (0–3 meters/0–10 feet deep)

Contaminants that can be addressed: Organic contaminants: Petroleum ●, Chlorinated Solvents, Pesticides ●, Nutrients: Nitrogen.

Not applicable for: POPs ●, Explosives ●, Radionuclides ■, Metals ■, Salts.

Degradation Bosques are used to target concentrated areas of soil contamination up to 3 meters (10 feet) below the surface. They treat organic contamination by breaking down the pollutant into smaller, less toxic substances in the root zone, stems or leaves of the plant, or by volatilizing the pollutant and releasing it into the air. Degradation Bosques are typically used to treat more recalcitrant organic compounds in soil that would not often be broken down by natural attenuation alone. Species selection can influence degradation rates. Each species releases different root exudates and the associated microbial profile will vary. Some root exudates have similar chemical compositions to organic contaminants, so plant species should be selected by the contaminants found on a given site.

Degradation Bosques can be successful in cleaning up petroleum, chlorinated solvents and pesticide spills deep in the soils which have not yet migrated to the groundwater. They can also be used to

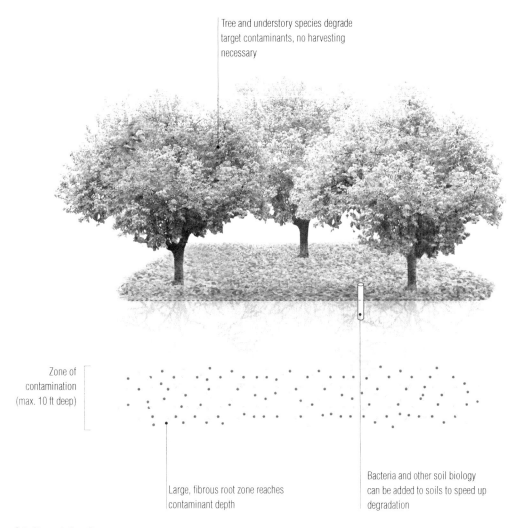

Tree and understory species degrade target contaminants, no harvesting necessary

Zone of contamination (max. 10 ft deep)

Large, fibrous root zone reaches contaminant depth

Bacteria and other soil biology can be added to soils to speed up degradation

Figure 4.7 Degradation Bosque

219

remediate high quantities of nitrogen found deep in the soil profile, by stimulating microbiology that volatilizes it into the air. In addition, some nitrogen will be metabolized by the plant and incorporated into the biomass.

Typical applications

- **Leaking underground storage tanks:** Where contaminants have not yet migrated to the groundwater, Degradation Bosques can target deeper contamination around LUSTs.
- **Fertilizer spills:** Where high concentrations of fertilizer exist deep in the soil profile, plants can be utilized to enhance the conversion of nitrogen in the soil to nitrogen gas by associated soil bacteria. In addition, plants can take up and use the nutrients, incorporating them into the plant's biomass. The contaminant is no longer toxic in its new organic form as part of the plant's biomass. This is referred to as phytometabolism.

Plant selection

1 *How are the best species for degradation chosen?* Plant species with certain root-exudate profiles and associated microbiology can be paired with contaminants to speed degradation rates. Degradation plant species are provided in Chapter 3, listed by contaminant.

2 *How deep is the contamination?* It is important to ensure that the root depth of a particular species will reach the location of a contaminant on site. The root-depth profile for a particular species should be carefully matched to the location of the pollutant in the soil.

Other design considerations

1 *Phytotoxicity:* 'Hot spots' can have concentrations of pollutants that prohibit plant growth. Amendments may have to be added to the soil to aid in plant establishment. Plants that can thrive in non-ideal soils have been utilized in past applications for their ease in plant establishment, rather than choosing plants to match root exudates or the soil-microbiology profile with the contaminant of concern. The idea is that any new oxygen and biological activity that are introduced into the soil via the roots of the plants will be a benefit to the contaminant breakdown.

Typology 4.8: Degradation Hedge and Living Fence

220

Description: Shrub species are planted to degrade contamination in the soil up to 4 feet deep. Contaminant removal is provided without harvesting the plant.

Primary mechanisms at work: Rhizodegradation , Phytodegradation , Phytovolatilization , Phytometabolism

Target: Surface soils (0–1.3 meters/0–4 feet deep)

Contaminants addressed: Organic contaminants: Petroleum ●; Chlorinated Solvents , Pesticides ●, Nutrients: Nitrogen .

Not applicable for: POPs ●, Explosives ●, Radionuclides ■, Metals ■, Salts .

Degradation Hedges utilize shrub species to define areas within sites, while also degrading contaminants in the surface soils. The degradation functioning of the plant is the same as described in the previous Degradation Bosque typology. The only difference here is that Degradation Bosques typically treat contamination deep in the soil profile and Degradation Hedges treat soils closer to the surface with dense, fibrous roots.

Popular in English gardens, Living Fences are often utilized to create boundaries between garden rooms. Almost always created out of willow species, Living Fences are typically used for aesthetic and screening purposes. However, as a phytotechnology typology, willow species can be selected to degrade organic contaminants such as Petroleum, Chlorinated Solvents and Pesticides, while still creating a functioning art form.

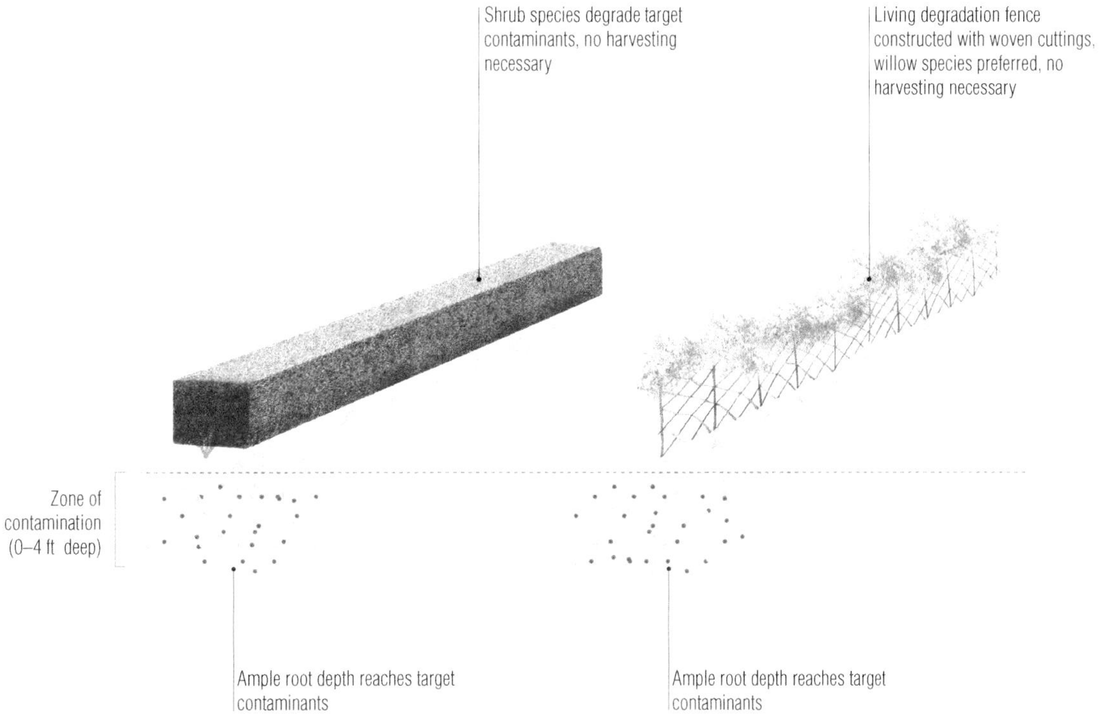

Shrub species degrade target
contaminants, no harvesting
necessary

Living degradation fence
constructed with woven cuttings,
willow species preferred, no
harvesting necessary

Zone of
contamination
(0–4 ft deep)

Ample root depth reaches target
contaminants

Ample root depth reaches target
contaminants

Figure 4.8 Degradation Hedge/Living Fence

221

To install a willow fence, dormant cuttings, typically ranging from 0.3 to 1.8 meters (1–6 feet) long, are inserted into the soil. As the cuttings grow, the shoots are woven into the fence to create a variety of shapes and patterns. Willow species have many ornamental benefits, and leaf and stem color can be selected to provide interest, even when choosing between degradation species.

Typical applications

- **Perimeters of gas stations, auto-repair shops, dry cleaners, urban industrial sites:** Similar to Interception Hedgerows (see 4.6), which target groundwater pollution, Degradation Hedges can be planted around the edges of properties to degrade organic pollutants within the soils. The buffer can serve aesthetic purposes to mitigate undesirable views and define the site boundary. Mixed species can be considered so as to provide other ecological functions such as habitat and wildlife corridors within the urban context.

- **Community gardens:** Living Fences can be planted around the edges of community garden sites to degrade organic pollutants within the soils and capture excessive nutrient and pesticide run-off. The fence can serve as a security and aesthetic boundary, in addition to its degradation capabilities.

Plant selection

1 *What species will degrade the site contaminants?* Plant species with certain root exudates and associated microbiology profiles can be paired with contaminants to speed degradation rates. Degradation plant species are provided in Chapter 3, listed by contaminant.

2 *How deep is the contamination?* It is important to ensure that the root depth of a particular species will reach the location of a contaminant on site. The root-depth profile for a particular species should be carefully matched to the location of the pollutant in the soil. Typically, the root systems of hedges and Living Fences do not grow past 1.3 meters (4 feet) in soil depth, and can be much less.

3 *How are Living Fence species selected?* The plant must be easily propagated by cuttings and have flexible young growth (shoots) to weave into a fence. Willow species can be selected, based on the target contaminant of concern. The cuttings are available only in the dormant season for installation.

Other design considerations

1 *Layering:* Degradation Hedges can be paired with Interception Hedgerows and other typologies around the perimeter of sites, to maximize the degradation of organic contaminants and nutrients in a small area.

2 *Pruning and maintenance:* Living Fences require continual maintenance and pruning to establish the shape. Typically, fences must be pruned and trained at least once per year.

3 *Irrigation of Living Fences:* As phreatophytes, willow species need water to grow. Irrigation is often required to establish willow species. Once established, many varieties are drought tolerant. Water requirements should be considered when selecting specific species.

Typology 4.9: Degradation Cover

Description: Degradation Covers utilize thick, deep-rooted herbaceous species to remove contaminants in surface soils up to 1.5 meters (5 feet) deep. Contaminant removal is provided without harvesting the plant.

Primary mechanisms at work: Rhizodegradation , Phytodegradation , Phytovolatilization , Phytometabolism

Target: Surface soils (0–1.5 meters/0–5 feet deep)

Contaminants addressed: Organic contaminants: Petroleum ●; Chlorinated Solvents ●, Pesticides ●, Nutrients: Nitrogen ●.

Not applicable for: POPs ●, Explosives ●, Radionuclides ■, Metals ■, Salts ▨.

Degradation Covers are typically utilized over large areas of land to speed up the natural attenuation and breakdown of organic pollutants such as petroleum compounds like Polycyclic Aromatic Hydrocarbons (PAHs). As the plant provides oxygen, sugars and other root exudates to the soil, the microbial environment is enhanced, leading to the degradation of targeted contaminants.

222

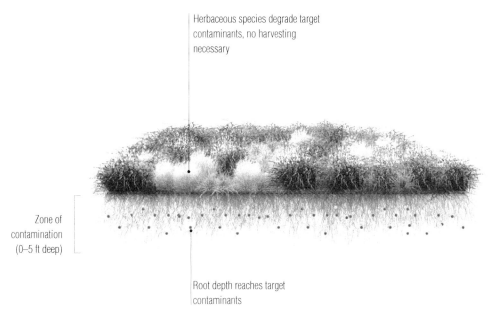

Herbaceous species degrade target
contaminants, no harvesting
necessary

Zone of
contamination
(0–5 ft deep)

Root depth reaches target
contaminants

Figure 4.9 Degradation Cover

Deep-rooted, drought-tolerant prairie grass species are often utilized. The thick, fibrous root zone of the grass stimulates activity in the rhizosphere and the plants can be easy to establish and maintain. Mixed-species covers have been found to be more effective than monocultures, encouraging a diverse microbial environment. Furthermore, the addition of legumes or other nitrogen-fixing species tends to benefit the system.

223

Typical applications

- **Military bases:** Surface soils at military bases are frequently impacted with petroleum compounds. Many of the heavier, recalcitrant portions of the petroleum can be difficult to break down. Degradation Covers that produce large volumes of below-ground biomass can be effective at speeding up the degradation process. These covers can often be installed while the base is still in operation, as the plant species are often low in height and generally unobtrusive.
- **Former industrial sites/salvage yards:** While formerly used brownfield sites with petroleum-impacted soils are sitting vacant, low-height Degradation Covers can be employed to help break down the heavier fractions present in soil.

Plant selection

1 *What are the best degradation plant species for cover?* Plant characteristics to look for in selecting Degradation Covers are deep-rooted, drought-tolerant, fibrous root-system species with an ability to spread well to cover large areas. Degradation plant species are provided in Chapter 3, listed by contaminant. Low groundcover species and prairie grasses are most often utilized for this typology.

Other design considerations

1 *Mixed contamination:* Many sites will have mixed contaminants. Petroleum and chlorinated solvents that can be more easily degraded are sometimes found with heavy metals and other inorganics that cannot be degraded. A mixed cover can be considered to combine the degradation functions described here for organics with the function of a the Stabilization Mat (see Figure 4.1, p. 202) or Extraction Plot (see 4.10, this page) for the inorganics.

Typology 4.10: Extraction Plots

Description: Hyperaccumulator plants or high-biomass crop species are used to extract inorganic pollutants or recalcitrant organic pollutants from the soil. The plants must be harvested to remove the contamination from the site.

Primary mechanisms at work: Phytoextraction , Phytometabolism

Target: Soils (0–1 meter/0–3 feet deep)

Contaminants addressed:

Shorter term: Some Metals : Arsenic, Selenium, Nickel

Long term: Metals : Cadmium, Zinc

At this time not applicable for: Cyanide; Metals : (Boron (B), Cobalt (Co), Copper (Cu), Iron (Fe), Manganese (Mn), Molybdenum (Mo), Chromium (Cr), Fluorine (F), Lead (Pb), Mercury (Hg), Aluminum (Al), Silver (Ag) and Gold (Au); Radionuclides ; Salts .

Not applicable for: Petroleum , Chlorinated Solvents , Pesticides , Explosives : All these groups can be treated with, degradation mechanisms, *not* extraction.

Extraction Plots utilize hyperaccumulators or fast-growing accumulator species to remove inorganic contaminants from the soil and groundwater. They are not used for organic contaminants because organic contaminants are typically degraded or volatilized and extraction and harvesting is unnecessary. Extraction Plots have been most effectively utilized for low levels of arsenic or selenium contamination on a site, or to remove nickel from the soil for phytomining. The harvested material should be tested for pollutant concentrations prior to disposal and must be properly disposed of in a hazardous waste facility if extracted concentrations are high enough. Most other inorganic contaminants (other than arsenic, selenium and nickel) are usually not taken up in concentrations high enough to assist in remediation in an acceptable time frame, though cadmium and zinc may be able to be extracted over longer periods of time, under the right circumstances.

A hyperaccumulator is a type of plant that takes up a particular element in concentrations 10–100 times higher than a typical species does. However, even though a plant is a hyperaccumulator, very often it is not able to remove a particular contaminant to below regulatory limits. Very often inorganic contaminants are so tightly bound to the soil, or soil chemistry makes contaminants unavailable, so

224

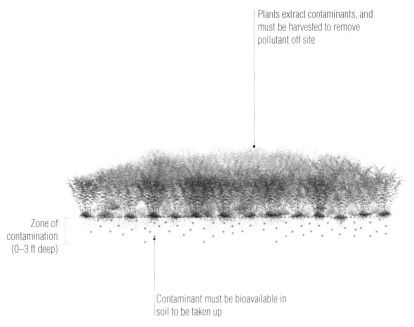

Plants extract contaminants, and must be harvested to remove pollutant off site

Zone of contamination (0–3 ft deep)

Contaminant must be bioavailable in soil to be taken up

Figure 4.10 Extraction Plots

that plants, even hyperaccumulators, will not be able to extract them. Plants may be used to take up the bioavailable fraction of metal contamination in some cases, but are often unable to remediate the soil to within regulatory limits because of the large fraction unavailable to biological processes.

For slightly elevated levels of metals in soil such as cadmium and zinc, Extraction Plots can be grown and harvested over a period of decades to remove the bioavailable fraction of the metal. The soil may still be contaminated, but the bioavailable amount would be depleted. This can be an effective strategy for removing bioavailable metals from fields where food crops are grown.

Typical applications

- **Residences with arsenic:** Urban and suburban residential sites can be contaminated with arsenic that has leached from older wood structures (such as decks, trellises and fences) that were built with lumber that was pressure treated with arsenic. In addition, arsenic was historically an ingredient in pesticides used for termite control. The Chinese Brake Fern (*Pteris vittata*) has been shown to effectively extract low concentrations of arsenic from field sites, remediating soils in as little as two years. The above-ground portions of the plant are harvested and tested to achieve contaminant removal on site, and sent to a hazardous waste facility if required. The challenge is that *Pteris vittata* and other similar hyperaccumulating tropical ferns are currently the only species found to hyperaccumulate arsenic at a rate that is viable for field remediation. These plants are not cold tolerant and are considered an annual in colder climates below US hardiness Zone 8. The fern must be planted as a plug rather than seed, and it becomes very costly in cold climates to replant the species every year. Thus, arsenic remediation via extraction is typically only effectively implemented in warmer climates (see Case study p. 149, Chapter 3)

- **Selenium:** Selenium occurs naturally in soils. In some areas it can be found in concentrations high enough to leach into water and negatively affect human health. Sometimes increased selenium concentration can be caused by mining and other intensive human land uses. Several plant species, such as *Stanleya pinnata*, have been identified to effectively extract selenium and even volatilize it into the air. In some cases the plants may not need to be harvested after the growing season, as most of the selenium has volatilized into the air. If harvested, the plants can often be recycled instead of treated as waste, since selenium is an essential micronutrient for animals. Selenium has been effectively extracted from contaminated sites with forage species that are harvested and fed to cattle as a supplement.

- **Nickel phytomining:** Nickel is one of the few metals for which hyperaccumulators have been used effectively to extract the metal from the soil. Because nickel is in such high market demand, there is the potential to mine nickel from the ground with plants. Dr. Rufus Chaney and other scientists have developed a patent to grow hyperaccumulating plants on nickel-rich soils, harvesting the plants and selling the bio-ore, which is burned to an ash and then used in smelting (see Case study, p. 163).

- **Long-term agricultural field remediation:** Agricultural fields can become contaminated with heavy metals from sources like continual manure application containing unknown levels of metals, nearby mining operations, intensive industrial land use, and naturally occurring high metal levels in soil. The bioavailable fraction of metals in soil may be a threat to food-crop plants, contaminating the food supply with elevated levels of the metals in consumed portions, or by inhibiting plant growth. In addition, livestock that eat the contaminated crops can bioaccumulate the metals, leading to even larger doses of metals exposure to humans when ingested. One potential long-term application of phytotechnologies is to extract the bioavailable fragment of metals in soils. The soil may remain contaminated, but the bioavailable fraction that previously entered the food system can be reduced or eliminated. Extraction Plots can be grown and harvested on the fields for years, then the land can be returned to growing food crops once the bioavailable fraction has been removed (see Case studies, pp. 159 and 171).

- Another use of Extraction Plots on agricultural lands is to entirely replace the food crops with hyperaccumulator species that will very slowly remediate the land over a period of decades, or even centuries, with consistent periodic harvesting. Even low levels of elevated cadmium can take a very long time to remove from a site with such techniques. Biomass crops such as grass species, willow and poplar are being evaluated as energy crops that, with continual harvesting, may slowly remediate the site over time (see Case studies, pp. 159 and 171).

Plant selection

1 *How are extraction plant species selected?* For arsenic-, selenium- and nickel- impacted soils, hyperaccumulating plants or high biomass 'accumulator' species can be considered for Extraction Plots (see plant species lists in Figures 3.38, 3.42, 3.45, 3.47 and 2.17). Other metals extraction should not be considered without very long timescales. A note of caution: Many plants may hyperaccumulate metals, but still cannot reduce concentrations low enough to provide site remediation. Very often site designers will look up hyperaccumulator plant lists and extrapolate that extraction can be completed

when, in truth, the metal is not bioavailable and is too strongly tied to the chemical composition of the soil to be extracted. Extraction Plots are generally not recommended for remediation except in the cases noted above with arsenic, nickel and selenium. Keeping the contaminant on site and removing the risk of human exposure is often the best phytotechnology-treatment option available for inorganic contamination. Stabilization Mats (4.1), Evapotranspiration Covers (4.2) and Groundwater Migration Tree Stands (4.5) can often work together to hold contaminants on site rather than removing them.

Other design considerations

1 *Bioavailability:* Many inorganic contaminants, such as metals and radionuclides, can exist in the soil but not in forms bioavailable to the plant. For a full explanation, see Chapter 3. A detailed analysis of bioavailability must be performed by a soil scientist on the site design team before phytoextraction can be proposed for a site.

2 *Harvesting:* Once an inorganic contaminant is extracted, the plants must be harvested in order for the contaminant to be removed. The contaminant concentrations in the harvested biomass can be tested to determine if the plant must be disposed of in a hazardous waste facility or if it is acceptable to dispose of it in a municipal landfill. In most plots extracting arsenic, selenium or nickel, disposal in a hazardous waste facility should be anticipated.

3 *Risk and bioaccumulation:* Unlike degradation typologies that completely remove a contaminant, Extraction Plots move the contamination into the above-ground parts of the plant, where it may become accessible for consumption by insects, animals and other predators. This mobilization in Extraction Plots can create new, unanticipated vectors for exposure. A careful analysis must be completed to determine if contaminant exposure is likely to occur, and if bioaccumulation is a risk factor.

227

Typology 4.11: Multi-Mechanism Mat

Description: A mixed herbaceous planting utilizing many, or even all, of the phytotechnology mechanisms. The objective is to provide the maximum amount of phytotechnology benefit over a large area with mixed contamination, using low-height species.

Primary mechanisms at work: Phytoextraction , Phytometabolism , Phytodegradation , Phytostabilization , Phytovolatilization

Target: Soils (0–1.5 meters/0–5 feet deep)

Contaminants that can be addressed: Any

Multi-Mechanism Mats are designed with all of the extraction, degradation and stabilization mechanisms in mind to create a low, herbaceous meadow-like planting that maximizes phytotechnology impact while minimizing exposure risk. Elements of Extraction Plots (4.10) Degradation Covers (4.9) and Planted Stabilization Mats (4.1) are combined to create a multifunctioning dense planting of working

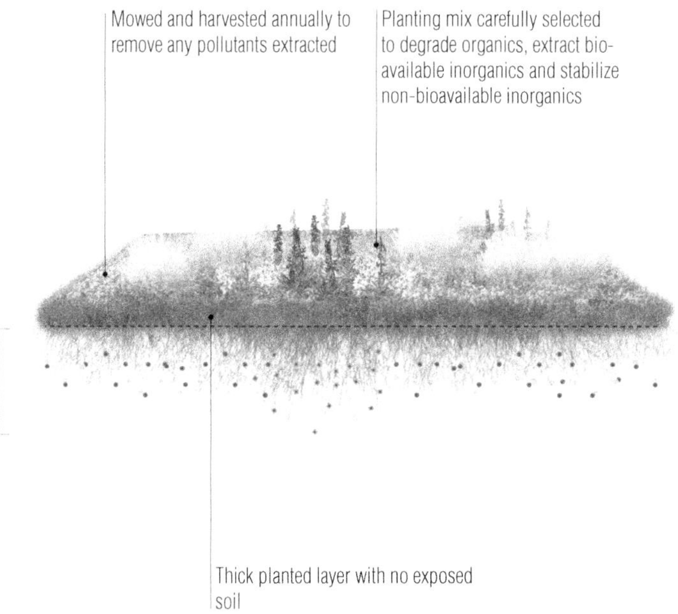

Mowed and harvested annually to remove any pollutants extracted

Planting mix carefully selected to degrade organics, extract bio-available inorganics and stabilize non-bioavailable inorganics

Zone of contamination (0–5 ft deep)

Thick planted layer with no exposed soil

Figure 4.11 Multi-Mechanism Mat

vegetation on a site. Multi-Mechanism Mats should be cut and harvested at the end of each growing season to remove the maximum amount of pollutants from a site.

Typical applications

- **Vacant lots:** While urban parcels remain vacant, Multi-Mechanism Mats can be planted as a holding strategy to provide some cleanup while the site waits for a future use. This typology is suitable in cases where there are no immediate plans for development of a vacant lot; yet, with some minimal care and maintenance, the planting can provide some remediation, wildlife and aesthetic benefits. Spontaneous vegetation typically dominates abandoned urban landscapes, but generally is considered unsightly. With minimal intervention, species selection can be more intentional, encouraging extraction of some metals where applicable, degradation of some organics and stabilization of dust and urban fill. Peter Del Tredici's *Wild Urban Plants of the Northeast* (Del Tredici, 2010) provides recommendations of spontaneous urban species adapted to high levels of disturbance from pedestrian and vehicular traffic. These plants are quick to reproduce and are adapted to the high pH levels often encountered in urban areas from road salt and leaching limestone from concrete. Del Tredici's recommended plant lists can be combined and compared with the phytotechnology plant lists in this book, to consider the creation of a successful and self-reliant Multi-Mechanism Mats. While some of Del Tredici's species are not known to have phytoremediation potential, these species may provide wildlife and ornamental values and help to stabilize the soil, while extraction or degradation species can be added to enhance remediation.
- **Railway and roadway corridors, underutilized industrial areas and other marginalized lands:** Seed mixes can be designed with extraction or degradation species to provide the benefits described

228

above, together with minimal maintenance. Annual harvesting and collection is still required if extraction is targeted.

- **Military bases and firing ranges:** Pollutants including explosives, petroleum, chlorinated solvents and metals are both historically present and continuously added to military landscapes with training activities. Low-height vegetation is required so as to maintain sight lines for firing ranges and use areas for training fields. Extraction and degradation species can be considered so as to increase the functionality of fields for remediation.

Plant selection

1 *Function:* Selection of plant species is based on the particular contaminants being targeted. See Chapter 3 for plant lists by contaminant.

Other design considerations

1 *Ecosystem services:* In addition to providing contaminant removal, Multi-Mechanism Mats can also prevent erosion, enhance wildlife and aesthetics and sequester carbon. The mixed composition of these buffers creates opportunities for species diversity and multi-level functioning.

2 *Biomass production:* If extraction of contaminants is targeted, species selection could also consider energy production, since plants must be cut annually.

229

Typology 4.12: Air-Flow Buffer

Description: The leaf surfaces of vegetation can physically intercept particulate matter from moving air, enhancing the air quality of areas downwind of the vegetation. Typically, no contaminant degradation is provided, and eventually the particulates will wash off the leaves and become a potential source of pollution in stormwater, for which other phytotypologies can provide remediation.

Target: Air

Primary mechanisms at work: Phytoaccumulation

Contaminants addressed: Air pollution particulate matter

Particulate matter is one portion of air pollution that can be removed with the introduction of vegetation (see Chapter 3 for discussion of the other components of air pollution). Where roadways exist, automobiles generate particulate matter. Studies have shown that the amount of particulate matter in air decreases with increasing distance from the roadway and an increase in vegetation cover.

Because the leaves physically filter the particulates out of the air, the particulates are only being sequestered on the leaf, and not degraded. During heavy rainfall or when deciduous tree leaves fall, the particulates are washed into the stormwater. For this reason, Stormwater Filters (see 4.15) should be considered with Air-Flow Buffers to prevent the particles from contaminating the stormwater run-off.

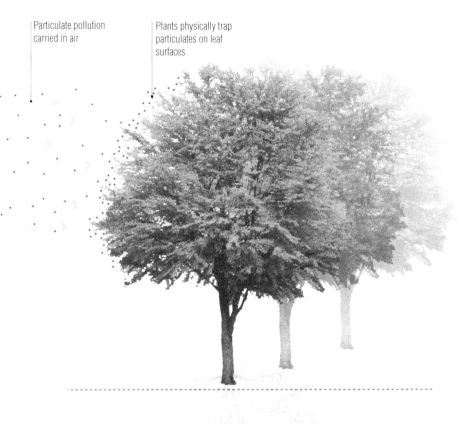

Particulate pollution carried in air

Plants physically trap particulates on leaf surfaces

230

Figure 4.12 Air-Flow Buffer

Typical applications

- **Street tree buffers:** Roadway buffers can be designed to maximize emissions contact with leaves, helping to collect particulates out of the air. Multi-layer buffers can be provided near land uses with human habitation, such as near residences, along roadways, parks and open spaces. Species selection can maximize particulate matter removal.

Plant selection

1 *How are species with good accumulation rates chosen?* More research needs to be completed to assess the connection between leaf surfaces and their impact on particulates removal. Species from preliminary studies are listed in Figure 3.60 on p. 196. Several tree species have been documented as accumulating particulate matter at higher rates than other species. This is primarily due to their leaf size (the greater the area, the more the removal) and the 'stickiness' of the leaf. Species with more leaf hairs and waxy surfaces tend to accumulate more particulate matter.

Other design considerations

1 *Canopy trapping:* Urban trees can potentially trap airborne contaminants at the street level, essentially forming a roof over streets and prohibiting the exchange of air between the atmosphere and the street environment. Where there are numerous pollutant sources below the canopy (e.g., automobiles), the street tree canopy could have a negative effect by minimizing the dispersion of the pollutants at ground level (Nowak, 2006), so placement of trees, wind direction and average wind speed should be carefully considered.

Typology 4.13: Green Wall

Primary mechanisms at work: Rhizodegradation , Phytosequestration , Phytometabolism , Rhizofitration

Target: Air or water

Contaminants addressed:

In air: Volatile Organic Compounds (VOCs) and particulate matter

In water: Nutrients and potentially other pathogens

Description: Green Walls are installations of plants on vertical surfaces, grown with or without soil. Soil biology living in the root zone of a Green Wall can degrade airborne Volatile Organic Contaminants. In order for this to be effective, air must be drawn through the root zone to degrade the contaminant.

231

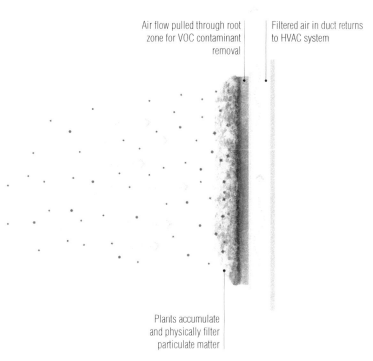

Air flow pulled through root zone for VOC contaminant removal

Filtered air in duct returns to HVAC system

Plants accumulate and physically filter particulate matter

Figure 4.13a Green Wall – Air Biofilter

Drip irrigation feeds wastewater to
vertical plants

Plants remove nutrients,
degrade organics and trap other
non-degradable contaminants in
root system

Figure 4.13b Green Wall – Air Biofilter

232

Particulate matter can also be removed via leaves of plants from accumulation, similar to Air-Flow Buffers (4.12). Water irrigated through the root zone may also have the potential to be cleaned. Green Walls can improve building insulation, mitigate temperature and have other sustainability benefits, but the data on their capacity to deliver human health benefits is mixed. Three categories of Green Walls can be considered.

1 *Vine Walls*: These are Green Walls comprised of climbing vines that are planted at the base or top of the wall.

2 *Living Walls:* The entire plant, including the root system, is integrated into a vertical growing system. The plant can be anchored in soil or another media affixed to a frame or other structure.

3 *Biofilter Walls:* A Biofilter Wall is a living wall where water or air flow is purposefully directed through the wall to have contact with the root zone of a plant for degradation or sequestration purposes.

Vine Walls and traditional Living Walls can be located in both indoor and outdoor environments. The air movement around these walls is not controlled. Typically, outdoor wind or traditional indoor ventilation passes by the static wall, and exposure of the leaf or root system to overall air volumes is minimal.

In indoor environments, VOCs are typically contaminants of concern. To achieve a decrease in VOCs in air, a large volume of air must pass through the leaves and root zone of a plant in order to significantly impact the contaminant concentrations in the surrounding environment. Plants themselves do not degrade VOC contaminants; the soil biology living in the root zones of plants completes the degradation. In most Vine Walls and Living Walls, the amount of air that would typically have contact with the microbial organisms in the root zone of a plant is minimal; therefore the impact of these walls on VOC removal can be quite small (see Chapter 3, p. 195) (Darlington, 2013).

However, if air is purposefully drawn through the wall, using the wall as a biofilter, the associated microbes in the root zones of the plant have greater exposure to the VOCs. The rate of VOC degradation in Green Walls has been shown to be effective when air is pulled through the root zone of a plant (see Case study, p. 197).

Polluted water can also be delivered through the root zone of plants in a vertical system to remove contaminants. Nutrients in wastewater have been the contaminant most commonly targeted for removal in these types of systems.

Typical applications

- **Large-building indoor air filter:** When air is drawn through a Green Wall, the wall becomes a biofilter and can effectively remediate VOCs. This technology can be used as a component of the ventilation system in large buildings.
- **Wastewater filters:** Green Walls have been utilized in schools and residential buildings to filter wastewater effluents by removing excessive nutrients.

Plant selection

1 *How are vertically grown plants selected?* Traditionally, plant selection in these systems has been based on aesthetics and the plant's ability to live in a vertical environment in the particular planting medium, water, light and temperature environments provided. Current research indicates that the specific plant species may not greatly affect the amount of contaminant removal, since most of the degradation occurs in the root biology of the plant. It is likely that providing more plant species diversity encourages greater microbial diversity and can increase degradation rates (Darlington, 2013).

Other design considerations

1 *Air and water flow:* As air and water are passed through the system, the rate and concentration of contamination in these vectors will affect the success of the contaminant removal. Careful study of precedents and collaboration with remediation scientists is required to obtain desired removal performance.

233

Typology 4.14: Multi-Mechanism Buffer

Description: A mixed planting targeted to utilize all of the phytotechnology mechanisms. The objective is to provide the maximum amount of phytotechnology benefit in a small footprint without the need to harvest any plant materials.

Primary mechanisms at work: Phytostabilization ⬚ ⬚, Phytohydraulics ⬚ ⬚, Rhizodegradation ⬚, Phytodegradation ⬚, Phytovolatilization ⬚ ⬚, Phytometabolism ⬚

Contaminants that can be addressed: Any

Multi-Mechanism Buffers are designed with all of the phytotechnology mechanisms in mind to maximize the impact of vegetation in minimizing exposure risks. They combine elements of all of the typologies presented in this chapter to degrade, stabilize and prevent contamination from spreading.

234

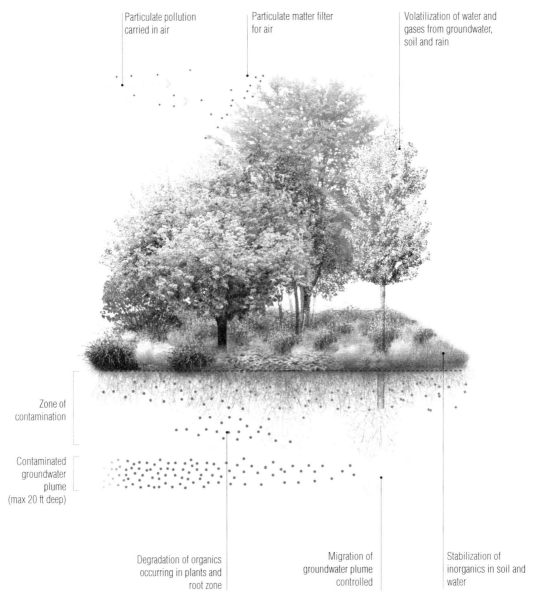

Particulate pollution carried in air

Particulate matter filter for air

Volatilization of water and gases from groundwater, soil and rain

Zone of contamination

Contaminated groundwater plume (max 20 ft deep)

Degradation of organics occurring in plants and root zone

Migration of groundwater plume controlled

Stabilization of inorganics in soil and water

Figure 4.14 Multi-Mechanism Buffer

Typical applications

- **Riparian buffers:** The ecological and remediation benefits of mixed-planting riparian buffers along rivers and water bodies is well documented. In addition to contaminant capture and removal, with careful plant selection these buffers can also greatly benefit wildlife and other ecological systems.
- **Corridor buffers for roadsides, railroads, industrial areas and agricultural plots:** Linear buffers can not only prevent water- and soil-borne contaminants from migrating, but also capture particulate matter in air pollution. Buffers close to the original sources of the contamination can contribute to minimizing the effort (and cost) required to remediate water resources downgradient.
- **Site perimeter buffers:** Plantings along the edges of sites can be mixed to include benefits of all the typologies previously described in this chapter in a minimal amount of space. Site perimeter buffers help keep the pollution impacts of site activities inside the site footprint, limiting impact on neighboring land uses.

Plant selection

1 *Function:* Selection of plant species is based on the particular contaminants being targeted. See Chapter 3 for plant lists by contaminant.

Other design considerations

1 *Ecosystem services:* The benefits of Multi-Mechanism Buffers have been acknowledged for decades. In addition to providing contaminant removal, they can also prevent erosion, enhance wildlife habitat and corridors, sequester carbon, increase real estate values and livability and enhance aesthetics and recreational opportunities. The mixed composition of these buffers creates opportunities for species diversity and multi-level functioning.

235

Typology 4.15: Stormwater Filter

Description: Plantings and soil remove and trap contaminants from stormwater. Organic pollutants may be degraded and nitrogen contamination in water may be converted into a gas and returned to the atmosphere. Inorganic contaminants may be immobilized and remain on site in the soil. The objective is to remove contaminants from stormwater at the source, before they spread to groundwater or other water bodies.

Target: Stormwater

Primary mechanisms at work: Rhizofiltration

Contaminants addressed:

Degraded/removed: Nitrogen , Petroleum ●, Chlorinated Solvents , Pesticides ●.

Pollutants mobilized in stormwater

If plants are harvested, some nutrients and extracted inorganics can be removed from site

Plants degrade organics from stormwater run-off

Inorganics trapped in soil matrix and plant roots

Nitrogen transformed to gas by bacteria

Figure 4.15 Stormwater Filter

236

Held in soil/plants: Metals ■, Phosphorus , POPs ●

Slowly extracted over time if plants are harvested: Some Metals ■, some Phosphorus , Nitrogen

Contaminants not effectively treated: Salts ▨

Contaminated stormwater is most often generated by rainfall on impervious surfaces such as roads and sidewalks. Debris on these surfaces is picked up by the rain and is mobilized. Stormwater Filters capture the run-off near the targeted impervious surface before the pollutants mobilize into other water bodies or groundwater.

Stormwater Filters are also referred to as bio-swales, vegetated swales, vegetated filter strips, rain gardens and detention basins. These various types of Stormwater Filters are solutions tailored to the particular contaminant(s), stormwater flow volume and speed, climate and space available for the installation. In addition, traditional grey-infrastructure engineering solutions are often paired with planted Stormwater Filter components to maximize treatment. Both the grey-infrastructure components and green Stormwater Filters are frequently referred to as Best Management Practices (BMPs). They are often combined to create a 'treatment train,' where each selected BMP targets a particular segment of the contamination and the BMPs are linked together to provide the best sequence of treatment in the space available.

Since Stormwater Filter design techniques and removal capacities have been greatly documented, they will not be covered in detail here. Much of the treatment success of these systems has to do with the design of the soil media and water retention times, in addition to selecting plants that will survive the stormwater and contaminants. A general overview is provided below.

Typical applications

- **Roadsides and parking lots:** Stormwater Filters are often used at the edges of impervious surfaces to collect and remove pollutants mobilized by the water. The most typical contaminants encountered include Nutrients (Phosphorus and Nitrogen) ◗, Petroleum PAHs ●, and Metals ▮.
- **Agricultural fields:** Due to heavy fertilizer and pesticide use, stormwater run-off from agricultural fields is a significant threat to adjacent water bodies and groundwater. Stormwater Filters can be applied at field edges to target removal of Nutrients and Pesticides ●.

Plant selection

Most Stormwater Filter plants are selected based on their survival characteristics and ability to withstand the amount of water coming into the system, periods of drought (since they are usually not irrigated) and general lack of maintenance activities. In addition, some further criteria can be considered.

1 *Degradation/Removal:* Organic contaminants such as petroleum, chlorinated solvents and pesticides can be degraded by plants and their associated microbes. In addition, nitrogen can be removed from the system by denitrifying bacteria that turn the nitrogen into a gas. In general, species that produce the greatest biomass, root depth and root growth will remove the most pollutants from the system (Read et al., 2009). If degradation of organics is a priority, maximize species diversity within a system and choose plants that maximize biomass production so as to achieve target goals. In addition, some specific species have been used to target some of the organic compounds that are more recalcitrant and difficult to break down.

2 *Stabilization:* In Stormwater Filters, inorganic contaminant removal generally occurs because contaminants are captured and held within the Stormwater Filter. The contaminant is not degraded or removed, but instead the system acts like a sponge, holding the contaminant while the water passes through. The filtration happens in two ways:

Physical: By controlling water speed and retention time, particles can be settled out of the system as sedimentation. This is one of the best mechanisms to catch inorganics. The sedimentation that results must be removed, so that as new water passes through, it is not re-suspended and released from the Stormwater Filter. Selecting plants that have a scrubbing, slowing effect on water velocity may help with physically removing inorganics from the stormwater.

Chemical: As water infiltrates into the soil, contaminants can chemically bind with soil particles. The media of Stormwater Filters can be manipulated to maximize the immobilization of inorganic contaminants. At some point the media 'fills up' and reaches its carrying capacity if no plants are provided. By integrating plants into the system, new receptors for binding contamination can continually be created, using the plants' organic matter and oxygen released through the roots. This is why systems with plants typically perform better over time

237

than systems without plants. Because most of the work is done in the rhizosphere of the plants, systems with the greatest diversity of soil biology tend to perform better, therefore systems thickly planted with a variety of species tend to perform better than monocultures.

3 *Extraction:* Inorganic contaminants, such as metals, can be extracted by plants in very small quantities. Plant species can be selected to remove metals more quickly than others; however, in order to remove the contaminants from the system, the plants must be harvested. Typically, even with hyperaccumulator plants, the amount of metals removal by the plants will be quite small, as compared to the total amount of metals in the system. For this reason, Stormwater Filters are generally designed to catch and stabilize metals, rather than extract them. To enhance pollutant removal, choose species that have deep and thick root systems (Read et al., 2009). However, if a particular target metal is of concern and annual harvesting can be accomplished, some systems may be designed with hyperaccumulator species to slowly remove some portion of the contaminant over time (see Chapter 3).

4 Nitrogen and phosphorus can be removed from the system with plant harvesting, though typically this is not utilized, as the amount extracted in the plant biomass is typically small in comparison to the amount removed by other mechanisms in a Stormwater Filter. For example, maximizing the denitrification process of soil bacteria turning nitrogen into a gas is a much more efficient mechanism than harvesting the plants. For phosphorus, maximizing stormwater contact with the soil for stabilization via infiltration is the best mechanism. The amount extracted in the plant biomass is so small in comparison to the amount removed via soil contact and infiltration that plant extraction and harvesting is generally not worth the effort. However, if even small portions of nitrogen or phosphorus are targeted for removal from the system, the plants should be harvested each year before they die back/decompose and release the nutrients back into the system. Since nitrogen and phosphorus are essential plant nutrients, no plant is considered a hyperaccumulator of these nutrients. The general rule is that if more biomass is produced, typically the nitrogen and phosphorus extraction is maximized, so fast-growing plants with lots of biomass production for harvesting are the most useful species. If harvest is considered, high-biomass species typically use more nitrogen and phosphorus, so nutrient removal is often greatest with high biomass-producing species.

Other design considerations

1 *Evapotranspiration:* Consider maximizing evapotranspiration within the system so as to have the capacity to treat greater volumes of water. Plants with high evapotranspiration rates can move a large amount of water out of the system through evapotranspiration. See plant list in Figure 2.18, p. 48, for plants with high evapotranspiration rates.

Typology 4.16: Surface-Flow Constructed Wetland

Description: Water is directed through a series of planted marshes and engineered soil media at varying depths to remove contaminants. The objective is to clean the water as it passes through the system.

Some organic contaminants and nitrogen can be removed/degraded completely and other inorganic contaminants can be filtered out and held in the soil.

Target: Stormwater, wastewater, groundwater

Primary mechanisms at work: Rhizofiltration

Contaminants addressed:

Degraded/removed: Nitrogen , Petroleum ●, Chlorinated Solvents , Pesticides ●, other organic contaminants of concern

Held in soil/plants: Explosives ●, Most Metals ■, Phosphorus , POPs ●

Slowly extracted over time, if plants are harvested: Some Metals ■, Phosphorus , Nitrogen

Contaminants not effectively treated: Salts

Surface-Flow Constructed Wetlands closely mimic the ecosystem of a natural wetland by utilizing plants to filter water through plant root zones, a planted medium and open water zones. They are highly engineered to obtain treatment capacities. Most of the treatment does not occur within the plants themselves, but rather, in the biofilm on the roots of the plants and within the biology and chemistry of the water and planted media. Often, separate wetland cells are provided with different types of media with or without oxygen (aerobic or anaerobic) in a treatment train to address the specific contaminants within the water. The remediation role of plants in these systems supports the microbial life and soil media in the wetland. The plants deliver organic matter, oxygen, nutrients, sugars and other root exudates to the system.

239

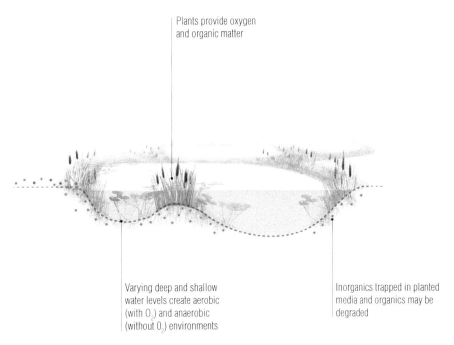

Plants provide oxygen and organic matter

Varying deep and shallow water levels create aerobic (with O_2) and anaerobic (without O_2) environments

Inorganics trapped in planted media and organics may be degraded

Figure 4.16 Surface-Flow Wetland

These systems have been thoroughly detailed in other publications and will be only briefly referenced here. The hydrology and media must be designed by an experienced constructed wetlands specialist to achieve required removal rates. Failures of constructed wetlands to date have been high, due to improper design, implementation or maintenance. It is critical to engage an experienced engineer when taking on these projects.

Typically the main treatment mechanisms in Surface-Flow Constructed Wetlands are similar to those described above in Stormwater Filters.

1 Many organic contaminants can be degraded.

2 Nitrogen is removed as a gas, via the denitrification process in anaerobic cells.

3 Inorganic contaminants are filtered out of the water and stabilized and held in the soil or in the plants themselves. They remain on site, but the water is cleaned.

4 Phytoextraction of inorganic contaminants is typically not an objective, since the plants would need to be harvested to obtain removal. Typically the amounts extracted are minimal in comparison to the amount of inorganics filtered out and held in the soil. In addition, phytoextraction of inorganics is often seen as a potential detriment, since animals can access the above-ground portions of the plants and an exposure pathway is created.

5 For all plants, phytometabolism of nitrogen and phosphorus into the plant biomass does occur each year. The amount of nitrogen and phosphorus in the plant biomass can be removed from the wetland if the plants are harvested each year, but it is typically not significant enough to render this useful.

240

Typical applications

- **Municipal and industrial wastewater:** Wastewater from municipal sewer systems and industrial applications has been successfully treated with Surface-Flow Constructed Wetlands. The contaminants treated range from excessive nutrients to temperature and heavy metals.
- **Landfill leachate:** Water leaking out of municipal and industrial landfills can be laden with contamination. This landfill leachate can be pumped into a constructed wetland and treated.
- **Stormwater wetlands:** Constructed wetlands can be effectively utilized as part of the treatment train for stormwater run-off.

Plant selection

1 *Specificity:* Since constructed wetlands can be designed to remove any number of contaminants, plant selection will respond to the specific media selected, hydrology and climate. Reference peer-reviewed publications on constructed wetlands to aid in plant selection (see Chapter 6). Emergent species as well as submerged aquatic or floating-leaved aquatic species should be considered.

Other design considerations

1 Consider pairing constructed wetland systems with upland Groundwater Mitigation Tree Stands or Phytoirrigation systems so as to maximize treatment capacity.

Typology 4.17: Subsurface Gravel Wetland

Description: Contaminated water is treated by pumping the water slowly through subsurface gravel beds, where it is filtered through plant root zones and soil media in a vertical or horizontal flow pattern. Some organic contaminants and nitrogen can be degraded/removed completely and other inorganic contaminants can be filtered out and held in the media.

Target: Stormwater, wastewater, groundwater

Primary mechanisms at work: Rhizofiltration , Phytostabilization

Contaminants that can be addressed

Degraded/removed: Nitrogen , Petroleum , Chlorinated Solvents , Pesticides and other organic contaminants of concern held in soil/plants; Metals , Phosphorus , POPs .

Slowly extracted over time if plants harvested: Some Metals , Phosphorus , Nitrogen .

Contaminants not effectively treated: Salts

Contaminated water is treated by pumping the water slowly through subsurface gravel beds, where it is filtered through plant root zones and soil media. The water flows 3–8 inches under the surface to prevent exposure to the water, mosquito breeding and odors. These systems require less land area than traditional Surface-Flow Constructed Wetlands and often can have better removal efficiencies for certain contaminants, but are typically more expensive to build and offer less habitat for waterfowl and other aquatic animals and organisms.

241

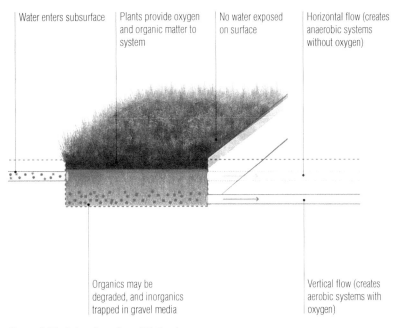

Water enters subsurface

Plants provide oxygen and organic matter to system

No water exposed on surface

Horizontal flow (creates anaerobic systems without oxygen)

Organics may be degraded, and inorganics trapped in gravel media

Vertical flow (creates aerobic systems with oxygen)

Figure 4.17 Subsurface Gravel Wetland

Water is pumped to flow horizontally or vertically through subsurface systems. The pattern of flow has a significant effect on the amount of oxygen in the system, which thereby drives contaminant removal efficiencies. Subsurface flow systems are often combined with other conventional treatment train add-ons such as aeration devices to obtain maximum removal efficiencies.

Typical applications

- **See Surface-Flow Constructed Wetlands (4.16):** Subsurface Gravel Wetlands can be used in the same applications as traditional surface-flow systems, but typically require much less space. They are often utilized in land-restricted environments where constructed wetland technology is desired for contaminant removal.

Plant selection

1 *See Surface-Flow Constructed Wetlands (4.16).*
2 *Rooting depth:* It is important to match the rooting depth of selected species to the depth of the gravel media. Only emergent wetland plant species with deeper rooting zones tend to be used in Subsurface Gravel Wetlands, as their roots will reach the water in the gravel beds.

Other design considerations

1 *Winter function:* Because the water is below the surface in these systems, contaminant removal through the winter months is often possible, since the system does not freeze. An additional mulch layer is sometimes added during cold temperatures to provide insulation so as to maintain treatment effectiveness in winter months.

Typology 4.18: Floating Wetland

Description: Plantings installed on structures are floated on existing water bodies to filter contaminants out of the water. Organic pollutants may be degraded and nitrogen contamination in water may be converted to a gas and removed. Inorganic contaminants are immobilized and remain attached to the roots or floating structure substrate or are extracted into plants. The Floating Wetlands may be harvested to remove any pollutants extracted from the water. The biomass produced is often rich in essential elements and may be composted and reused for nutrient recycling, if not found to be too toxic.

Target: Existing surface water bodies: lakes, rivers, ponds, streams and other basins of water

Primary mechanisms at work: Rhizofiltration

Contaminants addressed:

Degraded/removed: Nitrogen   , Petroleum ●, Chlorinated Solvents ●, Pesticides ●

Held in soil/plants: Explosives ●, Metals ■, Phosphorus   , POPs ●

Slowly extracted over time if plants are harvested: Some Metals ■, Phosphorus   , Nitrogen   .

Contaminants not effectively treated: Salts ▦

Existing contaminated water bodies may be treated with Floating Wetlands. Often local species can be utilized to provide other ecosystem services such as water cooling and habitat creation while removing contaminants.

Typical applications

- **Urban rivers:** Inundated with pollutants from both historic and current uses, urban rivers can often contain excessive levels of nutrients and metals in the water column. Floating Wetlands can be installed to filter out many pollutants as the water flows by. In addition to the pollutants covered in this book, living biological contamination, pathogens, viruses and emerging pollutants of concern like pharmaceuticals may also be able to be treated with proper sizing, filtration media and plant selection.
- **Canals, ponds and rivers connecting agricultural fields and golf course water hazards:** Due to heavy fertilizer and pesticide use, water bodies adjacent to agricultural fields and golf courses often

243

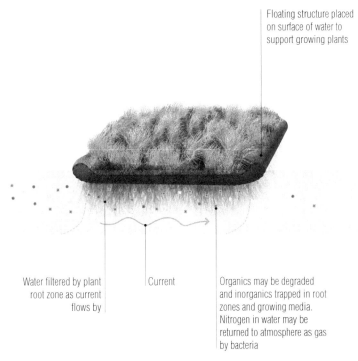

Floating structure placed
on surface of water to
support growing plants

Water filtered by plant
root zone as current
flows by

Current

Organics may be degraded
and inorganics trapped in root
zones and growing media.
Nitrogen in water may be
returned to atmosphere as gas
by bacteria

Figure 4.18 Floating Wetland

contain high levels of nutrients and pesticides. Floating Wetlands can be installed on the surface to target removal.

- **Wastewater treatment facilities:** Both municipal wastewater facilities for human waste and industrial food processing facilities have utilized Floating Wetlands, or some version of Floating Wetlands, for pollutant removal. Living machines and eco-machines are a combination of Floating Wetland and other constructed wetland technology (see 4.16 and 4.17) to treat water with a series of vegetated tanks that target both abiotic and biological pollutants (Todd, 2013).

Plant selection

Most Floating Wetland plants are selected based on the particular contaminants being targeted, their survival characteristics and ability to withstand the encountered water inundation, pH and general lack of maintenance activities. In addition, some further criteria can be considered.

1 *Degradation/removal:* Organic contaminants such as petroleum, chlorinated solvents and pesticides can be degraded by plants and their associated microbes. In addition, the root system is a home for beneficial bacteria and microbes that remove nitrogen from the water and turn it into a gas. In general, the more biomass and growth produced by the plants, the more degradation and removal will occur. If degradation of organics is a priority, maximizing species diversity within the system and choosing plants that maximize biomass production can help to achieve the desired goals. In addition, some specific species have been used to target some of the organic compounds that are more recalcitrant and difficult to break down.

2 *Stabilization and extraction:* In Floating Wetlands, inorganics removal generally occurs because contaminants are captured and held in the root system of the plant, or in the substrate of the Floating Wetland. In some limited cases, the inorganics may be transported into the above-ground portions of the plant. Floating Wetlands act as large filters that must eventually be removed to remove the pollutants from the system.

3 *Nutrients:* In addition to the conversion of nitrogen to gas by beneficial microbes, nitrogen and phosphorus can be removed from the system by plant growth and harvesting. The amount extracted in the plant biomass is typically small, but repeat crops may be cultivated in one growing season to maximize the amount of nutrients metabolized and harvested. The biomass produced may be collected, composted and harvested for use as organic fertilizer. In freshwater systems, the contaminant of concern is typically phosphorus, whereas in saltwater systems, nitrogen is more problematic.

Other design considerations

1 *Aeration:* Many Floating Wetland systems are combined with aeration to maximize the beneficial biological processes occurring in the root zones of the plants. Aeration is usually maximized if placed in the bottom of the water rather than on the surface.

2 *Habitat:* Floating Wetlands may provide enhanced cover to promote fish, turtle and insect populations, as well as creating a food source for waterfowl.

244

5: Site programs and land use

In the varied built environment, many different site programs are encountered that could potentially release contaminants. Site uses such as gas stations, dry cleaners and industrial manufacturing sites are obvious potential offenders, but there are other less obvious landscapes such as cemeteries and residential back yards that can host a range of pollutants.

This chapter introduces the application of the plants and planting typologies covered in previous chapters to a broad range of sites, land uses and contaminants. Sixteen land-use categories are illustrated to consider what pollutants can be anticipated on these types of sites. Planting typologies developed in Chapter 4 are then applied, and aesthetic composition and consideration of natural and cultural systems are integrated. The objective of this chapter is to consider where opportunities exist for phytotechnology integration into site design practice. This will assist the landscape architect, engineer or site owner in developing tools to address the cleanup of pollutants for the reuse and development of such sites.

Organization

This chapter provides an overview and description of sites as land-use programs and, in particular, describes the industrial or infrastructural activities that occur on the sites, followed by a listing of the contaminants found in the soils and groundwater commonly associated with these land uses. The anticipated use of phytotechnology planting methods to remediate potential contamination is illustrated.

The land uses identified are generally for landscapes and planting zones in North America; however, the principles developed on this short list of site programs can also be applied to other site programs that are found internationally. Through the use of site typologies and identification of prior land use, the site designer will be able to link the following three elements:

1 cleanup of commonly found pollutants on a range of landscape types
2 environmental engineering techniques such as stabilization, capping or degradation planting and their purpose and logic in application
3 landscape design layouts using plants and phytotechnology applications combined.

This will allow the landscape architect to collaboratively engage with other site engineering disciplines and integrate these approaches into design strategies for the site, rather than carrying out landscape design after the initial civil and environmental engineering design has taken place. In addition, these typologies allow landscape architects to be pre-emptive in their planting approaches, and to anticipate potential pollution that may arise from future land uses on site. In this way, design work can be projective, based on evolving conditions, ultimately able to be both pragmatic and visionary.

Site land-use programs

In order to review the complex and varied nature of the built environment, 16 distinct site programs are described and illustrated in this section. The land uses are as follows:

1 Roadways and parking lots
2 Parks, open spaces, lawns and golf courses
3 River corridors and greenways
4 Railroad corridors
5 Light industrial and manufacturing sites
6 Gas stations and auto-repair shops
7 Dry cleaners
8 Funeral homes and graveyards
9 Urban residences
10 Vacant lots
11 Community gardens
12 Agricultural fields
13 Suburban residences
14 Landfills
15 Former manufactured-gas plants
16 Military uses

Source control

In addressing these various site types, the most important issue is to control the source of the pollution. For example, if fertilizers and pesticides are being used on the site for maintenance purposes and are contributing contaminants to the environment, can organic maintenance practices instead be utilized? If bluegrass lawns continually need cutting, causing excessive use of lawn mowers that burn and leak fuel, can low-mow, low-maintenance lawns be considered instead to prevent the fuel contaminant release? Ways to minimize source pollutants should always be considered first, including organic landscape maintenance practices, low-maintenance plant variety selection, and low-emission equipment specifications, to name a few. Below is a review of the 16 land-use programs.

5.1 Roadways and parking lots

Roads, roadsides and the finished surfaces of parking areas are varied in context, width, vehicle speed and slope, as well as their location, development and the vegetation cover on their sides. They can be considered singular as well as linked together as a continuous seam across the country, acting as conveyance routes for pollutants as well as people. According to the Federal Highway Administration, in 2008 there were 2,734,102 miles of paved public roads in the United States and the majority of Americans used a motor vehicle as their primary, if not only, mode of transportation. Cars themselves generate much of the pollution that is found along roadways. This includes heavy metals released from the brake linings of cars, tire debris, gasoline, oil leaks and drips on the road surface, and emissions, as well as other depositions of atmospheric constituents and particulate material onto the roadside. Often, vehicles emit heavy metals and petroleum products from incomplete combustion. These emissions bind to form one component of particulate matter (PM) air pollution. After several days suspended in the air, particulates fall to the surface and pollute surrounding soil and water. In parking areas and roadside soils, the contaminants generated from cars can include organic compounds, petroleum hydrocarbons and VOCs. The soil can also contain inorganic metals such as aluminum, cadmium, chromium, copper, lead, mercury and zinc. Cars made before 1993 may contain chlorofluorocarbons (CFCs – a widely used chlorinated solvent) in their air conditioning systems. Other sources of contamination include loosened pieces of the road surface or roadway base preparation itself (including recalcitrant PAHs generated from asphalt wear), nutrients from atmospheric deposition adhering to sediments, discarded objects, heat from road surfaces, de-icing materials and herbicides from seasonal maintenance activities.

249

The ecological effects of roadways on the environment are profound. In their seminal book, *Road Ecology*, Forman et al. illustrate that roadway pollutants have elevated effects up to 100 meters away from roadway edges (Forman et al., 2003, p. 205). In addition, there is a direct correlation between air pollution and highway capacity: the greater the capacity, the greater the air pollution in a metropolitan area (US PIRG, 2004). Refer to Chapter 3, p. 189 for more information on air pollution and roadways.

The presence of existing vegetation in drainage swales in both parking areas and roads is of potential use to phytotechnology design, as is the introduction of additional strategies to allow for the removal of contaminants at the source, thereby decreasing the effects of the pollutants downgradient.

Figure 5.1a Roadways and Parking Lots: Sources of Contamination

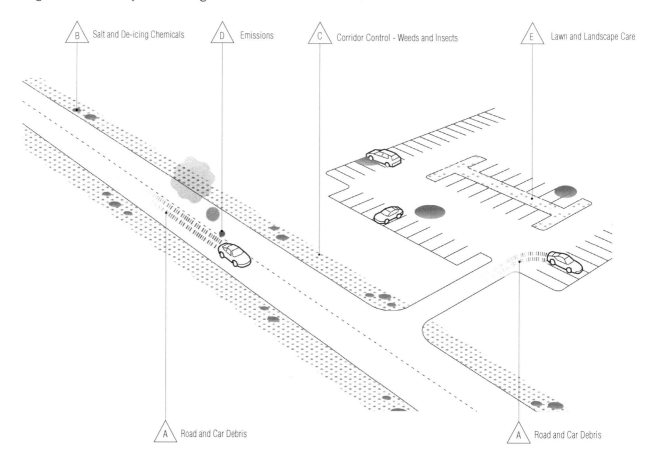

Key	Source	Description	Contaminants
A	Road and car debris	The following debris migrates to water, soil and air: • Tire debris: nutrients (phosphorus and nitrogen) and metals (especially mercury and zinc) picked up from atmospheric deposition and other sites • Road surface wear: petroleum • Brake lining and car parts: asbestos, nickel, copper, chromium • Metal plating and tires: metals including nickel, iron, copper, chromium, zinc, lead, cadmium, manganese • Fuel and oil: petroleum, salt, lead (formerly added to gasoline).	Nutrients (phosphorus and nitrogen): Ch. 3, p. 125 Metals: Ch. 3, p. 136 Petroleum: Ch. 3, p. 65 Air pollution: Ch. 3, p. 189
B	Salt and de-icing chemicals	Materials spread onto roadways, walkways and other surfaces to prevent ice formation.	Salt (sodium, chloride and other additives): Ch. 3, p. 179
C	Corridor control: weeds and insects	Pesticides, including insecticides and herbicides, are applied to control weed growth along roadsides and infestations of unwanted insects. These substances usually include salts and metals, which build up on sites over time. Historic applications may have deposited elevated levels of arsenic and lead or persistent organic pesticides such as DDT, DDE or Chlordane.	Pesticides and POPs: Ch. 3, p. 111 Salt (sodium, chloride and other additives): Ch. 3, p. 179 Metals: Ch. 3, p. 136 POPs: Ch. 3, p. 118
D	Emissions	Release of chemicals and particles into the air from automobiles. In addition, areas where gasoline emissions may have been historically concentrated may have lead contamination in soils.	Air pollution: Ch. 3, p. 189
E	Lawn and landscape care	Fertilizers and pesticides applied by site owners to ornamental landscapes can migrate into stormwater and groundwater. In addition, maintenance equipment for application of these products generates emissions and may leak fuel.	Pesticides and ● POPs: Ch. 3, pp. 111 and 118 Salt (sodium, chloride and other additives): Ch. 3, p. 179 Metals: Ch. 3, p. 136 Nutrients (phosphorus and nitrogen): Ch. 3, p. 125

Figure 5.1b Roadways and Parking Lots: Phytotechnologies to Address Contaminants

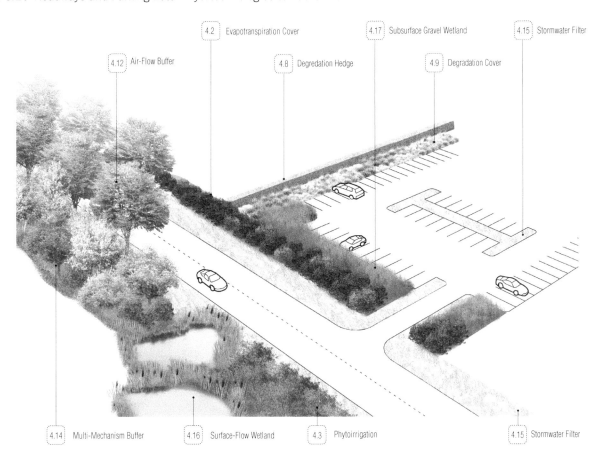

251

Key	Typology	Description	Addresses	Plant Lists
4.2 (p. 204)	Evapotrans-piration Cover	High evapotranspiration-rate species are planted to quickly transpire stormwater into the air, preventing contaminants from migrating off site. Where de-icing activities occur, salt-tolerant species must be utilized.	Within stormwater: Nutrients (phosphorus and nitrogen); ■ Metals; ● Petroleum	p. 48, 2.18 High evapotranspiration-rate species
4.3 (p. 207)	Phyto-irrigation	Stormwater is collected and irrigated onto plants to prevent contaminants from migrating off site. A solar-powered pump with drip tubing can be used for the irrigation system. Where de-icing activities occur, salt-tolerant species must be utilized.	Within stormwater: Nutrients (phosphorus and nitrogen); ■ Metals; ● Petroleum	p. 48, 2.18 High evapotranspiration-rate species
4.8 (p. 220)	Degradation Hedge	Plants with thick, fibrous root zones are used along the edges of the parking lot to intercept and degrade hydrocarbons in water run-off.	● Petroleum	p. 74, 3.5 Petroleum
4.9 (p. 222)	Degradation Cover	Deep-rooted perennials are used along the edges of the parking lot to degrade any hydrocarbons that might run off from the parking lot.	● Petroleum	p. 74, 3.5 Petroleum
4.12 (p. 229)	Air-Flow Buffer	Tree species that trap particulate matter from air in the canopy can be used as street trees to prevent pollution from migrating beyond the road corridor.	Air pollution: particulate matter	p. 196, 3.60 Air pollution, particulate matter
4.14 (p. 234)	Multi-Mechanism Buffer	Road pollutants have been shown to affect environments as far as 200 meters away from the road edge (Ministry of Environment, 2006: 9). Heavily vegetating this area with a mix of species can prevent the spread of pollutants. Inorganic contaminants may be held on site in the root zone and soil, while organic contaminants may be degraded. Particulate matter in the air may be captured and held on leaf surfaces.	Nutrients (phosphorus and nitrogen); ■ Metals: stabilized on site; ● Petroleum; Air pollution: particulate matter; ● Pesticides	p. 46, 2.17 Nutrients: high-biomass species p. 140, 3.37 Metals p. 74, 3.5 Petroleum p. 195, 3.59–3.60 Air pollution p. 113, 3.23 Pesticides
4.15 (p. 235)	Stormwater Filter	Plants and the soil filter out pollutants from the stormwater in a swale or linear filter strip. Where de-icing activities occur, salt-tolerant species must be utilized; salt is typically not removed in these systems.	Within stormwater: Nutrients (phosphorus and nitrogen); ■ Metals; ● Petroleum; ● Pesticides	Stormwater/Wetland species not included in this publication
4.16 (p. 238)	Surface-Flow Wetland	Plants and the soil filter out pollutants in a series of open water ponds and cells. Where de-icing activities occur, salt-tolerant species must be utilized; salt is typically not removed in these systems.	Within stormwater: Nutrients (phosphorus and nitrogen); ■ Metals; ● Petroleum; ● Pesticides	Stormwater/Wetland species not included in this publication
4.17 (p. 241)	Subsurface Gravel Wetland	Plants and associated media filter out pollutants in a series of gravel cells below the surface. No water is visible and the treatment footprint is typically smaller than in Surface-Flow Wetlands. Where de-icing activities occur, salt-tolerant species must be utilized; salt is typically not removed in these systems.	Within stormwater: Nutrients (phosphorus and nitrogen); ■ Metals; ● Petroleum; ● Pesticides	Stormwater/Wetland species not included in this publication

5.2 Parks, open spaces, lawns and golf courses

The land use and programs in this section – parks, open spaces, lawns and golf courses – are a broad range of designed landscape types that cover generally permeable, open and planted areas of varying scales from domestic and small-scale sites to larger, district-scale sites set within urban, suburban and rural contexts. These sites often contain graded topographic surfaces, roads, pathways, planting areas, modest public structures and storage areas, and water bodies ranging from vast lakes to smaller-scale pools and fountains. As such, they are the targets of multiple pollutants from ongoing maintenance activities involving excessive nutrients and pesticides in the soil, salt and de-icing solutions, and petroleum products from maintenance equipment such as industrial-scale grass mowers. Additional factors that contribute to pollution in parks include the previous or historic land use of the site, the construction materials used in building the park, atmospheric deposition and adjacent and nearby industrial land uses.

Parks inherently have a number of characteristics that prevent human exposure to contaminants that may exist below ground. For example, paved areas, dense turf grass and thickly mulched beds prevent direct exposure to users. However, the pesticides and fertilizers that are regularly and repeatedly applied to open spaces, especially golf courses, are a cause for concern. According to the Golf Course Superintendents of America, golf courses in the US spend over $8 billion each year on lawn chemicals and equipment (GCSAA, 2013). Many of the applied chemicals and nutrients migrate into the soil and groundwater, where they may eventually pollute nearby water bodies and drinking water. Planting schemes in parks are often performing some level of remediation, however, these systems can be enhanced and optimized to intercept these pollutants using phytotechnologies.

Pesticides

The US EPA permits the use of over 200 chemical pesticides to control weeds, fungus and insects, many of which are banned in the European Union because of health and ecological concerns. Nearly 80 million pounds of pesticides are used on US lawns every year, with the 30 most popular pesticides being used in over 90% of lawn-care treatment (Wargo et al., 2003). An estimated 5–10% of total applied pesticides are lost to run-off, most often ending up in local surface waters and aquifers (Haith and Rossi, 2003). While source control (preventing the application of these materials) is one way to mitigate these contaminants, remediation strategies include vegetated buffers and filters that prevent pesticides from leaching and traveling off site.

Fertilizers

Over-fertilization of lawns due to commercialized 'step' programs, social pressures and lack of knowledge is well documented (Spence et al., 2012). Nitrogen and phosphorus pollution occur mainly off site, once these nutrients leach away from the original application site, into water bodies and drinking-water sources. Nutrient overloading in water bodies causes algal blooms, eutrophication, lack of oxygen

252

and degradation of both native plant and animal species (Rosen and Horgan, 2013). The ecological effects of nutrient loading and eutrophication can be devastating to the ecosystem economies on which many people depend. Vegetation remediation efforts should be sited surrounding surface waters and along downward slopes to intercept moving water, and above areas of high groundwater tables where nutrients can easily reach the water below the surface.

Maintenance equipment

Motorized machines for lawn care and maintenance commonly use petroleum products, such as gasoline and lubricants. Wherever these machines are stored and refueled, some level of hydrocarbon contamination can be expected. Americans use about 800 million gallons of gasoline annually in power mowers, with an estimated 17 million gallons spilled during refueling (US EPA, 1996). In addition, contaminant spills when filling machinery with pesticides or fertilizers should also be considered. Planted buffers can potentially control and degrade these contaminants around maintenance areas.

Contaminant source control

To prevent pollutant generation in these environments, low-maintenance landscapes that require less ongoing care can be designed from the outset of the project. For those looking to replace the conventional turf grass environment, a host of new, low-growing, low-mow species composed of fescues and warm-season grasses such as Buffalo Grass (*Buchloe dactyloides*) require minimal water and mowing, significantly reducing run-off and chemical applications. In addition, minimizing irrigated water should be considered. High irrigation rates not only allow contaminant migration but prevent the development of robust plant root systems and often consume potable water supplies. Changing irrigation rates to water deeply and infrequently promotes deeper, more extensive root growth, with healthier plants that require fewer chemical treatments. These plants may then pre-emptively be ready to deal with small-scale spills with their large root biomass and associated root biology, rather than their highly irrigated counterparts, with their associated smaller root systems. Organic maintenance practices can additionally be mandated, and low fuel-consumption maintenance machinery can be specified.

253

Figure 5.2a Parks, Open Spaces, Lawns and Golf Courses: Sources of Contamination

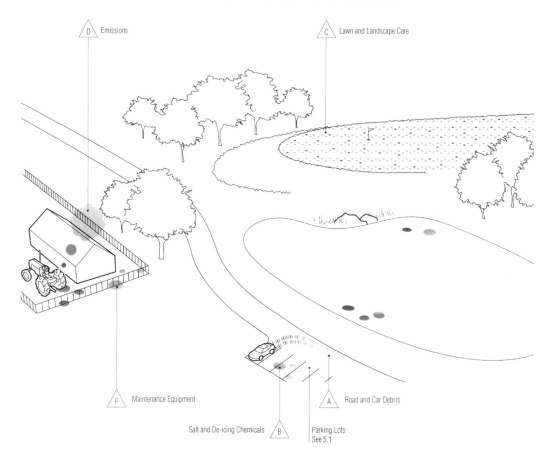

254

Key	Source	Description	Contaminants
A	Road and car debris	See 5.1: Roadways/Parking Lots	Nutrients (phosphorus and nitrogen): Ch. 3, p. 125 ■ Metals: Ch. 3, p. 136 ● Petroleum: Ch. 3, p. 65 Air pollution: Ch. 3, p. 189
B	Salt and de-icing chemicals	See 5.1: Roadways/Parking Lots	Salt (sodium, chloride and other additives): Ch. 3, p. 179
D	Emissions	Release of chemicals and particles into the air from maintenance equipment and automobiles. Emissions include particulate matter, partially combusted petroleum and metals. Emissions may settle out as solids in open water bodies.	Air pollution: Ch. 3, p. 189
E	Lawn and landscape care	Fertilizers and pesticides applied during maintenance can migrate into water supplies and build up in soil. Pesticides, including insecticides, fungicides and herbicides usually include metals and sometimes salts, which build up on sites over time. Applied fertilizers may quickly leach into soils and water, causing excessive nutrient loading in local surface-water bodies and groundwater.	● Pesticides and ● POPs: Ch. 3, pp. 111 and 118 ■ Metals: Ch. 3, p. 136 Nutrients (phosphorus and nitrogen): Ch. 3, p. 125
F	Maintenance equipment	Lawn maintenance equipment such as mowers, tractors and trailers generates emissions and fuel leaks. In addition, accidental spills of fertilizer and pesticides can occur when maintenance equipment is filled. Repeat occurrences of spills in the same area may create hot spots of pollutants.	● Petroleum: Ch. 3, p. 65 ● Pesticides and ● POPs: Ch. 3, pp. 111 and 118 ■ Metals: Ch. 3, p. 136 Nutrients (phosphorus and nitrogen): Ch. 3, p. 125

Figure 5.2b Parks, Open Spaces, Lawns and Golf Courses: Phytotypologies to Address Contaminants

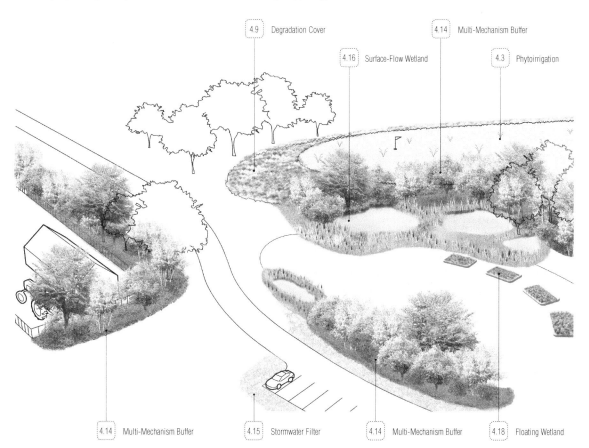

4.9   Degradation Cover   4.14   Multi-Mechanism Buffer

4.16   Surface-Flow Wetland   4.3   Phytoirrigation

4.14   Multi-Mechanism Buffer   4.15   Stormwater Filter   4.14   Multi-Mechanism Buffer   4.18   Floating Wetland

Key	Typology	Description	Addresses	Plant Lists
	Organically Maintained Landscape	Eliminate source pollutants wherever possible. Organic pesticides tend to be less toxic, and slow-release fertilizers may prevent migration into surrounding water and soils.	Nutrients; ● Pesticides; ■ Metals	
4.3 (p. 207)	Phytoirrigation	Groundwater around parks and golf courses can become polluted with excessive levels of nutrients. This nutrient-rich water can be pumped up to the surface and irrigated onto existing golf courses to provide fertilizer-enriched water. The grasses use some of the nutrients in the irrigated water, allowing cleaner water to pass back down to the groundwater. Solar-powered pumps for irrigation can be considered to reduce energy consumption for the irrigation operation.	Within groundwater: Nutrients (phosphorus and nitrogen)	Irrigate onto any plant species preferring high macro-nutrient levels
4.9 (p. 222)	Degradation Cover	Perennials with thick, fibrous root systems can be used along the edges of the golf course to trap and remove excess nutrients and pesticides before they migrate into stormwater or groundwater. High biomass-producing species tend to remediate nutrients at the highest rates.	Nutrients (phosphorus and nitrogen); ● Pesticides	p. 46, 2.17 Nutrients: high-biomass species p. 113, 3.23 Pesticides
4.14 (p. 234)	Multi-Mechanism Buffer	The highest concentrations of pesticides and nutrient run-off from sites occurs during the first rain event after application or during prolonged periods of irrigation (Smith and Bridges, 1996). Heavily vegetated buffers around golf courses and other lawn areas may prevent the spread of pollutants. Dust particles from pesticides may be intercepted with these buffers.	Nutrients (phosphorus and nitrogen); ■ Metals; ● Petroleum; Air pollution: particulate matter	p. 46, 2.17 Nutrients: high-biomass species p. 113, 3.23 Pesticides p. 195, 3.60 Air pollution
4.15 (p. 235)	Stormwater Filter	Around impervious surfaces where stormwater becomes contaminated run-off, provide stormwater filters. Where de-icing activities occur, salt-tolerant species must be utilized; salt is typically not removed in these systems.	Within stormwater: Nutrients (phosphorus and nitrogen); ■ Metals; ● Petroleum	Stormwater/Wetland species not included in this publication
4.16 (p. 238)	Surface-Flow Wetland	Where open water bodies exist on site, create Surface-Flow Wetlands at the edges to filter out pollutants in a series of open water ponds and cells. With proper design, nitrogen may be returned to the atmosphere as a gas and phosphorus can be bound to soils.	Within water: Nutrients (phosphorus and nitrogen); ● Pesticides; ● Petroleum; ■ Metals	Stormwater/Wetland species not included in this publication
4.18 (p. 242)	Floating Wetland	Where open water bodies exist on site, Floating Wetlands can be placed on the surface to help extract and degrade pollutants that have leached into the water body.	Within water: Nutrients (phosphorus and nitrogen); ● Pesticides; ● Petroleum; ■ Metals	Stormwater/Wetland species not included in this publication

5.3 River corridors and greenways

The river corridor and corresponding greenway is a special type of landscape element that is found as part of the living infrastructure of regions, cities and local landscapes. The typology consists of linear sections of planted soft banks and grassed edges, and transitions from land to a watercourse of varying width, depth and character. A greater variety of vegetation may thrive along these corridors, as compared to the landscapes beyond. This variety may range from park-like swatches of grass-covered banks dotted with mature canopy trees and shrubs, to river-edge buffers and emergent plants.

Consistent with every river corridor is the role that the watercourse plays in channeling pollutants that are discharged from local industries, roadways, overflow pipes, illegal discharges of chemicals and waste products, and general atmospheric emissions. Contamination may be found in the water bodies themselves through discharges of polluted groundwater and from adjacent seeping soils. Many older communities contain combined sewer overflow systems (CSO) that can discharge human waste materials into rivers during storm events. Continual stormwater discharges create an influx of phosphorus, nitrogen, heavy metals, hydrocarbons and pesticides from roads and adjoining land uses including agricultural fields. The corridors are, unfortunately, also the venue of continuous illegal waste dumping along their length that can include metals such as lead, paint and construction waste materials and hydrocarbons. Finally, the range of emissions that can be adjacent to river corridors includes airborne particulate matter and metals and all six of the priority air pollutants identified by the US EPA (see Chapter 3, p. 189). Atmospheric deposition can cause these pollutants to appear within the water column of the river corridor and in the surrounding greenway.

256

Contaminant source control

The most important consideration for cleanup in these environments is to stop the point source and non-point source pollutants entering the corridor. Disconnecting CSOs from local waterways, or at least minimizing the amount of stormwater entering these systems, is a critical improvement. Industries must be regulated and dumping activities monitored to ensure prevention of future pollutant releases.

5.4 Railroad corridors

Collectively, railroad corridors make up a significant amount of the post-industrial land area found in urban and ex-urban locations. As a linear open space system of passenger and freight train lines and storage yards, all threaded through varied landscapes, railroad corridors connect other post-industrial sites along their routes, creating a network of adjoining smaller sites of differing land uses, histories and pollutant paths. A typical railroad corridor or right of way (ROW) consists of three major elements in the width and extent of the corridor.

The first is the flat, layered rail bed, comprising a lower ballast with top coat of aggregate which support a surface system of precast concrete or wooden ties upon which lie the continuous metal rail sections. Historically, wooden railroad ties (also called sleepers) were preserved with creosote, a form of coal tar that is a difficult form of petroleum to degrade, or with copper arsenate, which leaches arsenic into the soil surrounding the sleepers. The ballast, created from gravel or urban fill, can contain heavy metals and PAH hydrocarbons such as coal ash.

The second element is the adjacent sloped banks on either side, containing drainage swales and a variety of planted ground surfaces, gradient changes, edge conditions, pathways and boundary fences and walls. These areas often contain traces of adjacent land uses where loading or unloading of raw materials or manufactured goods may have occurred.

The third element is volunteer vegetation and support soils. Historically, herbicides have been used to maintain a clear, safe corridor for the passage of trains, and traces of heavy metals and salts can be found in the wake of their use. Railroad corridors are also supported in the ROW by a large number of switching boxes, electrical panels, signals, viaducts, bridges, crossings and rail station infrastructure, which can leach POPs and PCBs (from transformers), chlorinated solvents, petroleum products and heavy metals.

The most common pollutant in railway corridors is PAHs (Ciabotti, 2004). These are often generated by both the creosote, noted before, and the train operation itself, with most PAHs being derived from the incomplete burning of fuel sources such as coal, oil or wood. As an example, sooty exhaust from diesel engines can settle along the tracks, creating PAH deposits. Lead and mercury contamination are also often found in historic emissions from diesel combustion. Contamination from emissions is correlated to the distance from the track, with the most heavily contaminated zone being 10 meters (32 feet) or less away from the track, a moderate pollution zone 10–50 meters (32–165 feet) away and a slightly polluted zone 50–100 meters (165–330 feet) away (Ma et al., 2009). The trains also emit heavy metals from operations and from old, flaking paint surfaces (Pb, Cd, Cu, Zn, Hg, Fe, Co, Cr, Mo), combined with the PAHs in the form of fuel oils, combustion and exhaust, lubricant oils, condenser fluids, machine grease and transformers oils (Wilkomirski et al., 2011). Concentrations of contaminants tend to be higher in rail yards and stations, where cars sit idle and slow leaks can occur over time (Wilkomirski et al., 2011).

Outside of the ROW, over time a continuous series of complementary land uses will have grown up along the railroad corridor. A heterogeneous mix of contaminants can exist where transfer siding tracks were present, not to mention the potential for contamination from the adjacent land uses themselves.

The railroad corridor offers a range of opportunities for individual phytotechnology site applications, as well as repetitive remediation opportunities along its continuous length. While the typology in this section illustrates still-active corridors, planting typologies for cleanup can also be integrated into railroad corridor recreation-conversion projects, such as rails to trails.

257

Figure 5.3a River Corridors and Greenways: Sources of Contamination

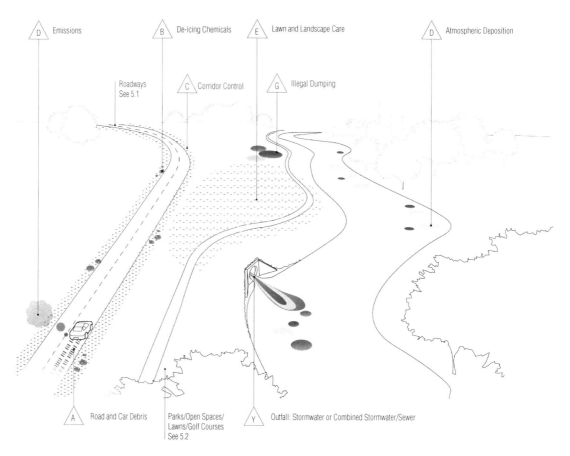

258

Key	Source	Description	Contaminants
A	Road and car debris	See 5.1: Roadways/Parking Lots	Nutrients: Ch. 3, p. 125 ■ Metals: Ch. 3, p. 136 ● Petroleum: Ch. 3, p. 65 Air pollution: Ch. 3, p. 189
B	Salt and de-icing chemicals	See 5.1: Roadways/Parking Lots	Salt (sodium, chloride and other additives): Ch. 3, p. 179
D	Emissions/atmospheric deposition	Release of chemicals and particles into the air from automobiles and industrial sources. Rivers can be sinks for contaminated air particles that have settled out and deposited onto water surfaces.	Air pollution: Ch. 3, p. 189
E	Lawn and landscape care	Maintenance of riverside pathways and adjacent parklands, See 5.2: Parks/Open Spaces/Lawns/Golf Courses.	● Pesticides and ● POPs: Ch. 3, pp. 111 and 118 ■ Metals: Ch. 3, p. 136 Nutrients: Ch. 3, p. 125
G	Illegal dumping/buried debris and waste	River corridors are often back yards for current or former polluting industrial operations. Illegal dumping and industry overspills should be expected, although pinpointing hot spots of contamination may be challenging.	All contaminants possible – see Ch. 3
Y	Outfall: stormwater or combined sewer/stormwater (CSO)	Many stormwater drains from adjacent roadways and parcels directly discharge to river corridors. Combined stormwater and sewer systems are common in older urban areas and may overflow into adjacent waterways during large storm events. Significant releases of nutrients, metals and other biological pollutants are common with CSOs.	Nutrients Ch. 3, p. 125 ■ Metals: Ch. 3, p. 136 ● Petroleum: Ch. 3, p. 65 Air pollution: Ch. 3, p. 189 Total Suspended Solids (TSS) Bacteria, BOD and living organisms

Figure 5.3b River Corridors and Greenways: Phytotypologies to Address Contaminants

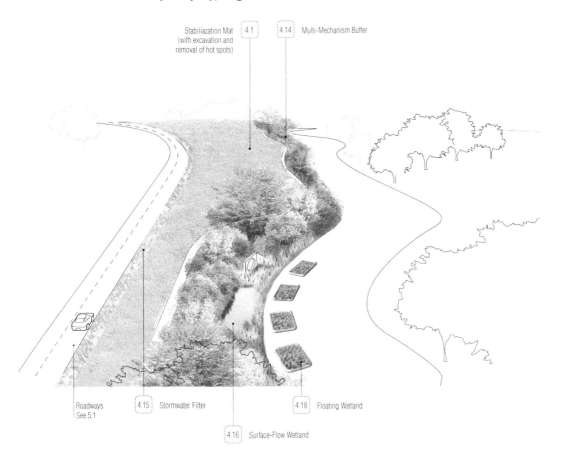

Stabiliazation Mat
(with excavation and
removal of hot spots)   [4.1]   [4.14]   Multi-Mechanism Buffer

Roadways   [4.15]   Stormwater Filter   [4.18]   Floating Wetland
See 5.1

[4.16]   Surface-Flow Wetland

259

Key	Typology	Description	Addresses	Plant Lists
4.1 (p. 202)	Stabilization Mat	Due to the unknown nature of river-bank soils and buried materials, vegetative capping to hold pollutants on site, away from risk of exposure to humans and wildlife, can be an effective option. Where hot spots of contamination are known, excavation and removal of heavily polluted soils should occur before remaining soils are stabilized.	■ Metals Air pollution (deposited)	p. 140, 3.37 Metals excluders
4.14 (p. 234)	Multi-Mechanism Buffer	Riparian Buffers along river corridors prevent pollution migration from adjacent land uses. In addition to contamination removal, these vegetated areas can serve important wildlife corridor functions.	All: see Ch. 3 ◐ ● ■ ● ● ● ■ ■	All: see Ch. 3 (Select plants to address specific adjacent land uses)
4.15 (p. 235)	Stormwater Filter	On the downhill side of paved surfaces and compacted landscapes, stormwater filters can be used to filter out the pollutants closer to the source before they compound in river systems.	Within stormwater: Nutrients (phosphorus and nitrogen); ■ Metals; ● Petroleum	Stormwater/Wetland species not included in this publication
4.16 (p. 238)	Surface-Flow Wetland	Polluted outfall waters and contaminated river water can be directed through Surface-Flow Wetlands constructed along the edges of rivers to filter out pollutants.	Within water: Nutrients (phosphorus and nitrogen); ● Pesticides; ● Petroleum; ■ Metals; BOD/Bacteria	Stormwater/Wetland species not included in this publication
4.18 (p. 242)	Floating Wetland	Place Floating Wetlands on the surface of rivers to act as filters as polluted water passes through. In addition, plants can be harvested from these systems, composted and nutrients recycled.	Within water: Nutrients (phosphorus and nitrogen); ● Pesticides; ● Petroleum; ■ Metals	Stormwater/Wetland species not included in this publication

Figure 5.4a Railroad Corridors: Sources of Contamination

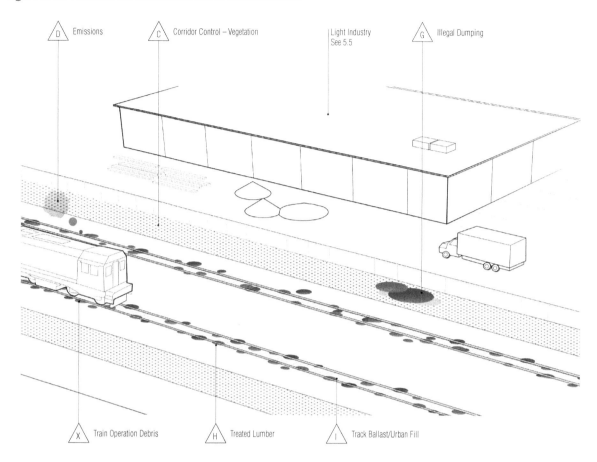

Key	Source	Description	Contaminants
C	Corridor control: vegetation	Herbicides are often utilized along rail corridors to control vegetation growth. They often include metals (most typically lead and arsenic) and salts, which build up on sites over time in addition to the pesticides themselves.	● Pesticides and ● POPs: Ch. 3, pp. 111 and 118 ■ Metals: Ch. 3, p. 136 ▨ Salts: Ch. 3, p. 179
D	Emissions	Release of emissions into air from train engines and partial fuel combustion.	Air pollution: Ch. 3, p. 189
G	Illegal dumping/buried debris and waste	Rail corridors often link heavily polluting industrial operations, and overtopping of freight cars and accidental spills can result in buried debris and waste. Adjacent land-use activities where on- and off-loading occurred can often provide insight into what contaminants may be encountered.	● Petroleum: Ch. 3, p. 65 ● POPs: Ch. 3, p. 118 ■ Metals: Ch. 3, p. 136 ▫ Nutrients Ch. 3, p. 125
H	Treated lumber	Railroad ties, historically treated with creosote or copper arsenate to prevent rotting, may be found under current or historic tracks. Many historic railroad beds still have creosote ties or debris present, and arsenic levels 10x natural background levels (Ma et al., 2009).	● Petroleum: creosote: Ch. 3, p. 65 ■ Metals: arsenic and copper, Ch. 3, p. 136
I	Track ballast/urban fill	The base of track areas is often built up with urban fill, often including coal ash. Ballast may contain metals.	● Petroleum, especially PAHs: Ch. 3, p. 65 ■ Metals: Ch. 3, p. 136
X	Train operation debris	Operation of train engines and cars, especially brake dust, can release petroleum products, lubricants and metals.	● Petroleum-PAHs: Ch. 3, p. 65 ■ Metals: Ch. 3, p. 136

Figure 5.4b Railroad Corridors: Phytotypologies to Address Contaminants

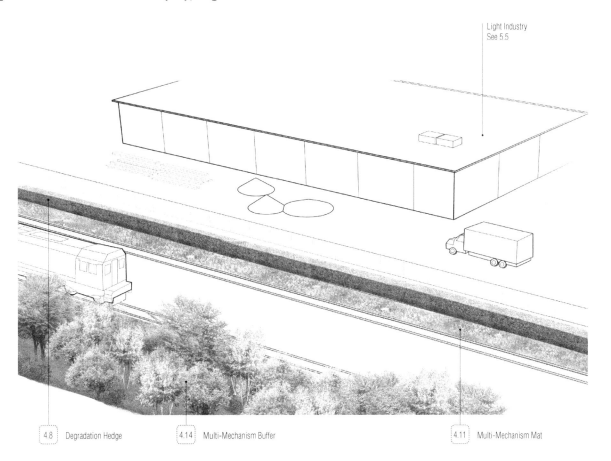

Light Industry
See 5.5

4.8 Degradation Hedge

4.14 Multi-Mechanism Buffer

4.11 Multi-Mechanism Mat

261

Key	Typology	Description	Addresses	Plant Lists
4.8 (p. 220)	Degradation Hedge	To buffer adjacent land uses from surface soil contaminants in rail ROWs, add degradation hedges to remove hydrocarbons or chlorinated solvents. Since only a thin line of vegetation may be possible, the concentration of contaminants may not be totally degraded, but some remediation may be possible. Hedges may be pruned or left to grow naturally.	● Petroleum; ● Chlorinated solvents	p. 74, 3.5 Petroleum p. 97, 3.13 Chlorinated solvents
4.11 (p. 227)	Multi-Mechanism Mat	Between and around tracks it may be possible to install low-growing vegetation to stabilize, degrade or even extract existing pollutants on site. Degradation of PAHs should be provided, potentially some extraction/ harvesting for arsenic and stabilization of other non-bioavailable contaminants by creating a thick vegetation mat to control pollution dispersal via erosion. A yearly mowing regime where clippings are collected and removed from site can slowly remove bioavailable metals from sites over time. A research program called "Green Tracks" is being developed in Berlin, in which tolerant plant species are planted between rails to capture and hold and potentially degrade pollutants as they are emitted by trains (Gorbachevskaya et al., 2010).	● Petroleum: PAHs; ● Pesticides: herbicides; ■ Metals; Air pollution (deposited)	p. 140, 3.37 Metals excluders p. 74, 3.5 PAHs: degradation p. 146, 3.38 Arsenic: extraction
4.14 (p. 234)	Multi-Mechanism Buffer	Heavily vegetated buffers along rail corridors can prevent pollution migration to adjacent land uses. Organic contaminants such as prevalent PAHs can potentially be degraded, while inorganics such as lead and arsenic are captured and held in soils. Particulate matter in air pollution may be captured on leaf surfaces, buffering adjacent lands. In addition to contamination remediation, these vegetated areas can serve important wildlife corridor functions.	All: see Ch. 3 ●●■ ●●●■	All: see Ch. 3 (Select plants to address specific adjacent land uses)

5.5 Light industrial and manufacturing sites

Light industrial sites contain a complex mixture of building structures, in-ground infrastructure, storage structures, haulage areas and fuel areas to support a full range of manufacturing activities. Light industrial sites differ from heavy industrial locations which have on-site activities such as steel-making, shipbuilding and metal-refining factories, by using partially processed materials that are shipped to the site to be further worked and assembled. Finished goods such as clothing, prepared food, household appliances, furniture and electronics for domestic and overseas markets are produced on light industrial sites. In addition, these sites have a less intense impact on the environment than do heavy industrial activities, which often cause significant groundwater and waste contamination over longer periods of time. Light industrial sites require a relatively small amount of raw materials, factory size and power usage. While light industry can cause relatively little pollution, particularly when compared to heavy industries, some light industrial sites can cause significant risk of contamination through their soils, sediments and groundwater. Electronics manufacturing can create potentially harmful levels of lead and other metal or chemical wastes in soil, due to improper handling of solder and waste products such as cleaning and degreasing agents used for machinery and factory equipment. The types of activities on light industrial sites change often, as manufacturing cycles change. It is therefore likely that over time a site in this category has had multiple occupancies with a range of often unrelated manufacturing activities with varying types and levels of contaminants.

262

Some common contaminants on many of these sites are leaky underground or above-ground storage tanks with fuels (petroleum) or chlorinated solvents; air conditioning or heating units (chlorinated solvents); transformers leaking PCBs; road debris from shipping trucks (see 5.1); and emissions.

5.6 Gas stations and auto-repair shops

Gas stations are ubiquitous across the North American landscape, creating the physical network required to support an American lifestyle heavily reliant on the automobile. Consequently, gas stations occupy key real estate at highly visible street corners and transportation intersections. Because of their location, as well as the high probability of their contamination, abandoned gas stations are both highly visible and highly contaminated sites whose remediation costs can hinder redevelopment.

Internationally, there have been efforts to reduce contamination from gas stations either through the process of modifying the gas station infrastructure or through legislation regulating fuel additives such as lead and MTBE. Most filling stations, wherever they are located in the world, are built in a similar manner, with most of the fueling infrastructure underground and pumping machines in the forecourt (the part of the filling station where vehicles are refueled). Many of the facilities also house garage and auto-repair facilities with storage and disposal of degreasers, waste oils and air conditioning fluids.

According to the US EPA, the three main sources of contaminant release on gas station sites are "product delivery piping failures, corrosion of unprotected tanks, and spills and overfills" (US EPA Office of Underground Storage Tanks and Office of Brownfields and Land Revitalization, 2010). The US EPA warns that gasoline vapors getting into the air is also a major concern, resulting from both normal operation and spills that vaporize.

Fuel storage

Single or multiple fuel storage tanks are typically located underground. Pre-1998, the majority of gas station contamination events in the US were due to Leaky Underground Storage Tanks (LUSTs). Underground storage tanks (USTs) were required to comply with new federal US standards to provide more durable corrosion resistant tanks by 1998. For this reason, many smaller locally owned gas stations closed in the late 1990s, due to their inability to comply with the new double-lined tank and other environmental regulations. Many of the old underground tanks are still in place and potentially leaking. Even those USTs up to federal standards still experience leaks. As of 2013, 436,406 of 514,123 reported LUSTs have been cleaned up, with 77,717 releases remaining (US EPA, 2013).

Fuel transfer and pumping

Fuel is usually offloaded from a tanker truck into the tanks through a valve located on the filling station's perimeter. Fuel spills in this location can occur during filling. Fuel from the storage tanks travels to the retail pumps through underground pipes located in a service channel that usually connects the forecourt and the tanks. Fuel tanks, dispensers and nozzles used to fill car tanks often employ vapor-recovery systems which prevent releases of vapor into the atmosphere. The area around the fuel dispensers has a drainage system, since fuel sometimes spills on the ground. Any liquids present on the forecourt will flow into a channel drain before they enter a petroleum interceptor, which is designed to capture any pollutants and filter them from rain-water.

Gas additives

Leaded gas was used in gas stations until the 1970s, before being phased out, due to concerns about lead poisoning, as well as the advent of the catalytic converter and new gasoline additives. Leaded gasoline historically moved into soils and groundwater from spills and leaks in USTs, and also was released into the atmosphere during combustion, as emissions. High lead concentrations can still be found in soils around gas stations, toll booths and other areas where cars repeatedly idled and emitted fumes before the 1970s. Currently MTBE and other additives which replaced lead in gasoline are of concern, since they can quickly migrate into groundwater when spilled.

In addition, buried debris and waste disposal from any service garages on site is of concern. Auto-repair shops are associated with a number of chlorinated solvents such as methylene chloride and tricholoroethylene, which are used as paint removers, brake oil cleaner and degreasers, as well as CFC recharge for air conditioners. Shop owners or workers may release chlorinated solvents into soil and waterways through illegal dumping or accidental spills.

263

Figure 5.5a Light Industry: Sources of Contamination

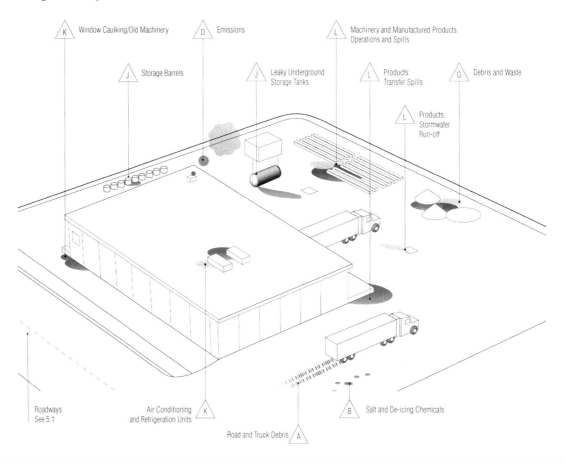

Key	Source	Description	Contaminants
A	Road and truck debris	See 5.1: Roadways/Parking Lots	Nutrients: Ch. 3, p. 125 ■ Metals: Ch. 3, p. 136 ● Petroleum: Ch. 3, p. 65 Air pollution: Ch. 3, p. 189
B	Salt and de-icing chemicals	See 5.1: Roadways/Parking Lots	Salt (sodium, chloride and other additives): Ch. 3, p. 179
D	Emissions	Release of chemicals and particles into the air from industrial operations and trucking.	Air pollution: Ch. 3, p. 189
G	Illegal dumping/buried (or unburied) debris and waste/trash overspills	Both accidental and intentional spills, buried debris and waste can be found on light industrial sites. The historic and current industries on site should be investigated to determine what contaminants may be encountered.	Anything is possible and depends upon the current and past industrial use. The most common include the following: ● Petroleum: Ch. 3, p. 65 Chlorinated solvents: Ch. 3, p. 94 ■ Metals: Ch. 3, p. 136
J	Storage tanks	Any underground or above-ground storage tank or container used to store fuel or industrial products can leak or rupture.	● Petroleum: Ch. 3, p. 65 Chlorinated solvents: Ch. 3, p. 94
K	Air conditioning and refrigeration units/window caulking/old machines	Air conditioning units or refrigeration units, especially older or decommissioned units, can leak coolants. Historic caulking on buildings or old transformers and machines can be laden with PCBs.	Chlorinated solvents (CFCs and freon): Ch. 3, p. 94 ● POPs (PCBs): Ch. 3, p. 118
L	Machinery and manufactured products: operations and spills	On-site machinery and fabrication lines can leak fuel, lubricants, coolants and solvents. In addition, if the machines require chemical inputs, spills and overtopping can occur during the transfer. Areas directly under machinery and areas where chemicals are transferred or disposed during the industrial process should be considered as potentially polluted sites.	Anything is possible and depends upon the current and past industrial use. The most common include the following: ● Petroleum: Ch. 3, p. 65 Chlorinated solvents: Ch. 3, p. 94 ■ Metals: Ch. 3, p. 136

Figure 5.5b Light Industry: Phytotypologies to Address Contaminants

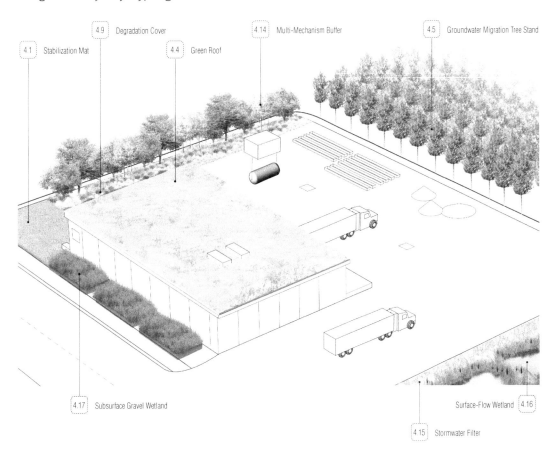

4.9	Degradation Cover
4.1	Stabilization Mat
4.4	Green Roof
4.14	Multi-Mechanism Buffer
4.5	Groundwater Migration Tree Stand
4.17	Subsurface Gravel Wetland
4.16	Surface-Flow Wetland
4.15	Stormwater Filter

265

Key	Typology	Description	Addresses	Plant Lists
4.1 (p. 202)	Stabilization Mat	Where highly recalcitrant pollutants exist, vegetation with thick, fibrous root systems can be used to hold pollutants on site. Thickly established vegetation may prevent erosion and migration of pollution.	● POPs, including PCBs; ■ Metals	p. 140, 3.37 Metals excluders
4.4 (p. 210)	Green Roof	Large, flat, industrial buildings are ideal locations for Green Roofs. The primary purpose is to evapotranspire water, preventing generation of stormwater that would otherwise run through the site, and pick up contaminants.	Stormwater vector	Green Roof species not included in this publication.
4.5 (p. 213)	Groundwater Migration Tree Stand	If contaminants leach into groundwater, consider tree stands downgradient of the plume to naturally pump up the water, degrading organic contaminants and/or filtering out inorganic pollutants. A detailed Water Mass Balance must be completed by an engineer to calculate how many trees will be needed to make the plume capture effective.	Groundwater vector: especially ● Petroleum and ◐ Chlorinated solvents	p. 45, 2.15 Phreatophytes
4.9 (p. 222)	Degradation Cover	Petroleum and chlorinated solvents may be able to be degraded through targeted plantings, especially when done soon after spills. This may be effective especially where old fuel tanks and barrels of organic liquids were stored.	● Petroleum; ◐ Chlorinated solvents	p. 74, 3.5 Petroleum p. 97, 3.13
4.14 (p. 227)	Multi-Mechanism Buffer	Heavily vegetated buffers along property lines can prevent pollution migration to adjacent land uses. Organic contaminants such as fuels and chlorinated solvents can potentially be degraded, while inorganics such as metals can be captured and held in soils. Particulate matter in air pollution can be filtered out onto leaf surfaces of select species. These vegetated areas can additionally serve important wildlife corridor functions.	All: see Ch. 3 ●◐■ ●●◐■◐	All: see Ch. 3 (Select plants to address specific adjacent land uses)
4.15 (p. 235)	Stormwater Filter	Impervious surfaces prevail on industrial sites, and stormwater filters can be installed downgradient to filter out pollutants. Wherever possible, stormwater generated from industrial uses should be disconnected from city collection systems and treated on site. In stormwater filters, inorganic contaminants are filtered in the soil media, while organic contaminants may be degraded. Where de-icing activities occur, salt-tolerant species must be utilized; salt is not typically removed in these systems.	Within stormwater: Nutrients; ■ Metals; ● Petroleum; ◐ Chlorinated solvents	Stormwater/ Wetland species not included in this publication
4.16 and 4.17 (pp. 241 and 242)	Surface-Flow and Subsurface Gravel Wetland	Both manufacturing wastewater and site stormwater can be directed into Surface-Flow (open water) and Subsurface (gravel) Constructed Wetlands to remove pollutants. With proper design, contaminants may be mitigated. Where de-icing activities occur, salt-tolerant species must be utilized; salt is not typically removed in these systems.	Within wastewater and stormwater: Nutrients; ■ Metals; ● Petroleum; ◐ Chlorinated solvents	Stormwater/ Wetland species not included in this publication

Figure 5.6a Gas Stations and Auto-Repair Shops: Sources of Contamination

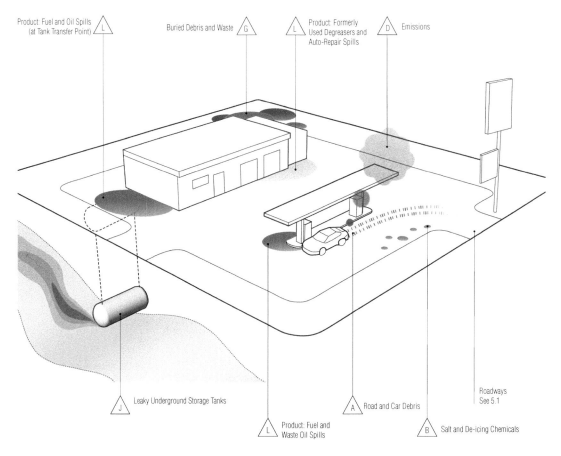

Product: Fuel and Oil Spills
(at Tank Transfer Point) L

Buried Debris and Waste G

Product: Formerly
Used Degreasers and L
Auto-Repair Spills

Emissions D

266

Leaky Underground Storage Tanks J

Product: Fuel and L
Waste Oil Spills

Road and Car Debris A

Salt and De-icing Chemicals B

Roadways
See 5.1

Key	Source	Description	Contaminants
A	Road and car debris	See 5.1: Roadways/Parking Lots	Nutrients: Ch. 3, p. 125 ■ Metals: Ch. 3, p. 136 ● Petroleum: Ch. 3, p. 65 Air pollution: Ch. 3, p. 189
B	Salt and de-icing chemicals	See 5.1: Roadways/Parking Lots	Salt (sodium chloride and other additives): Ch. 3, p. 179
D	Emissions	Release of chemicals and particles into air from automobiles. Lead in surrounding soils is common at gas stations, due to historic leaded gas emissions.	Air pollution: Ch. 3, p. 189
G	Illegal dumping/buried (or unburied) debris and waste	On older gas station sites, buried debris and waste can sometimes be found in back corners of sites, behind the buildings, including lead batteries, old car parts and paint-stripping and degreasing products made from chlorinated solvents.	● Petroleum: Ch. 3, p. 65 ■ Metals: Ch. 3, p. 136 Chlorinated solvents (paint stripping/air conditioning): Ch. 3, p. 94
J	Leaky underground storage tanks (LUSTs)	LUSTs are the most common source of contamination releases at gas stations and can impact around water.	● Petroleum: Ch. 3, p. 65
L	Product: fuel and oil spills, formerly used degreasers	Fuel spills are frequent at transfer points for delivery trucks and retail fueling pumps. In addition, auto-repair shops generate spills, including oil and air conditioning fluids. Historic locations of spills may be polluted with lead from past use of leaded gasoline. In addition, chlorinated solvents were historically often used as degreasers in autobody repair shops.	● Petroleum: Ch. 3, p. 65 Chlorinated solvents: Ch. 3, p. 94 ■ Lead: Ch. 3, p. 172

Figure 5.6b Gas Stations and Auto-Repair Shops: Phytotypologies to Address Contaminants

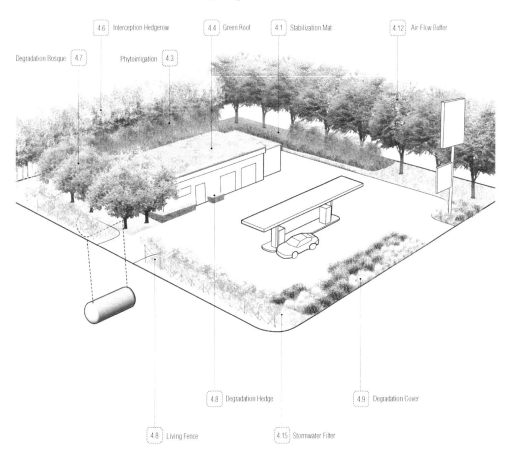

4.6 Interception Hedgerow   4.4 Green Roof   4.1 Stabilization Mat   4.12 Air-Flow Buffer

Degradation Bosque 4.7   Phytoirrigation 4.3

4.8 Degradation Hedge   4.9 Degradation Cover

4.8 Living Fence   4.15 Stormwater Filter

Key	Typology	Description	Addresses	Plant Lists
4.1 (p. 202)	Stabilization Mat	Where old buried waste exists, vegetation can be used to hold pollutants on site, away from risk of exposure. Thickly established vegetation can prevent erosion and migration of pollutants.	■ Metals	p. 140, 3.37 Metals excluders p. 86, 3.5 Petroleum tolerant
4.3 (p. 207)	Phyto-irrigation	Polluted groundwater at gas stations can be pumped up and irrigated onto gas station plantings targeted to degrade petroleum/chlorinated solvents. Solar-powered pumps for irrigation can be considered to reduce energy consumption for the irrigation operation.	● Petroleum; ◐ Chlorinated solvents	p. 74, 3.5 Petroleum p. 97, 3.13 Chlorinated solvents
4.4 (p. 210)	Green Roof	Green Roofs can be installed to evapotranspire water, preventing generation of stormwater that would otherwise run over paved areas and migrate contaminants.	Stormwater vector	Green Roof species not included in this publication.
4.6 (p. 216)	Interception Hedgerow	Downgradient of LUSTs and auto-repair service bays, Interception Hedgerows can be created to degrade gasoline in groundwater. Since only a thin line of vegetation along the property line may be possible, the concentration of contaminants may not be totally degraded, but some remediation may be possible.	● Petroleum	p. 74, 3.5 Petroleum p. 97, 3.13 Chlorinated solvents
4.7 (p. 218)	Degradation Bosque	Degradation Bosques can be installed around transfer points to break down freshly spilled fuel before it contaminates groundwater.	● Petroleum	p. 74, 3.5 Petroleum
4.8 (p. 220)	Degradation Hedge	Around edges of auto-repair shops, Degradation Hedges can break down fuel and chlorinated solvent spills that may leach out from the inside of the garage.	● Petroleum; ◐ Chlorinated solvents	p. 74, 3.5 Petroleum p. 97, 3.13 Chlorinated solvents
4.8 (p. 220)	Living Fence	At the property line, attractive living fences can be installed to break down fuel spills generated uphill of the plantings.	● Petroleum; ◐ Chlorinated solvents	Use *Salix spp.*
4.9 (p. 222)	Degradation Cover	Fuel spills washed into adjacent landscape areas can be broken down with targeted plantings. Many ornamental grasses with large root systems and petroleum-degradation capabilities can create attractive entry landscapes. Vegetation must be thickly planted and maintained with no visible mulch, to maximize root-zone coverage.	● Petroleum; Nitrogen	p. 74, 3.5 Petroleum
4.12 (p. 229)	Air-Flow Buffer	Heavily vegetated buffers along property lines can prevent migration of pollution from adjacent land uses. Particulate matter in air pollution can be filtered out and trapped on leaf surfaces of select species to keep pollutants on site.	Air pollution	p. 195, 3.59–3.60 Air pollution
4.15 (p. 235)	Stormwater Filter	Stormwater filters can be installed downgradient of impervious surfaces to remove pollutants. Wherever possible, stormwater generated from gas stations should be disconnected from city collection systems and treated naturally on site.	Within stormwater: Nutrients; ■ Metals; ● Petroleum; ◐ Chlorinated solvents	Stormwater/Wetland species not included in this publication

5.7 Dry cleaners

Modern dry cleaning uses non-water-based solvents to remove soil and stains from clothing and fabrics. Early dry cleaners used a greater variety of solvents, including gasoline and kerosene. This required a larger industrial-complex building structure and infrastructure to be constructed, with storage, delivery and waste disposal areas for chemicals, oils and solvents. Since World War II, the use of chlorinated solvents carbon tetrachloride and trichloroethylene (TCE) have given way to perchloroethylene (Perc, or PCE), which became the predominating solvent of choice within the industry. Perc systems required smaller equipment, less floor space and could be installed in retail locations. As a result of this innovation, today the majority of clothes are cleaned by Perc. Some dry cleaners promoting 'green' methods typically substitute glycol ethers for the Perc. The primary source of contamination in dry-cleaning facilities is equipment failures and equipment operations, including removal or replacement of parts. Together these sources account for about two-thirds of the cases of contamination (Linn et al., 2004).

Perc can quickly pollute groundwater when it is spilled and leached into soils on a site, and can quickly volatize into the air, releasing VOCs that can negatively affect air quality. When TCE or PCE are in groundwater below the surface of buildings, these chlorinated solvents continually migrate from the groundwater into a gas form that can penetrate building floors, affecting indoor air quality. The US EPA has required elimination of any ground-level Perc dry cleaners within predominantly residential buildings by 2020 because of air quality concerns in residential units.

268

5.8 Funeral homes and graveyards

These land uses are one of the least-considered sites for potential contamination, although they traditionally have substantial prospective pollutants. Together they represent a largely invisible source of pollution and yet, because of varied cultural norms and the delicate nature socially and culturally of the land use, this has not been a topic for discussion and research. Funeral homes, as a light industry, use potentially carcinogenic 'occupational liquids' such as formaldehyde and embalming materials. Graveyards function less as resting grounds than as specialized landfills for the materials of embalming and encasement. The typical cemetery ground, for example, contains enough coffin wood to construct more than 40 homes, 900-plus tons of casket steel and 20,000 tons of vault concrete (Harris, 2008). Add to that enough embalming fluid to fill a small swimming pool and pesticides and herbicides to keep the graveyard preternaturally green, and you have a pollution mix similar to those found on other light industrial and manufacturing sites.

In older cemeteries, arsenic may be the longest-enduring contaminant. A highly toxic and powerful preservative, arsenic was a mainstay of early embalming solutions in the pre- and post-Civil War years. By 1910, so many embalmers had perished from their efforts to preserve the dead with arsenic that the federal government stepped in and banned its use in embalming solutions. Arsenic is less likely to taint the environs of newer graveyards. Elevated concentrations of copper, lead, zinc and iron, the metals used in casket construction, can still be discovered. In addition, formaldehyde is the main ingredient of practically all embalming solutions on the market today. Formaldehyde is nonetheless a human carcinogen, and because of its potentially toxic effect when released into the environment, the US EPA regulates it as a hazardous waste. The funeral industry, however, legally buries over three gallons of formaldehyde-based 'formalin' embalming solution every time it inters an embalmed body. Like the contents of any landfill, the embalmed body's toxic cache escapes from its host and eventually leaches into the environment, tainting surrounding soil and groundwaters. Cemeteries bear the chemical legacy of their embalmed dead well after the graves have been filled.

5.9 Urban residences

Another commonly overlooked site with pollutants in its exterior spaces is the urban residence. These landscape spaces, whether private gardens, paved yards, storage areas, parking areas or planting beds, are in direct daily contact with families, including children and seniors. Urban residences can be separate single-family structures or multiple apartment buildings standing side by side on smaller lots. While there is great variation of building types nationally and internationally, depending on location, climate, materials and cultural needs, the types of elements found on site are rarely as varied. The issue of the presence of lead in household paint was brought to light in the 1970s and its use was banned in 1978 in the United States. Areas of typical lead contamination in urban lots include the drip line around the edge of residences to about 3 feet (1 meter) from the face of the building. The soils here can contain flakes or particles of lead paint in the top horizons of the soils, up to 18 inches (45 centimeters) in depth. They may be mixed in with exterior wall shingles or old caulking that has hardened and fallen from doorjambs and windows. Older caulking contained PCBs, and asbestos in shingles is common in older homes. Regulatory authorities for removal and disposal. Finally, within the ground conditions of an urban lot the ground fill can contain lead, arsenic and PAHs as well as old buried domestic oil tanks that may be corroded or ruptured.

269

Figure 5.7a Dry Cleaners: Sources of Contamination

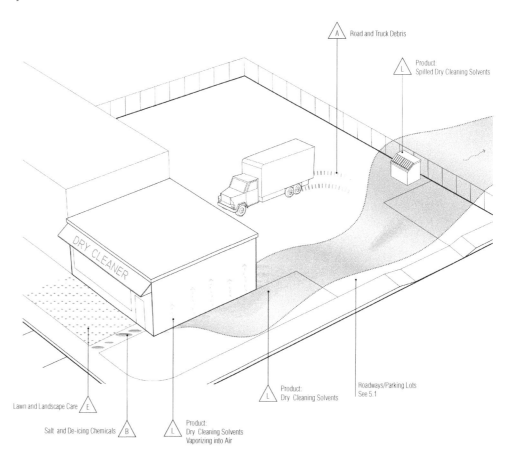

270

Key	Source	Description	Contaminants
A	Road and truck debris	See 5.1: Roadways/Parking Lots	Nutrients: Ch. 3, p. 125 ■ Metals: Ch. 3, p. 136 ● Petroleum: Ch. 3, p. 65 Air pollution: Ch. 3, p. 189
B	Salt and de-icing chemicals	See 5.1: Roadways/Parking Lots	Salt (sodium, chloride and other additives): Ch. 3, p. 179
E	Lawn and landscape care	See 5.2: Parks/Open Spaces/Lawns/Golf Courses	Nutrients: Ch. 3, p. 125 ● Pesticides: Ch. 3, p. 111 ■ Metals: Ch. 3, p. 136
L	Product: dry-cleaning solvents	Spilled dry-cleaning solvents can migrate quickly into groundwater or volatilize into harmful VOC air pollutants. Solvent contamination can often be found near the rear entry of dry cleaners, where delivery spills or historic 'dumping out the back door' occurred. In addition, areas around dumpsters where solvents were improperly discarded may be contaminated.	Chlorinated solvents: Ch. 3, p. 94

Figure 5.7b Dry Cleaners: Phytotypologies to Address Contaminants

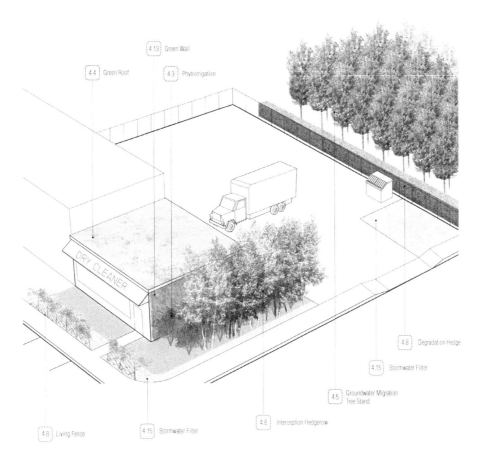

271

Key	Typology	Description	Addresses	Plant Lists
4.3 (p. 207)	Phytoirrigation	Polluted groundwater can be pumped up and irrigated onto dry cleaner plantings targeted to degrade chlorinated solvents. Solar-powered pumps for irrigation can be considered to reduce energy consumption for the irrigation operation.	Chlorinated solvents	p. 48, 2.18 High evapo-transpiration rate species
4.4 (p. 210)	Green Roof	Green Roofs can be planted to evapotranspire water, preventing generation of stormwater that could further mobilize dry-cleaning solvents in groundwater. More importantly, dry-cleaning solvents released into the air as VOCs may be able to be degraded in the root systems of rooftop plants. As the VOCs pass through the roof structure, interaction with the plants' root microbiology may break down the VOCs.	Stormwater vector: Chlorinated solvents	Green Roof species not included in this publication.
4.5 (p. 213)	Groundwater Migration Tree Stand	When solvents leach into groundwater, stands of trees can be placed downgradient of the plume to naturally pump up the water, degrading dry-cleaning products. A detailed Water Mass Balance must be completed by an engineer to calculate how many trees will be needed to make the plume capture effective.	Groundwater vector: Chlorinated solvents	p. 45, 2.15 Phreatophytes
4.6 (p. 216)	Interception Hedgerow	Interception Hedgerows can be installed along property lines to degrade migrating solvents. Since only a thin line of vegetation may be possible, the concentration of contaminants may not totally be degraded, but some remediation may be possible.	Chlorinated solvents	p. 97, 3.13 Chlorinated solvents
4.8 (p. 220)	Degradation Hedge	At edges of pavement and lot lines, hedges with thick, fibrous roots can be installed to mitigate potential stormwater run-off.	Within stormwater: Nutrients (phosphorus and nitrogen); ● Petroleum; Chlorinated solvents	p. 97, 3.13 Chlorinated solvents
4.8 (p. 220)	Living Fence	At the property line, attractive Living Fences can be installed in conjunction with Stormwater Filters to address pollutants in site run-off. In addition, Living Fences may also address adjacent shallow groundwater plumes contaminated with dry-cleaning solvents.	Chlorinated solvents	Use *Salix spp.*
4.13 (p. 231)	Green Wall	Dry-cleaning solvents that are dissolved in air as VOCs may be able to be degraded in the root systems of plants. As the VOCs pass through Green Walls, any interaction with the plants' root microbiology may break down the VOCs. These systems can also be incorporated into interior HVAC systems to improve air quality, as long as the air is drawn through the root zone, rather than just being passively exposed to leaf and soil surfaces.	VOCs in air	Green Wall species not included in this publication.
4.15 (p. 235)	Stormwater Filter	Stormwater Filters can be installed downgradient of impervious surfaces, such as sidewalks, parking lots and traditionally maintained lawn areas to remove pollutants. Where de-icing activities occur, salt-tolerant species must be utilized; salt is not typically removed in these systems.	Within stormwater: Nutrients (phosphorus and nitrogen); ■ Metals; ● Petroleum; Chlorinated solvents	Stormwater/Wetland species not included in this publication

Figure 5.8a Funeral Homes: Sources of Contamination

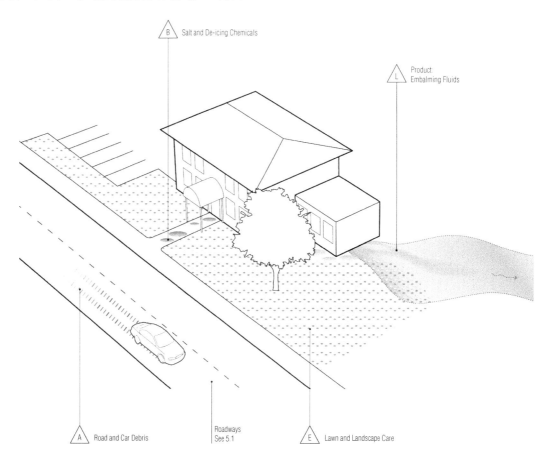

272

Key	Source	Description	Contaminants
A	Road and truck debris	See 5.1: Roadways/Parking Lots	Nutrients: Ch. 3, p. 125 ■ Metals: Ch. 3, p. 136 ● Petroleum: Ch. 3, p. 65 Air pollution: Ch. 3, p. 189
B	Salt and de-icing chemicals	See 5.1: Roadways/Parking Lots	Salt (sodium, chloride and other additives): Ch. 3, p. 179
E	Lawn and landscape care	See 5.2: Parks/Open Spaces/Lawns/Golf Courses	Nutrients: Ch. 3, p. 125 ● Pesticides and ● POPs: Ch. 3, pp. 111 and 118 ■ Metals: Ch. 3, p. 136
L	Product: embalming fluids	During preparation processes at funeral homes, embalming fluids released or spilled can migrate quickly into groundwater. Contamination may be found near the tools used for the embalming process or at delivery points.	Other contaminants of concern, embalming fluids: Ch. 3, p. 124

Figure 5.8b Funeral Homes: Phytotypologies to Address Contaminants

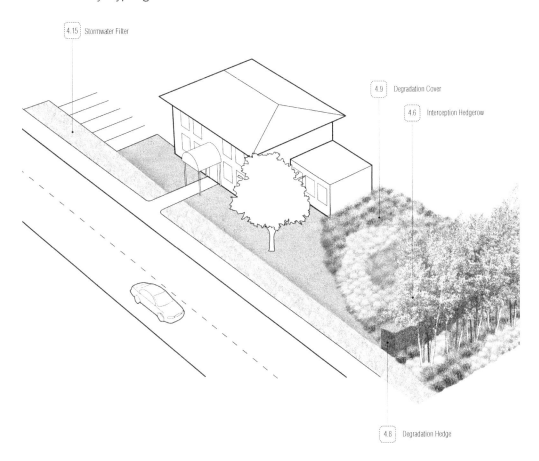

4.15 ┊ Stormwater Filter

4.9 ┊ Degradation Cover

4.6 ┊ Interception Hedgerow

4.8 ┊ Degradation Hedge

273

Key	Typology	Description	Addresses	Plant Lists
4.6 (p. 216)	Interception Hedgerow	Interception Hedgerows can be installed along property lines to degrade migrating fluids downgradient of the building. Since only a thin line of vegetation may be possible, the concentration of contaminants may not be totally degraded, but some remediation may be possible.	Other organic pollutants of concern	p. 45, 2.15 Phreatophytes
4.8 (p. 220)	Degradation Hedge	At lot lines or building edges, deep-rooting hedges may degrade potential embalming fluid spills.	Other organic pollutants of concern	p. 46, 2.17 High-biomass species
4.9 (p. 222)	Degradation Cover	Around the edges of the embalming building where processing occurs, ornamental grasses with degradation capabilities can be installed in traditional landscape swaths. Vegetation must be thickly planted and maintained to maximize root-zone coverage.	Other organic pollutants of concern	p. 46, 2.17 High-biomass species
4.15 (p. 235)	Stormwater Filter	Install Stormwater Filters downgradient of impervious surfaces, such as sidewalks, roadways, parking lots and traditionally maintained lawn areas to remove pollutants. Where de-icing activities occur, salt-tolerant species must be utilized; salt is typically not removed in these systems.	Within storm water: Nutrients (phosphorus and nitrogen); ■ Metals; ● Petroleum; Other organic pollutants of concern	Stormwater/Wetland species not included in this publication

Figure 5.8c Graveyards: Sources of Contamination

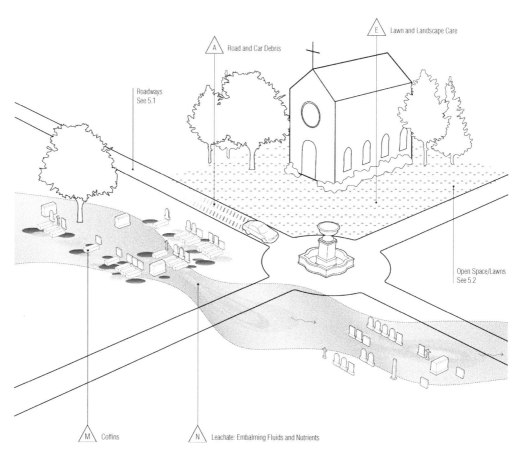

Key	Source	Description	Contaminants
A	Road and car debris	See 5.1: Roadways/Parking Lots	Nutrients: Ch. 3, p. 125 ■ Metals: Ch. 3, p. 136 ● Petroleum: Ch. 3, p. 65 Air pollution: Ch. 3, p. 189
E	Lawn and landscape care	See 5.2: Parks/Open Spaces/Lawns/Golf Courses	Nutrients: Ch. 3, p. 125 ● Pesticides and ● POPs: Ch. 3, pp. 111 and 118 ■ Metals: Ch. 3, p. 136
M	Coffins	Metals are often used in coffin construction and can affect nearby soils and water. In addition, prior to 1900 arsenic was used in the embalming process.	■ Metals: Ch. 3, p. 136
N	Leachate: embalming fluids and nutrients	As bodies decompose, embalming fluids and nutrients can leach into groundwater.	Nutrients: Ch. 3, p. 125 Other organic contaminants of concern: Ch. 3, p. 124 ■ Metals: Ch. 3, p. 136

Figure 5.8d *Graveyards: Phytotypologies to Address Contaminants*

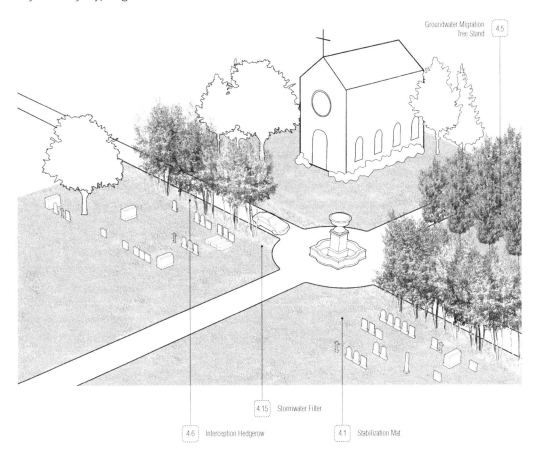

Groundwater Migration
Tree Stand   4.5

4.15   Stormwater Filter

4.6   Interception Hedgerow   4.1   Stabilization Mat

275

Key	Typology	Description	Addresses	Plant Lists
	Organically Maintained Landscape	Eliminate source pollutants wherever possible. Organic pesticides tend to be less toxic, and slow-release fertilizers may prevent migration into surrounding water and soils.	Nutrients; ● Pesticides and ● POPs; ■ Metals	
4.1 (p. 202)	Stabilization Mat	The thick turfgrass typically covering graveyards can function well to stabilize interred metals. The vegetation cover should be maintained as thickly as possible. Grass clippings may contain higher levels of pollutants; care should be taken if grass clippings are composted for reuse.	■ Metals	p. 140, 3.37 Metals excluders
4.5 (p. 213)	Groundwater Migration Tree Stand	Tree stands can be placed downgradient of graves to help intercept and clean any leachate generated from decomposition. A detailed Water Mass Balance must be completed by an engineer to calculate how many trees will be needed to make the plume capture effective.	Groundwater vector: ◍ Chlorinated solvents	p. 45, 2.15 Phreatophytes
4.6 (p. 216)	Interception Hedgerow	Interception Hedgerows can be installed as street plantings along roadways to intercept leachates.	◍ Chlorinated solvents; Nutrients	p. 45, 2.15 Phreatophytes
4.15 (p. 235)	Stormwater Filter	Stormwater Filters can be installed downgradient of roadways and traditionally maintained lawn areas to remove pollutants.	Within stormwater: Nutrients (phosphorus and nitrogen); ■ Metals; ● Petroleum; ◍ Chlorinated solvents	Stormwater/Wetland species not included in this publication

Figure 5.9a Urban Residences: Sources of Contamination

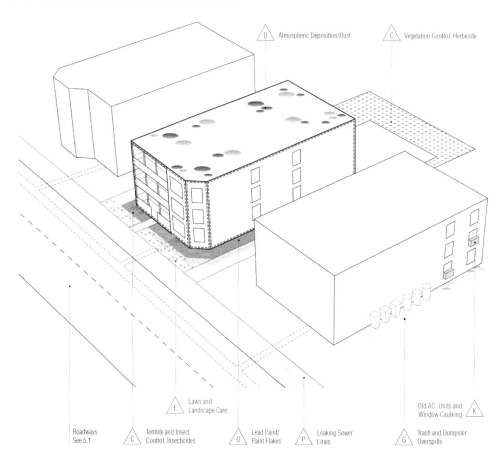

276

Key	Source	Description	Contaminants
C	Termite and insect control: insecticides Vegetation control: herbicides	Both historic and present spraying for termites, roaches, ants, wasps and other insects can leave pesticide and POPs residues. Until 1988, Chlordane, a carcinogenic POP, was used for termite control in wood-built residences and can still frequently be found in soils around wooden buildings. In addition, arsenic was a common component of historic pesticides. Where vegetation, especially invasive plants, is found along back property lines, herbicides may have been utilized.	● Pesticides: Ch. 3, p. 111 ● POPs: Ch. 3, p. 118 ■ Metals: Ch. 3, p. 136 Nutrients: Ch. 3, p. 125
D	Atmospheric deposition/ dust	Chemicals and particles released into the air from automobiles and industrial sources can settle out onto urban rooftops. When it rains, these contaminants can get picked up in the stormwater.	Air pollution: Ch. 3, p. 189
E	Lawn and landscape care	Traditional gardening and lawn care with fertilizers and pesticides can also generate excess nutrients and leave chemical residues in soil and groundwater. See 5.2: Parks/ Open Spaces/Lawns/Golf Courses	Nutrients: Ch. 3, p. 125 ● Pesticides and ● POPs: Ch. 3, pp. 111 and 118 ■ Metals: Ch. 3, p. 136
G	Trash overspills	Trash overspills in and around dumpster and garbage areas have the potential to leach contents. Decomposing organics create nutrients.	Nutrients: Ch. 3, p. 125
K	Air conditioning units/ window caulking	Air conditioning units or refrigeration units, especially older or decommissioned units, can leak coolants. Historic caulking on buildings often contains PCBs.	● Chlorinated solvents (CFCs and freon): Ch. 3, p. 94 ● POPs (PCBs): Ch. 3, p. 118
O	Lead paint/paint flakes and asbestos	Wooden homes painted before 1978 contain lead in old paint on the structure and paint flakes within soils. Even when old paint is no longer on a home, soils within the drip line of the building (about 3 feet from the building face) can be highly contaminated with lead from old flakes and previous removal activities. In addition, asbestos shingles are still commonly found on older residences.	■ Lead: Ch. 3, p. 172 Asbestos
P	Leaking sewer lines	Leaking sewer pipes leach raw, untreated sewage into soils and groundwater.	Nutrients: Ch. 3, p. 125 Bacteria, BOD and living organisms Other organic contaminants of concern: pharmaceuticals: Ch. 3, p. 124 ■ Metals: Ch. 3, p. 136

Figure 5.9b Urban Residences: Phytotypologies to Address Contaminants

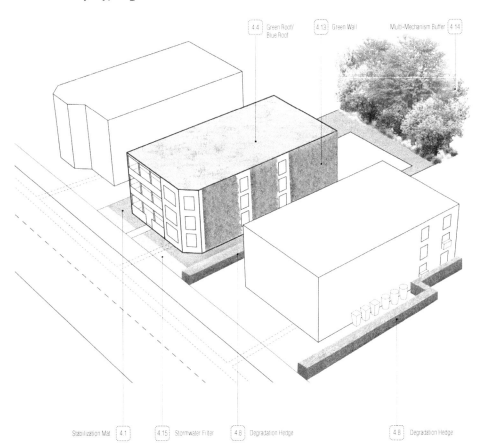

Key	Typology	Description	Addresses	Plant Lists
4.1 (p. 202)	Stabilization Mat	Around wooden structures previously painted with lead paint or shingled with asbestos, thick vegetation may prevent soil exposure. This essentially caps the site with vegetation. Alternatively, impervious pavements can be used to prevent human contact with the impacted soils. It is especially important to ensure that children are not exposed to this soil. Lead cannot be taken up and remediated with plants, so minimizing risk of exposure is the best management practice.	■ Lead, asbestos, arsenic	p. 140, 3.37 Metals excluders
4.4 (p. 210)	Green Roof/ Blue Roof	Green Roofs/Blue Roofs can be installed to evapotranspire water, preventing generation of stormwater that could further mobilize dust particles accumulating on rooftops and ground-level pollutants.	Stormwater vector: ■ Metals and ● POPs settled from air/dust	Green Roof species not included in this publication
4.8 (p. 220)	Degradation Hedge	Around sewer lines and trash areas, where nutrient releases can be found, consider deep-rooting hedges.	Nutrients	p. 46, 2.17 Nutrients: high-biomass species
4.13 (p. 231)	Green Wall	In some innovative communities, Green Walls are being used to filter out pollutants from grey wastewater along interior and exterior building walls. The water, once filtered by the plants, can be reused on site for irrigation and other non-potable uses. Wastewater is run through the planted walls to remove excess nutrients, BOD and emerging contaminants.	Nutrients (phosphorus and nitrogen): Ch. 3, p. 000; Bacteria, BOD and living organisms; Emerging contaminants: including pharmaceuticals	Green Wall species not included in this publication
4.14 (p. 234)	Multi-Mechanism Buffer	Heavily vegetated buffers along shared lot lines can mitigate on-site-generated pollutants. Organic contaminants, such as some herbicides and pesticides, can potentially be degraded, while inorganics such as lead and arsenic are captured and held in soils. Particulate matter in air pollution may be able to be captured on leaf surfaces, buffering adjacent lands. In addition to remediating contamination, these vegetated areas can serve important wildlife corridor functions.	All: see Ch. 3 ◐ ● ■ ● ● ● ■ ●	All: see Ch. 3 (Select plants to address specific adjacent land uses)
4.15 (p. 235)	Stormwater Filter	Stormwater Filters can be installed downgradient of impervious surfaces such as sidewalks and traditionally maintained lawn areas to remove pollutants.	Within stormwater: Nutrients (phosphorus and nitrogen); ■ Metals; ● Petroleum; ◌ Chlorinated solvents	Stormwater/Wetland species not included in this publication

5.10 Vacant lots

An increasing number of sites, by virtue of their abandonment and idle nature, are now considered vacant. This is a temporary state, but likely to have an impact on the concentrations and perpetuation of soil and groundwater pollution. Many examples of this land use can be found in the shrinking city of Detroit, MI, with 25 square miles of vacant land, comprising 19 square miles of purely empty land, 5 square miles with vacant residential structures and 1 square mile of underutilized industrial land (Detroit Future City, 2012). Vacant land has mainly had previous occupancies by residential, some form of manufacturing, extraction industries or waste storage. This is likely to have caused a range of contamination, including the full range of petroleum products, oils and greasing agents, as well as chemicals used in industrial processes, such as solvents and PCBs. Vacant sites also have the potential for ongoing illegal dumping of waste materials on the ground surface. This can generate a range of conditions from piles of urban fill and construction rubble to the illegal disposal of chemical wastes (often at night), creating a 'cocktail' of pollutants in the upper layer of soils and into the groundwater.

By virtue of shifting economic markets or the shrinkage of cities, many sites have vacant residential structures in a dilapidated state, simply standing empty and slowly weathering. The buildings themselves provide sources of contaminants while breaking down, particularly lead, copper and zinc, as well as products from construction materials such as lead paint. Add in buried and forgotten domestic oil tanks, abandoned cars, and asbestos in older residential construction, and you often have an uncertain mixture of contaminants.

278

5.11 Community gardens

Community gardens in North America are typically constructed in urban areas and promote flower and vegetable growing through a collective community group. They are now often part of an open space network of green spaces/corridors that were formerly abandoned sites (see 5.10 Vacant lots). In other countries, these sites are also set within residential neighborhoods or close to community facilities such as schools. Many governmental agencies have specific organizations and guides available for addressing urban food gardening in contaminated soils, and these should be referred to when taking on urban agriculture projects.

Many of the contaminants found on these sites are the result of existing urban fill materials and can contain ashes, lead, arsenic, metals and PAHs. Former railroad ties used to construct low walls and planter areas can contain creosote and coal tar, and pressure-treated lumber treated with arsenic may have been utilized if these were constructed before 2004. Metals and pesticides may exit from compost and plant debris, and fertilizer and pesticides from gardening activities. Lastly, lead-paint flakes from adjacent and previous structures can be a significant contaminant of concern, as well as lead from leaded gasoline and zinc from tire debris.

5.12 Agricultural fields

In agriculture, a field is an area of arable land, enclosed or otherwise, used for agricultural purposes such as cultivating crops, as an enclosure for livestock or left to lie fallow for future use. The presence of pollutants from agricultural practices in food production and water management on agricultural fields of all scales and types is increasingly global. Pollution arises from the application of fertilizers, composts (including animal manure), herbicides and pesticides, as well as a range of localized contamination arising from the storage of chemicals and the use of heavy machinery in crop production. This includes all manner of petroleum products, oils, lubricants and solvents. Over one-third of the world's workers are employed in the agriculture industry, although the percentages of agricultural workers in developed countries have decreased significantly over the past several centuries, due to mechanization. Modern agronomy, plant breeding, agrochemicals and technological improvements have sharply increased yields from cultivation, but at the same time have caused the widespread presence of pollutants, leading to ecological damage and negative human health effects. Selective breeding and modern practices in animal husbandry have similarly increased the output of meat, but have raised concerns about the health effects and environmental disposal of the antibiotics, growth hormones and other chemicals commonly used in industrial meat production.

5.13 Suburban residences

279

The suburban residential lot has fewer pollutants than are found in the urban residence typology built on city fill (see 5.9 Urban residences). Past agricultural uses prior to suburban development, however, may have left traces of pesticides and arsenic in the soils through the spraying of orchards or other agricultural practices. Growth pressures from urban areas often led to the use of agricultural lands for suburban residences, as well as the recycling of available abandoned lands such as former landfills, quarries, mining sites and abandoned military training and munitions-proving grounds on the urban fringes. It should be noted that although these prior uses are scattered throughout the landscape, the suburban residence is more likely to be located on former agricultural sites.

The layout of the suburban residence may be comprised of a detached dwelling with adjacent outlying structures such as storage and a garage, a significant amount of lawn and planted areas, including potential canopy trees as well as in-ground septic systems. Contaminants from individual septic systems that support housing development are significant sources of concern, releasing large amounts of unregulated nutrients, pharmaceuticals and other emerging contaminants of concern. If an individual well for water supply exists on the property, it can potentially be impacted by groundwater pollutants. For example, arsenic contamination in drinking-water wells is common in areas where the bedrock is naturally high in arsenic, and pesticides can leach into drinking water supplies. Adjacent to the main house structure can be a range of balcony, deck, patio and overhead timber structures. Pressure-treated wood containing arsenic may have been used prior to 2004, and areas under existing decks may be contaminated with the residue in the upper soils.

Figure 5.10a Vacant Lots: Sources of Contamination

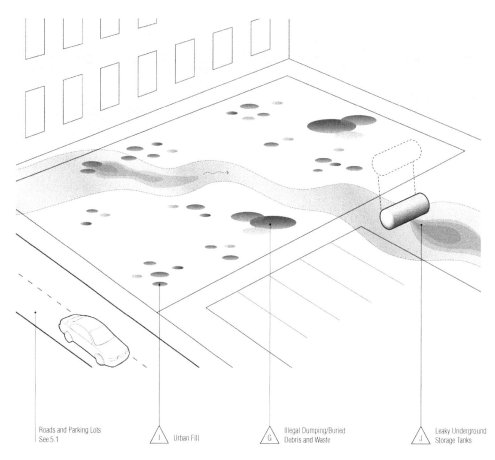

Roads and Parking Lots
See 5.1

I Urban Fill

G Illegal Dumping/Buried
Debris and Waste

J Leaky Underground
Storage Tanks

280

Key	Source	Description	Contaminants
G	Illegal dumping/buried (or unburied) debris and waste	Illegal dumping is a common occurrence on unmonitored vacant sites. Buried debris and waste can include almost anything, but difficult or expensive to dispose of building materials, such as asbestos shingles, lead- and arsenic-impacted soils and creosote-impregnated timber are common.	● Petroleum: Ch. 3, p. 65 ■ Metals: Ch. 3, p. 136 ● Chlorinated solvents: Ch. 3, p. 94
J	Leaky underground storage tanks (LUSTs)/ above-ground storage tanks and barrels	Abandoned underground storage tanks are commonly found on vacant lots. They are especially prevalent where oil fueled previous buildings.	● Petroleum: Ch. 3, p. 65
I	Urban fill	Almost any contaminant can be found in urban areas that were historically landfilled. The more difficult-to-degrade contaminants may be found in these soils, including coal ash and other PAHs, metals and POPs. Lead from old paint and arsenic and Chlordane from old pesticides are also common.	● Petroleum-PAHs: Ch. 3, p. 65 ■ Metals: Ch. 3, p. 136 ● POPs: Ch. 3, p. 118

Figure 5.10b Vacant Lots: Phytotypologies to Address Contaminants

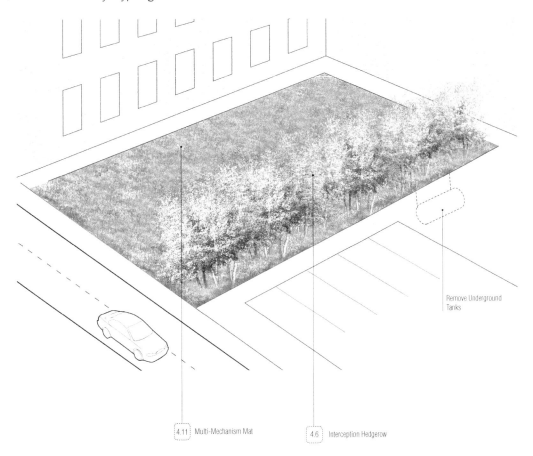

Remove Underground Tanks

4.11 Multi-Mechanism Mat

4.6 Interception Hedgerow

281

Key	Typology	Description	Addresses	Plant Lists
4.6 (p. 216)	Interception Hedgerow	Interception Hedgerows can be installed as street plantings along the property edges to intercept contaminated groundwater plumes from leaky underground storage tanks or buried waste on site.	● Petroleum	p. 45, 2.15 Phreatophytes
4.11 (p. 227)	Multi-Mechanism Mat	While vacant sites remain unused, Multi-Mechanism Mats can be installed as a holding strategy to start remediating pollutants. A low-maintenance, urban meadow mix can be designed to stabilize, degrade or even extract existing pollutants on site. If pollutants on site are not known, some assumptions about potential contaminants can be made. Degradation of petroleum and pesticides should be integrated, potentially some extraction/harvesting for arsenic and stabilization of other non-bioavailable contaminants, including Chlordane, POPs and lead. Vegetation should create a thick mat by which pollution dispersal via erosion is controlled. A yearly mowing regime where clippings are collected and removed from site can slowly remove bioavailable metals over time.	● Petroleum; ● Pesticides; ■ Metals; Air pollution (deposited in soils); ● POPs	p. 140, 3.37 Metals excluders p. 74, 3.5 Petroleum p. 97, 3.13 Chlorinated solvents p. 113, 3.23 Pesticides: degradation p. 146, 3.38, 3.42, 3.43, 3.45, 3.47 Arsenic, cadmium and zinc, nickel, selenium: extraction

Figure 5.11a Community Gardens: Sources of Contamination

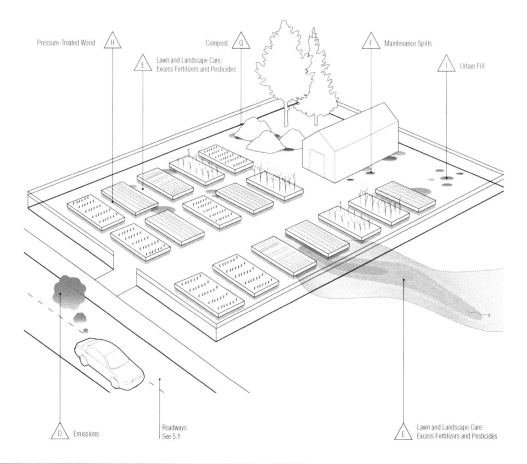

Pressure-Treated Wood H

Compost Q

F Maintenance Spills

E Lawn and Landscape Care:
Excess Fertilizers and Pesticides

I Urban Fill

D Emissions

Roadways
See 5.1

E Lawn and Landscape Care:
Excess Fertilizers and Pesticides

Key	Source	Description	Contaminants
D	Emissions	Release of chemicals and particles into the air from automobiles. Protect gardens from this air pollution source wherever possible.	Air pollution: Ch. 3, p. 189
E	Lawn and landscape care	Excess fertilizer and pesticide applications in community gardens can leach into groundwater and contaminate local soils.	● Pesticides: Ch. 3, p. 111 ■ Metals: Ch. 3, p. 136 Nutrients: Ch. 3, p. 125
F	Maintenance/storage spills	In the structure where community gardens store shared tools and equipment, product spills can occur, including fuel for mowers and other gas-powered equipment, fertilizers, herbicides, insecticides and fungicides.	● Petroleum: Ch. 3, p. 65 ● Pesticides: Ch. 3, p. 111 ■ Metals: Ch. 3, p. 136 Nutrients (phosphorus and nitrogen): Ch. 3, p. 125
H	Treated lumber	Landscape timbers used to create gardening beds can be treated with anti-rot chemicals that can migrate into soils. In pressure-treated lumber made before 2004, arsenic is a common contaminant that leaches from the copper-arsenate preservation material. In some gardens, old creosote-impregnated railroad ties can be found	■ Metals: arsenic Ch. 3, p. 143 ● Creosote: petroleum PAHs: Ch. 3, p. 65
I	Urban fill	Community gardens are often built on vacant lots, where urban fill is prevalent in soils. Coal ash and other PAHs are common as well as metals and POPs. Lead from old paint and arsenic and Chlordane from old pesticides are frequently found.	● Petroleum: PAHs: Ch. 3, p. 65 ■ Metals: Ch. 3, p. 136 ● POPs: Ch. 3, p. 118
Q	Compost	Some pollutants will not break down during composting of garden wastes. Where pesticides and man-made fertilizers have been used, metals and salts can sometimes be compounded over time. In addition, use of human biosolid compost should be avoided in community gardens, due to its likely metal content.	■ Metals: Ch. 3, p. 136

Figure 5.11b Community Gardens: Phytotypologies to Address Contaminants

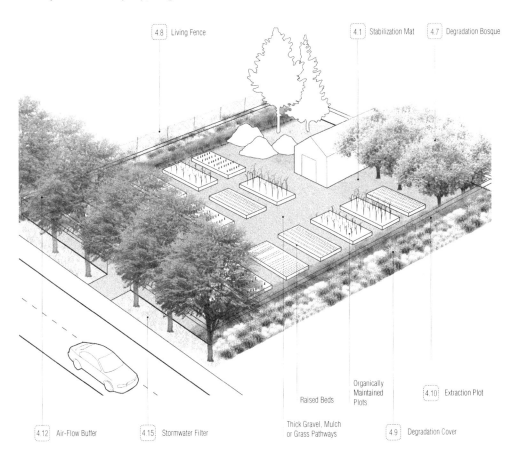

4.8   Living Fence

4.1   Stabilization Mat

4.7   Degradation Bosque

Organically
Maintained
Plots

4.10   Extraction Plot

Raised Beds

Thick Gravel, Mulch
or Grass Pathways

4.9   Degradation Cover

4.12   Air-Flow Buffer

4.15   Stormwater Filter

283

Key	Typology	Description	Addresses	Plant Lists
	Organically Maintained Gardens	Eliminate source pollutants wherever possible. Organic pesticides tend to be less toxic and slow-release fertilizer less migratory.	Nutrients; ● Pesticides and ● POPs; ■ Metals	
	Raised Beds	Where existing soils are impacted with lead, arsenic or other contaminants that can endanger edible food crops, construct raised beds, a minimum of 12 inches deep and fill with clean soils for growing edibles.	■ ● All	
	Thick Gravel, Mulch or Grass Pathways	Minimize all potential contact with existing soils by providing a minimum 3-inch thick layer of gravel, mulch or grass over all walking surfaces or other areas where existing soils could be exposed. Minimize dust generation as much as possible.	■ ● All	
4.1 (p. 202)	Stabilization Mat	Where urban fill with potential lead and arsenic impact exists, vegetation can be used to hold pollutants on site, away from risk of exposure.	● POPs including PCBs; ■ Metals	p. 140, 3.37 Metals excluders
4.7 (p. 218)	Degradation Bosque	Near storage and maintenance areas, install degradation trees and shrubs to break down spilled fuel, pesticides and excess nutrients.	● Petroleum; Nutrients; ● Pesticides	p. 74, 3.5 Petroleum p. 113, 3.23 Pesticides p. 46, 2.17 Nutrients: high-biomass species
4.8 (p. 220)	Living Fence	At the property line, attractive Living Fences can be installed to break down any excess fertilizer or pesticides that may be present within the garden.	Nutrients; ● Pesticides and ●POPs	Use *Salix spp.*
4.9 (p. 222)	Degradation Cover	Degradation Covers can complement Living Fences at the property line, creating attractive, non-edible borders that buffer on-site-generated contaminants. They can also help to break down any bioavailable PAH petroleum that may be in soils from urban fill. Cut flowers can be integrated into these borders to provide attractive neighborhood edges.	Nutrients; ● Pesticides and ● POPs	p. 113, 3.23 Pesticides p. 46, 2.17 Nutrients: high-biomass species
4.10 (p. 224)	Extraction Plots	In a few cases of low-level contamination with highly bioavailable metals, plants may be used to extract, harvest and remove these metals. This may be effective if polluted soils are piled along the edges of a site and extraction plots are grown and harvested over many years to slowly clean the piled soil over time. Soil chemistry, plant selection, metals concentration and long-term maintenance will greatly affect the outcome.	■ Highly bioavailable metals: arsenic, nickel, selenium, cadmium, zinc	p. 146, 3.38 Arsenic p. 162, 3.45 Nickel p. 168, 3.47 Selenium p. 154, 3.42–3.43 Cadmium and zinc
4.12 (p. 229)	Air-Flow Buffer	A tree buffer along the street edge can prevent migration of air pollution from adjacent roadways onto the site.	Air pollution	p. 195, 3.59–3.60 Air pollution
4.15 (p. 235)	Stormwater Filter	Install Stormwater Filters on the downhill side of the garden to trap excess fertilizers and pesticides that may run off during irrigation. Vegetable gardens tend to generate significant excess nutrients when conventional fertilizers are used. Over-irrigation quickly leaches these fertilizers into water run-off.	Within stormwater: Nutrients; ■ Metals; ● Petroleum	Stormwater/Wetland species not included in this publication

Figure 5.12a Agricultural Sites: Sources of Contamination

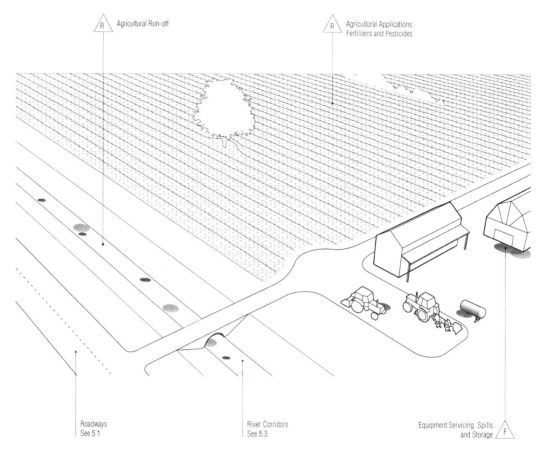

284

Key	Source	Description	Contaminants
⟁F	Equipment servicing, spills and storage	Where farm equipment, trucks, fertilizers and pesticides are stored and loaded, product spills can occur. This includes fuel for equipment, fertilizers, herbicides, insecticides and fungicides.	● Petroleum: Ch. 3, p. 65 ● Pesticides: Ch. 3, p. 111 ■ Metals: Ch. 3, p. 136 ○ Nutrients: Ch. 3, p. 125 ● POPs: Ch. 3, p. 118
⟁R	Agricultural applications/ run-off	Excess fertilizer, manure and pesticide applied in agricultural production can leach into groundwater and contaminate local soils. In addition, it can collect in field-side ditches and become more concentrated in local streams and rivers. Alga blooms from excess phosphorus leaching are common in adjacent waterways and the effects become compounded downstream.	● Petroleum: Ch. 3, p. 65 ● Pesticides: Ch. 3, p. 111 ■ Metals: Ch. 3, p. 136 ○ Nutrients: Ch. 3, p. 125

Figure 5.12b Agricultural Sites: Phytotypologies to Address Contaminants

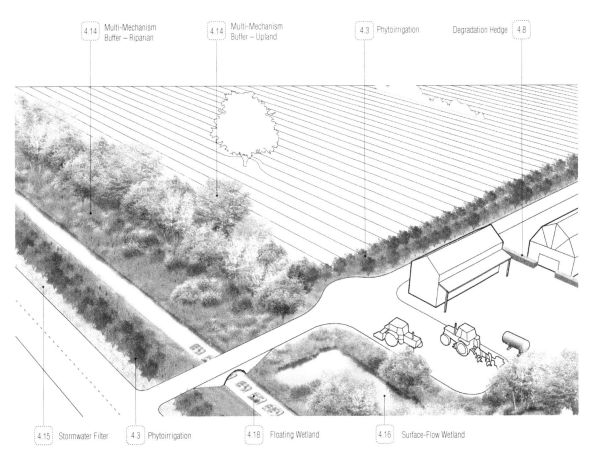

4.14 Multi-Mechanism Buffer – Riparian

4.14 Multi-Mechanism Buffer – Upland

4.3 Phytoirrigation

Degradation Hedge 4.8

4.15 Stormwater Filter

4.3 Phytoirrigation

4.18 Floating Wetland

4.16 Surface-Flow Wetland

285

Key	Typology	Description	Addresses	Plant Lists
	Organic Agricultural Practices, Integrated Pest Management	Eliminate source pollutants wherever possible. Carry out soil testing and add nutrients only when levels are low and it is required. If conventional products are utilized, continual monitoring of pests, moisture and soil fertility can allow for more accurate application of fertilizers and pesticides. This reduces both contamination and the product cost to farmers.	Nutrients; ● Pesticides; ■ Metals	
4.3 (p. 207)	Phyto-irrigation	Groundwater impacted with excess nitrogen or phosphorus can be pumped up to the surface and watered onto buffer plantings, providing nutrient-enriched water. The water will promote fast growth and the irrigated crop can be sold or used for biofuel production. Willow can be effectively used for this application because it can be cut and baled every few years, and pelletized for fuel to heat farm buildings. A solar-powered pump drip can be used for the irrigation system. Nutrient-rich surface waters may also be pumped out of ditches and used for this purpose.	Nutrients (phosphorus and nitrogen); ■ Metals; ● Petroleum	p. 48, 2.18 High evapotranspiration-rate species
4.8 (p. 220)	Degradation Hedge	Around edges of barns, maintenance and storage structures, deep-rooting hedges can be installed to degrade potential product and fuel spills.	Nutrients (phosphorus and nitrogen); ● Pesticides and ● POPs; ● Petroleum	p. 74, 3.5 Petroleum p. 113, 3.23 Pesticides p. 46, 2.17 High Biomass Species
4.14 (p. 234)	Multi-Mechanism Buffer	Vegetated buffers as little as 20 feet wide installed between agricultural fields and waterways can be beneficial in removing excess nutrients and pesticides. These buffers can be designed as mixed-species wildlife corridors, or a single species like willow can be installed to provide an alternative buffer cash crop for biofuel production. Buffers ideally should be greater than 50 feet wide.	Nutrients (phosphorus and nitrogen); ■ Metals; ● Petroleum; Air pollution: particulate matter	All: see Ch. 3
4.15 (p. 235)	Stormwater Filter	These systems can be installed along roadways and agricultural fields, wherever run-off is found. Where de-icing activities occur, salt-tolerant species must be utilized; salt is typically not removed in these systems.	Within stormwater: Nutrients ; ■ Metals; ● Petroleum; ● Pesticides	Stormwater/Wetland species not included in this publication
4.16 (p. 238)	Surface-Flow Wetland	Remediation wetlands can be installed along waterways to help filter out pollutants. In addition, these wetlands can serve as temporary holding and sedimentation ponds for farm run-off before it enters surface-water bodies.	● Petroleum	Stormwater/Wetland species not included in this publication
4.18 (p. 242)	Floating Wetland	In surface-water bodies and ditches, floating wetlands can be placed on the surface to help extract and degrade pollutants that have run off from fields. In addition, each year at the end of the season, the plants can be harvested from the floating structures and composted. The compost generated can then be applied to fields for nutrient recycling, both cleaning up the river and feeding agricultural soils.	Within water: Nutrients ; ● Pesticides; ● Petroleum ; ■ Metals	Stormwater/Wetland species not included in this publication

Figure 5.13a Suburban Residence: Sources of Contamination

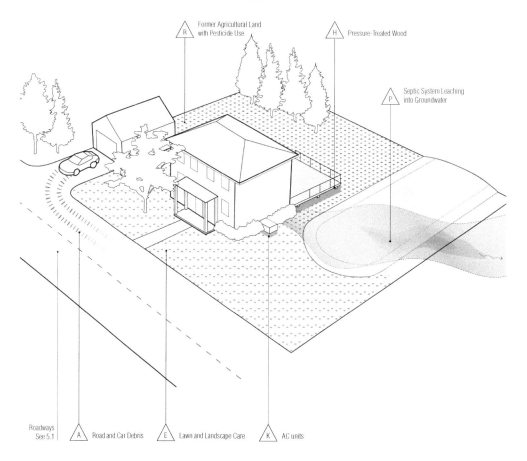

286

Key	Source	Description	Contaminants
A	Road and car debris	See 5.1: Roadways/Parking Lots	Nutrients: Ch. 3, p. 125 ■ Metals: Ch. 3, p. 136 ● Petroleum: Ch. 3, p. 65 Air pollution: Ch. 3, p. 189
E	Lawn and landscape care	See 5.2: Parks/Open Spaces/Lawns/Golf Courses	● Pesticides and ● POPs: Ch. 3, pp. 111 and 118 ■ Metals: Ch. 3, p. 136 Nutrients: Ch. 3, p. 125
H	Wood decks: treated lumber	Timber that has been pressure treated with anti-rot chemicals is often used for decks. Arsenic contamination is common beneath any pressure-treated wood deck constructed before 2004.	■ Metals: arsenic, Ch. 3, p. 143
K	Air conditioning units	Air conditioning units can leak coolants.	● Chlorinated solvents (CFCs and freon): Ch. 3, p. 94
P	Septic systems	Wastewater in many suburban homes, especially in the Northeastern US, is treated with on-site septic systems and leach fields. While BOD and pathogens are removed from the liquids before they migrate to groundwater, nutrients (nitrogen and phosphorus) are not removed. In addition, contaminants like pharmaceuticals are often not removed. These constituents can compound with many nearby septic systems and can greatly affect groundwater and drinking water.	Nutrients: Ch. 3, p. 125 Bacteria, BOD and living organisms Other Organic contaminants of concern: pharmaceuticals: Ch. 3, p. 124 ■ Metals: Ch. 3, p. 136
R	Former agricultural use/orchards with pesticide use	Many suburban homes are built on land that was once used for agriculture, including orchards. Old pesticides may remain on these sites, including lead and arsenic, two common additives to historic pesticides. It is especially important to consider this where children utilize the back yard areas and may be exposed to these metals.	● Pesticides and ● POPs: Ch. 3, pp. 111 and 118 ■ Metals: Ch. 3, p. 136 Nutrients: Ch. 3, p. 125

Figure 5.13b Suburban Residence: Phytotypologies to Address Contaminants

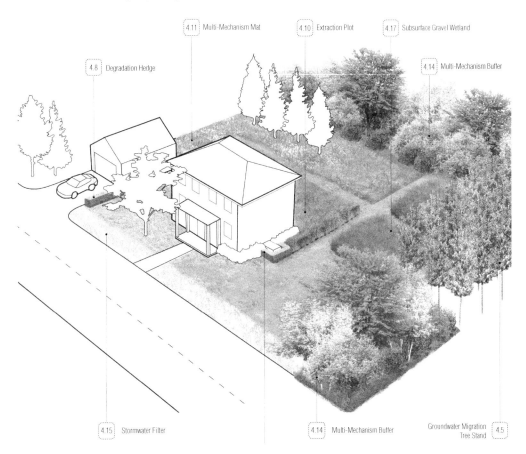

287

Key	Typology	Description	Addresses	Plant Lists
	Organically Maintained	Eliminate source pollutants wherever possible. Use non-toxic pesticides to address insect problems in the home.	Nutrients; ● Pesticides and ● POPs; ■ Metals	
4.5 (p. 213)	Groundwater Migration Tree Stand	Tree stands can be placed downgradient of septic systems to tap the groundwater and leachate and naturally transform the excess nitrogen in water back into atmospheric nitrogen. In addition, the tree stand may be able to help degrade other emerging contaminants such as pharmaceuticals.	Groundwater vector: Nitrogen	p. 45, 2.15 Phreatophytes
4.8 (p. 220)	Degradation Hedge	Around garages or sheds where lawn mowers, cars, fuel, or landscape products may be stored, Degradation Hedges can be installed to quickly break down any spills that may occur.	Nutrients (phosphorus and nitrogen); ● Pesticides; ● Petroleum	p. 74, 3.5 Petroleum p. 113, 3.23 Pesticides p. 46, 2.17 High Biomass Species
4.10 (p. 224)	Extraction Plots	Arsenic-hyperaccumulating ferns can be planted where old decks existed (or under decks) to extract the metal. Once harvested, the ferns may need to be landfilled at a hazardous waste site if the arsenic concentration in the ferns is above regulatory limits.	■ Arsenic	p. 146, 3.38 Arsenic
4.11 (p. 227)	Multi-Mechanism Mat	Where agricultural land uses may have existed previously, typically, the main contaminants of concern will be arsenic, lead and POPs once used in pesticides. The primary objective should be to stabilize the soil with a thick vegetative mat, preventing soil exposure. Second, extraction species may be able to be designed into the planting mix to slowly extract the arsenic over time. These plants would need to be cut and harvested at the end of each growing season.	● Pesticides: herbicides; ■ Metals; ● POPs	p. 140, 3.37 Lead and POPs: stabilization p. 146, 3.38 Arsenic: extraction
4.14 (p. 234)	Multi-Mechanism Buffer	Heavily vegetated buffers along property lines can help to treat excess fertilizers and septic contaminants on site, before they migrate. In addition to remediating contamination, these vegetated areas can serve important privacy and wildlife corridor functions.	All: see Ch. 3 ●●■ ●●●■	All: see Ch. 3
4.15 (p. 235)	Stormwater Filter	Install Stormwater Filters on the downhill side of roadways, driveways and lawns to trap road run-off and excess fertilizers and pesticides.	Within stormwater: Nutrients (phosphorus and nitrogen); ■ Metals; ● Petroleum	Stormwater/Wetland species not included in this publication
4.17 (p. 241)	Subsurface Gravel Wetland	Subsurface Gravel Wetlands can be designed to treat single-family wastewater loads, to amend traditional engineered septic systems. The advantage is that more contaminants can be treated and the area required for these subsurface systems is often small. Plant dormancy may affect functionality in winter months.	Within water: Nutrients; ● Petroleum; ■ Metals; BOD/Bacteria; Pharmaceuticals	Wetland species not included in this publication

5.14 Landfill

Landfills are one of the most commonly used methods around the world of managing and disposing of the many forms of waste generated through human settlements and industrial processes, including daily municipal household wastes as well as hazardous wastes and demolition and construction wastes. Landfills are also part of the larger system of collection, transport, processing or disposal, management and monitoring of waste materials. The waste-disposal process is generally undertaken to reduce the effect of wastes on health, the environment or aesthetics. One land use that is most closely associated with these sites of transformation and landscape-design work is the varying array of municipal solid-waste landfills derived from household and business waste, featuring nonhazardous materials including all manner of food scraps, paper, cardboard, clothing, packaging. These range in scale from the small, local town 'dump site' on the fringes of built-up areas or in suburban districts, to larger landfill landscapes that serve entire cities and conurbations.

Waste-management practices can differ significantly for developed and developing nations, for urban and rural areas and for residential and industrial producers. The US currently has 3,034 active landfills (US EPA, 2014d) and over 10,000 closed municipal landfills (US EPA 1988). Before 1960, however, every town (and many businesses and factories) had its own dump, creating many smaller landfills with unknown locations from that time period. Landfills were often established in abandoned or unused quarries, mining voids and borrow pits, or in low-lying wetlands and marshlands. A well-designed and well-managed landfill can still be a hygienic and relatively inexpensive method of disposing of waste materials, but in time even these landfills and their liquid leachate and air emissions can become hazardous.

Landfills are typically composed of a set of common elements including landfill 'cells' or defined mounds of daily trash disposed in cumulative layers, a circulation network of access and truck-haul roads to the cells and out again, a leachate collection and on-site treatment plant and facilities for security, control of vehicles and supporting structures for personnel and storage. Disposal of waste in a landfill involves burying the waste with 'day covers' of soil, where deposited waste is compacted daily to increase its density and stability and covered with a soil cover to prevent the attraction of vermin. The site is eventually capped with a 'final cover' of some combination of soil, clay or bituthene. This process remains the common practice in most countries.

Common throughout all landfills, irrespective of size, location or age, are two types of engineering and construction practices which are significant for the success of phytotechnology installations. Older landfills, especially those started before 1980, were unlined, meaning that the base of the waste fill was located on the existing ground surface, allowing the free passage of liquids and wastes down into the subsurface and groundwater. Furthermore, historic landfills used a final closing treatment of the waste cells of a simple earth cover. While unsafe by current standard landfill practices, this allows a good opportunity for retrofitting these structures with Evapotranspiration Covers on the landfill surface to prevent water from entering the waste pile, as well as the potential use of Phytoirrigation to remediate leachate from the landfill. Landfills after 1980 were lined with a range of clay, bituthene and mechanical products. In these situations, phytotechnology applications are limited, due to the engineering constraints of the liner and the avoidance of root penetration. In these cases planting can be carried out only at the edges of the landfill.

The design characteristics of a modern landfill include methods to contain leachate, such as clay or a plastic lining material. All landfills will likely eventually fail and leak leachate into ground and surface water. State-of-the-art plastic (HDPE) landfill liners 100ml thick and plastic pipes allow chemicals and gases to pass through their membranes, but will still become brittle, swell and break down in time. Lined landfills leak in very narrow plumes, whereas unlined landfills will produce wide plumes of leachate. Both lined and unlined landfills are often located next to water bodies such as rivers, lakes and ponds, making leakage detection and remediation difficult. Plume detection by monitoring wells can also be very difficult.

Another common product of landfills is gas composed of methane, carbon dioxide and leachate. The gas that is produced as organic wastes are anaerobically digested can kill surface vegetation. Many landfills have landfill gas-extraction systems installed. Gas is pumped out of the landfill using perforated pipes and flared off or burned in a gas engine to generate electricity. Wastes with high moisture content or which receive artificial irrigation or rain-water, surface or groundwater infiltration produce both leachate and methane gas at a significantly increased rate.

289

Finally, while all landfills could require remediation, landfills built in the last 60 years will often require a thorough cleanup, due to the disposal of highly toxic chemicals manufactured and sold since the 1940s.

5.15 Former manufactured-gas plants

Manufactured fuel-gas utilities were founded first in England, and then in the rest of Europe and in North America in the 1820s. From the late nineteenth century to the mid twentieth century, hundreds of manufactured-gas plants (MGPs) in North America, Europe and in urban centers internationally supplied fuel gases to homes and industry for domestic heating, cooking and lighting, and for public street lighting and power. Coal gas was produced through the distillation of bituminous coal in heated, anaerobic vessels called retorts. In this process, coal is broken down into its volatile components through the action of heat in a nearly oxygen-free environment. The fuel gases generated were mixtures of a number of chemical substances, including hydrogen, methane, carbon monoxide and ethylene. Coal gas also contained significant quantities of unwanted sulfur and ammonia compounds, as well as heavy hydrocarbons. Gases were drawn off from the retort and some of the vapors were converted to liquids consisting of water and coal tar, while others remained in a gaseous state. Sources estimate that between 1880 and 1950 approximately 11 billion gallons of coal tar were generated by the manufactured-gas industry in the US (Lee et al., 1992). The coal gas, however, still contained impurities, primarily gaseous ammonia and sulfur compounds. These were removed by cleaning the gas in water and by running the gas through beds of moist lime or moist iron oxides. One solution to disposing of these impurities was to dump them in

Figure 5.14a Landfill: Sources of Contamination

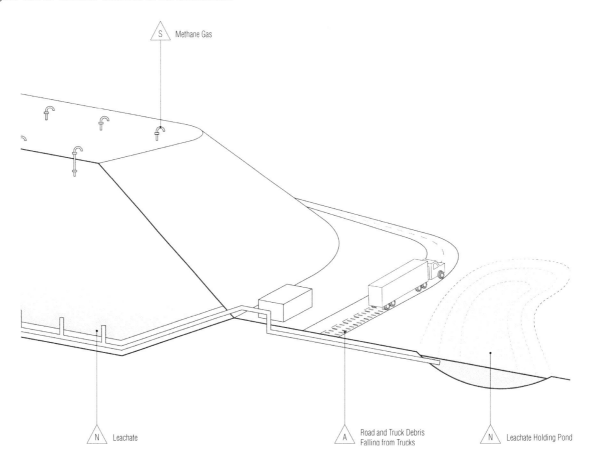

Key	Source	Description	Contaminants
△A	Road and truck debris	See 5.1: Roadways. Garbage trucks tend to track more pollutants than the average car or truck, due to the dumping of trash and potential leaking of pollutants.	Nutrients: Ch. 3, p. 125 ■ Metals: Ch. 3, p. 136 ● Petroleum: Ch. 3, p. 65 Air pollution: Ch. 3, p. 189
△N	Leachate	When rain-water runs through buried waste, the water picks up pollutants along the way, leaching out sideways and at the base. This polluted water is referred to as leachate, and is commonly collected through a piping system and directed into a holding pond or tanks, from where it is pumped and taken to a hazardous waste facility. The most common pollutants in municipal landfill leachate are nitrogen (usually in the form of ammonia), salt and metals; however, any pollutant that is soluble in water can become mobilized in the leachate.	Most common: Nitrogen: Ch. 3, p. 125 ▨ Salt (sodium, chloride and other additives): Ch. 3, p. 179 ■ Metals: Ch. 3, p. 136 All pollutants possible
△S	Methane gas	As landfills decompose, they release combustible methane gas that is usually collected and vented to the air through pipes and control valves. Occasionally, the gas is collected, purified and used as a local energy source.	Methane gas

Figure 5.14b Landfill: Phytotypologies to Address Contaminants

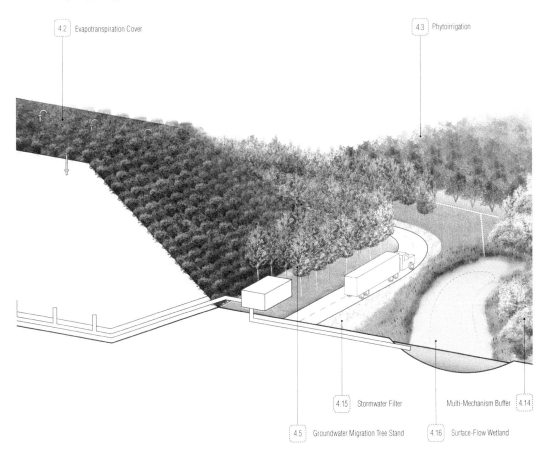

4.2 Evapotranspiration Cover

4.3 Phytoirrigation

4.15 Stormwater Filter

Multi-Mechanism Buffer 4.14

4.5 Groundwater Migration Tree Stand

4.16 Surface-Flow Wetland

291

Key	Typology	Description	Addresses	Plant Lists
4.2 (p. 204)	Evapotran-spiration Cover	Where landfills are unlined, high evapotranspiration-rate species can be planted on the surface of the landfill to quickly transpire into the air any rain that falls on the landfill. This prevents the generation of leachate, preventing contaminants from migrating off site. In addition, if species with a high leaf-area index are utilized (see Figure 4.2a), the canopy of leaves can also help prevent the water from infiltrating into the soil.	Leachate	p. 48, 2.18 High evapotranspiration-rate species
4.3 (p. 207)	Phyto-irrigation	Leachate from landfills can be collected in holding ponds and irrigated onto plantings to both remove the pollutants from the leachate and produce an economic crop. For example, fast-growing poplars or willows may be grown to produce biofuels or hardwood. Some metals and salts may also be able to be removed during the irrigation process. The salts and metals are bound to the soils and roots, while the nitrogen is transformed back into organic nitrogen or atmospheric nitrogen, removing it from the water. Phytoirrigation species can be planted on top of the landfill so that irrigated water is recycled and reused in a closed system.	Within leachate: Nutrients (phosphorus and nitrogen); Some ■ metals and ░ salt	p. 48, 2.18 High evapotranspiration-rate species
4.5 (p. 213)	Groundwater Migration Tree Stand	In lined landfills that are cracking and allowing small amounts of leachate to be generated, or in non-lined landfills where leachate is not being effectively collected and controlled, Groundwater Migration Tree Stands may be installed to naturally pump up and degrade the contaminated leachate. This system works best when nitrogen is the main contaminant of concern. A detailed Mass Water Balance must be conducted by an engineer to determine if this is possible and how many trees are needed to stop the migrating plume.	Leachate: nitrogen	p. 48, 2.18 High evapotranspiration-rate species
4.14 (p. 234)	Multi-Mechanism Buffer	Heavily vegetated buffers around the edges of the landfill can help mitigate airborne and groundwater-borne contaminants before they migrate. In addition to remediating contamination, these vegetated areas can serve important screening and wildlife corridor functions.	Nutrients (phosphorus and nitrogen); ■ Metals; ● Petroleum; Air pollution: particulate matter	All: see Ch. 3
4.15 (p. 235)	Stormwater Filter	Along access roads and near paved areas, plants and associated media filter out pollutants from the stormwater in a swale or linear filter strip. Where de-icing activities occur, salt-tolerant species must be utilized; salt is typically not removed in these systems.	Within stormwater: Nutrients (phosphorus and nitrogen); ■ Metals; ● Petroleum	Stormwater/Wetland species not included in this publication
4.16 (p. 238)	Surface-Flow Wetland	Vegetation may be able to be added to leachate-holding ponds to transform and degrade contaminants. Often, leachate-holding ponds are too toxic for plant growth. A series of wetland cells may be constructed to remove pollutants in a stepped system.	Within leachate: all pollutants	Stormwater/Wetland species not included in this publication

Figure 5.15a Former Manufactured-Gas Plants: Sources of Contamination

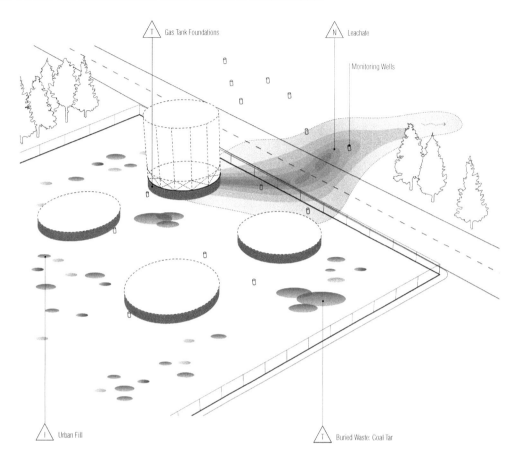

Key	Source	Description	Contaminants
⚠ I	Urban fill	MGPs were often constructed at river edges where sites were filled with debris to create usable land. Almost any contaminant can be found in urban fill areas. The more difficult-to-degrade contaminants are what typically remain in these soils, including coal ash and other PAHs, metals and POPs.	● Petroleum-PAHs: Ch. 3, p. 65 ■ Metals: Ch. 3, p. 136 ● POPs: Ch. 3, p. 118 All pollutants possible
⚠ N	Leachate	When rain-water or groundwater runs through buried waste at old MGP sites, the water can pick up pollutants along the way and generate a leachate plume. Over time, this plume usually becomes less concentrated because most of the PAHs remaining in aged soils after long periods of time will not mobilize in water. The most common pollutants in MGP leachate are petroleum; however, any pollutant that is soluble in water can become mobilized in the leachate.	Most common: ● Petroleum: Ch. 3, p. 65
⚠ T	Gas tank foundations/ buried coal tar	Old tank foundations and various areas of buried waste around former MGP sites typically contain coal tar, a highly recalcitrant sticky, black petroleum that persists on site. In addition, heavy metals such as arsenic and cyanide may be mixed in with the coal tar.	● Petroleum: Ch. 3, p. 65 ■ Metals: Ch. 3, p. 136

Figure 5.15b Former Manufactured-Gas Plants: Phytotypologies to Address Contaminants

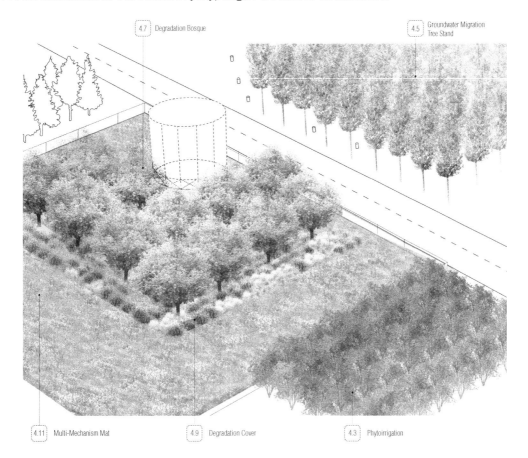

4.7 Degradation Bosque

4.5 Groundwater Migration Tree Stand

4.11 Multi-Mechanism Mat

4.9 Degradation Cover

4.3 Phytoirrigation

293

Key	Typology	Description	Addresses	Plant Lists
4.3 (p. 207)	Phytoirrigation	Leachate around MGP sites can be pumped up and irrigated onto plantings to potentially remove the pollutants from the leachate and prevent the plume from migrating. Solar-powered pumps can be considered.	Within leachate: ● Petroleum; some ■ metals	p. 48, 2.18 High evapotranspiration-rate species
4.5 (p. 213)	Groundwater Migration Tree Stand	Trees can be installed to tap into polluted groundwater and plumes, take up the water and degrade the petroleum. A detailed Mass Water Balance must be conducted by an engineer to determine if this is possible and how many trees are needed to stop the migrating plume.	Leachate: ● Petroleum	p. 48, 2.18 High evapotrans-piration-rate species
4.7 (p. 218)	Degradation Bosque	Within old MGP foundations where coal tar is likely, Degradation Bosques can be installed to break up the petroleum and slowly degrade it over time. In addition, this same strategy can be used where buried coal tar waste is found.	● Petroleum	p. 74, 3.5 Petroleum-PAH
4.9 (p. 222)	Degradation Cover	Under bosques of trees, or where open sight lines must be maintained, shorter plants can be used to create a Degradation Cover to remediate petroleum found in surface soils.	● Petroleum	p. 74, 3.5 Petroleum-PAH
4.11 (p. 227)	Multi-Mechanism Mat	Multi-Mechanism Mats installed on MGP sites can be designed to stabilize non-extractable metals, while slowly degrading tough PAH petroleum.	● Petroleum: PAHs; ■ Metals	p. 140, 3.37 Metals excluders p. 74, 3.5 Petroleum: degradation

waterways on site, while another solution was to dispose of them in vast pits or holding ponds at or near the gas plant. The sheer volume of waste products generated soon overwhelmed on-site storage capacity at most plants, but the stored contaminants often remain to this day.

Pipelines from natural gas fields were eventually built in the 1880s, linked to cities, and natural gas was used to supplement manufactured fuel-gas supplies, eventually completely displacing it. Manufactured gas ceased to be made in North America by the mid 1960s, but continued in Europe until the 1980s.

Today, many communities are home to these former manufactured-gas plants, long since abandoned yet still highly toxic with subsurface coal tars, creosote and heavy metals, often buried in site pits or in adjacent landfills; yet few in the surrounding community know of this potential hazard. Abandoned gas-works were demolished and the external steel skeletons surrounding the gas storage holders were removed for scrap metal, while new facilities were constructed on the sites, such as electrical substations built by utility companies. This left MGP sites with a complex mixture of soil and subgrade pollutants: coal tar lagoons often buried and not visible, as well as a mixture of old infrastructure from the gas-plant period, including railroad lines, gantries, cranes, coal storage and gasification equipment as well as potential pollutants from new industries located on the former MGP lands. More than 50,000 gas-works operated in the US at various times during that period. During the life span of the industry, billions of gallons of extremely hazardous wastes were generated and stored.

294

Coal tar and its associated wastes are extremely resistant to biodegradation. The chemicals and compounds that comprise this waste are extremely persistent and long lasting and the hazards presented by manufactured-gas wastes remain on many urban sites, hidden below the ground.

5.16 Military uses

Land that was previously or is currently occupied by military activities may contain a range of contamination arising from industrial processes, including intense areas of pollutants from fire-training exercises, munitions storage, proving grounds for test-firing weapons and the disposal of munitions and wastes. In very limited cases radionuclides will be present, usually within landfill areas. There are over 9,800 sites in the United States that have been reviewed by the DOD for evidence of contamination; over 2,650 of these properties were determined to be in need of environmental cleanup and restoration, at an estimated cost of $18 billion (Albright, 2013). The large areas and remoteness of DOD lands give the subject of site contamination an added dimension. Starting with ordinary types of activities, DOD sites contain all forms of manufacturing and industrial processes related to their active mission. These range from repair shops to road-construction bays and temporary housing for large-scale equipment including excavators, bulldozers and specialized troop vehicles. These are all supported by metal and fabrication shops employing the full range of oils, lubricants, coolants and refrigeration liquids. Many of these products can find their way into the groundwater if they are not disposed of according to regulatory practices. Bearing in mind that many of these sites have been continuously occupied since the 1930s and regulatory practices are not nearly as old as this, a build-up of pollutants is likely to have occurred in soils and groundwater, down to a considerable depth.

Specialized activities such as fire training and munitions and explosives testing have left large tracts of land contaminated with lead and explosives pollution, such as RDX and TNT. RDX is significantly mobile in water, and the very nature of an explosives test means that the contaminant is often injected almost directly into groundwater. Furthermore, the presence of airfields and runways brings concerns of groundwater contamination from fuel spills, as well as chemicals and de-icing solutions. Elsewhere on some sites, military cemeteries offer all of the same contaminants as are possible on other graveyards (see 5.8 Graveyards). Many of these sites function much like small cities, with all of the potential contamination of such urban systems.

Summary

This chapter has demonstrated that in urban and suburban environments multiple land uses are encountered that can release pollutants into the environment. After considering specific contaminants and their corresponding species types and phytotechnologies in Chapters 3 and 4, the objective is to not create a series of planting templates for particular site programs, but to instead think about where there may be opportunities for integrating phytotechnology applications into day-to-day design practice. The landscape architect, site designer, engineer or owner can develop tools to address the cleanup of pollutants that are also integrated with planting solutions for the reuse and development of such sites. Larger opportunities for the use of phytotechnologies on contaminated lands are likely to become available to the design professions in the coming years. These opportunities will arise from the application of phytotechnologies on sites internationally to mitigate climate change and its expected modifications to plant zones and the growth patterns of vegetation, the land banking of polluted land and the continued pressures of urbanization on contaminated sites.

1 With the growing globalization of design and planning services related to contaminated sites, the application of phytotechnologies will increase in a broader scope of environments, within a range of legal and regulatory conditions and a wide set of climatic conditions. The authors conclude that this will be an important professional design and planning opportunity for landscape architects and site designers in the coming years.

2 The modification of temperature and planting zones through climate change and the increase of temperatures and lengths of growing seasons will not only affect the range of species and types of plants that can be used in phytotechnology projects but also reduce the periods of dormancy in northern climes when installations will not be operating. While the authors believe this will not lead to significant changes in phytotechnology projects in the coming decade, it will give designers the ability to use a larger palette of plants, which will support the expansion of plant-based remediation.

3 Land banking, or the accumulation of contaminated sites by local authorities and private entities to aggregate adjacent smaller polluted sites, such as along railway corridors or in docklands, provides

Figure 5.16a Military Uses: Sources of Contamination

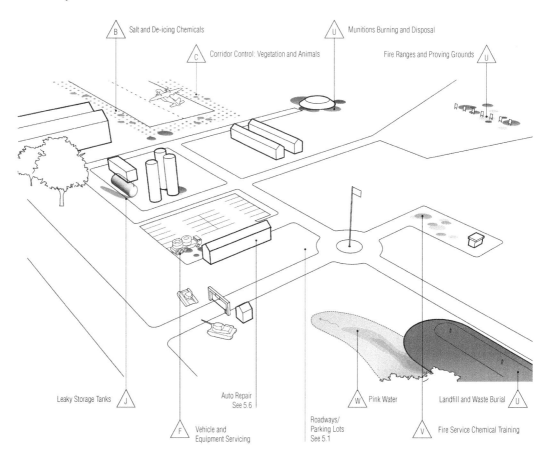

B Salt and De-icing Chemicals

U Munitions Burning and Disposal

C Corridor Control: Vegetation and Animals

Fire Ranges and Proving Grounds U

Leaky Storage Tanks J

Auto Repair
See 5.6

W Pink Water

Landfill and Waste Burial U

F Vehicle and
Equipment Servicing

Roadways/
Parking Lots
See 5.1

V Fire Service Chemical Training

Key	Source	Description	Contaminants
A	Roadways: road and truck debris	See 5.1: Roadways/Parking Lots	Nutrients: Ch. 3, p. 125 Metals: Ch. 3, p. 136 Petroleum: Ch. 3, p. 65 Air pollution: Ch. 3, p. 189
B	Salt and de-icing chemicals	Salt and de-icing chemicals are frequently used on airfields during cold winter months. These pollutants can quickly leach into groundwater if not controlled.	Salt (sodium, chloride and other additives): Ch. 3, p. 124 De-icing – other organic contaminants of concern: Ch. 3, p. 124
J	Leaking underground storage tanks (LUSTs)/above-ground storage tanks and barrels	Any underground or above-ground storage tank or container used to store fuel or solvents used in military activities may leak over time.	Petroleum: Ch. 3, p. 65 Chlorinated solvents: Ch. 3, p. 94
L	Machinery and trucks: operations and servicing	See 5.6: Gas Stations and Auto-Repair Shops. Truck and equipment servicing can leak fuel, as well as other fluids.	Petroleum: Ch. 3, p. 65 Chlorinated solvents: Ch. 3, p. 94
U	Munitions: burning and disposal/landfill and waste burial and firing ranges	When unused or unexploded munitions are discarded, they may be disassembled and landfilled on site. Munitions testing often leaves unexploded remnants which remain in soils and groundwater.	Chlorinated solvents: Ch. 3, p. 94 Explosives: Ch. 3, p. 103 Radionuclides: Ch. 3, p. 182 Metals: Ch. 3, p. 136
V	Fire and chemical training	Drills for fire, chemical fires and chemical warfare control may be conducted at military bases. TCE was historically used as a common fire retardant.	Chlorinated solvents: Ch. 3, p. 94
W	Pink water	Explosives and radionuclides that have leached into water are called 'pink water' since the color of the leachate is a bright pink. Pink water can be generated from landfilled munitions or unexploded ordnance or unintentionally buried munitions.	Chlorinated solvents: Ch. 3, p. 94 Explosives: Ch. 3, p. 103 Radionuclides: Ch. 3, p. 182 Metals: Ch. 3, p. 136

Figure 5.16b Military Uses: Phytotypologies to Address Contaminants

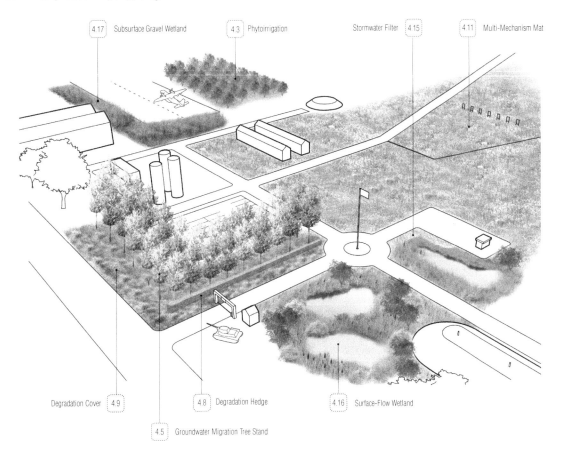

4.17 Subsurface Gravel Wetland   4.3 Phytoirrigation   Stormwater Filter 4.15   4.11 Multi-Mechanism Mat

Degradation Cover 4.9   4.8 Degradation Hedge   4.16 Surface-Flow Wetland

4.5 Groundwater Migration Tree Stand

297

Key	Typology	Description	Addresses	Plant Lists
4.3 (p. 207)	Phytoirrigation	Excess run-off and polluted water generated from de-icing activities can be collected in holding ponds and irrigated onto plants for degradation of the de-icing fluids.	Within water: ● Petroleum	p. 48, 2.18 High evapotrans-piration-rate species
4.5 (p. 213)	Groundwater Migration Tree Stand	Military sites often need to remain spare of trees for training and defense purposes. However, on the downgradient side of bases, where the military site meets the public, trees can be installed to intercept polluted groundwater and provide privacy screening. RDX and HMX explosives and propellants used on military sites can quickly travel and migrate in groundwater. Having an established tree stand around active sites creates the potential for intercepting any new contaminant that may be released. The trees act as pumps, controlling the water and potentially degrading the RDX and HMX. The metabolites produced in the degradation process must be carefully monitored, as they may still be toxic. A detailed Mass Water Balance must be conducted by an engineer to determine how many trees are needed to stop the migrating plume, and if it is possible.	Leachate: ● Petroleum	p. 48, 2.18 High evapotrans-piration-rate species
4.11 (p. 227)	Multi-Mechanism Mat	Where training grounds and firing ranges need to be left open, low-growing species can be installed to help stabilize contaminants. At this time, degradation of RDX and HMX has been shown to be possible in low-growing grass species; however, there are significant challenges with applying this in the field. Usually the soil is also contaminated with TNT or other substances that prevent plant growth. In addition, the metabolites that the RDX and HMX are broken down into can still be toxic. Radionuclides, if present, cannot be extracted, only stabilized on site. Plantings may be designed to create a thick vegetated mat to help cap pollutants on site.	◐ Chlorinated solvents; ● Explosives; ■ Radionuclides; ■ Metals; ● Petroleum	p. 140, 3.37 Metals Excluders p. 106, 3.19 Explosives
4.16 (p. 238)	Surface-Flow Wetland	Pink water has been shown to be effectively remediated with carefully designed Surface-Flow Wetlands. A series of ponds/cells is created to break down and trap the explosives, producing filtered water.	● Explosives	Wetland species not included in this publication
4.17 (p. 241)	Subsurface Gravel Wetland	De-icing chemicals can be successfully broken down in wetland systems. However, open-water Surface-Flow Wetlands are not desired near runways, since they tend to attract birds. In these scenarios Subsurface Gravel Wetlands can be used instead to collect and break down the de-icing fluids. Subsurface Gravel Wetlands have no open water and therefore are not as attractive to birds and wildlife.	De-icing chemicals	Wetland species not included in this publication

phytotechnology projects with increased scale and opportunity for phased installation/harvesting. The authors suggest that this offers an important area for landscape architects and site designers to plan and implement larger phytotechnology installations over time.

4 The evolution of the applied science, installation protocols and maintenance of phytotechnology will advance in coming years, despite the restrictions in funding mentioned earlier in the book. Other technologies to address groundwater and soils pollution are currently being developed using a wide range of methods, including bioremediation, thermal and electrical techniques. The authors consider that the growing interest, invention and investment in remediation technologies in general will assist the development of phytotechnology, which has an inherent ability to be twinned with other emerging methods in remediation treatment trains.

For the landscape architect and site designer there remains a need for a level of support via resources on phytotechnology in printed and digital media. Chapter 6, which follows, will outline a range of useful resources for the professional or student engaged in phytotechnology research or projects.

298

6: Additional resources

As has been noted in the Preface of this book and in subsequent chapters, the field of phytotechnology still requires the further development and documentation of a significant amount of applied research. The further evolution of phytotechnology installations and their monitoring need to be tested out in a variety of site conditions. This will continually change and will require access to largely dispersed information both nationally and internationally. In this chapter a list of resources from which to source information about phytotechnologies is provided for readers who may be pursuing their own research or attempting to carry out phytotechnology applications as part of design and development projects. These resources are divided into the following three areas.

1 *Organizations:* A list of governmental, academic, private and non-profit entities supporting phytotechnology research and innovation.
2 *Documents:* A summary of documents, books or journals that may be particularly helpful in specific areas of the phytotechnology field.
3 *Plant lists:* Resources to find additional plants for phytotechnologies, not listed in this book.

1   Organizations

International Phytotechnology Society (IPS)

http://www.phytosociety.org

This is the most important and relevant non-profit organization for the field. It is a worldwide professional society comprised of individuals and institutions engaged in the science and application of using plants to deal with environmental problems. Every year, the IPS holds an annual conference bringing together the latest scientific researchers and consultants, and all members of the public are welcome.

US Environmental Protection Agency (US EPA)

http://www.cluin.org

The US EPA was actively involved in funding research and investigating opportunities for phytotechnologies in the 1990s. This US government regulatory agency continues to provide some outreach and support to the field. Its website provides links to overview documents on the science and a project database of over 165 installed phytoremediation field sites.

Offshoots, Inc.

http://www.offshootsinc.com

Kate Kennen, co-author of this book, founded Offshoots, Inc. as a landscape architecture practice dedicated to consulting on phytotechnology projects. The studio is based in Boston and assists other designers and engineers in completing phytotechnology work nationally. Information on phytotechnologies is continually updated on the firm's website.

Greenland Project

http://www.greenland-project.eu

This is a collaboration of scientists working in Europe to share research on the gentle remediation of lands contaminated with metals. (See Chapter 3, pp. 159 and 171 for additional information on two case studies participating in this project.) The objective is to develop plant-based approaches to remediate large areas of polluted soils at low cost and without significant negative effects for the environment.

302

European Union COST (Cooperation in Science and Technology) Groups: Cost Action 837 (Phytoremediation), 859 (Phyto and Food Safety) and FA0901 (Putting Halophytes to Work – From Genes to Ecosystems)

http://lbewww.epfl.ch/COST837, http://w3.gre.ac.uk/cost859/ and *http://www.cost.eu/domains_actions/ fa/Actions/FA0901*

These three groups funded by the European Union consist of scientists collaborating to research phytoremediation issues specifically affecting Europe. Although these groups are no longer active, their notes and research are available online, as well as links to international resources.

Association for Environmental Health and Sciences Foundation

http://www.aehsfoundation.org

Dedicated to the challenge of soil, sediment and water cleanup and protection, this organization is a professional society of scientists and consultants investigating all kinds of sustainable remediation practices, not just phytotechnologies. It is a larger, more broad organization than IPS that addresses the field of remediation in general, and this non-profit holds two annual conferences each year.

2 Documents

International Journal of Phytoremediation

http://www.tandfonline.com

This journal is strictly dedicated to the field of phytoremediation. The latest peer-reviewed science and literature on plant-based systems can be found here. It is published on a quarterly basis.

Additional publications are listed by subject below. Many additional books are written on the subject of phytotechnologies; however, most are very scientific and quite technical. Included here are only overview materials that may be more accessible for designers. For a thorough list of applicable references, refer to the Bibliography on page 313.

A Upland, land-based phytotechnology systems

ITRC – PHYTO 3 (Interstate Technology and Regulatory Council, Phytotechnology Technical and Regulatory Guidance and Decision Trees, Revised)

http://www.itrcweb.org

This free, online 'how to' document is a step-by-step practical guide that details how to go about designing an upland phytotechnology installation. It lists critical regulatory, site analysis and site planning considerations, and additionally provides a good overview of the phyto field in general.

B Groundwater

Introduction to Phytoremediation of Contaminated Groundwater: Historical Foundation, Hydrologic Control, and Contaminant Remediation

Book by James E. Landmeyer (Springer, 2012 edition)

This book is a comprehensive text for understanding and evaluating the design, implementation and monitoring of phytoremediation projects for sites with contaminated groundwater.

C Constructed wetland systems

Treatment Wetlands, 2nd edition

Book by R. H. Kadlec and S. D. Wallace (Boca Raton, FL: CRC Press, Lewis Publishers, 2009)

This book is the seminal text on constructed wetlands, often referred to by regulatory agencies for its documentation of contaminant removal rates and case studies.

303

ITRC – WTLND-1 (Interstate Technology and Regulatory Council, Technical and Regulatory Guidance Document for Constructed Treatment Wetlands)

http://www.itrcweb.org

This free, online 'how to' document is a step-by-step practical guide that details the regulatory and technical processes for creating constructed wetlands.

ITRC – WTLND-2 (Interstate Technology and Regulatory Council, Characterization, Design, Construction, and Monitoring of Mitigation Wetlands)

http://www.itrcweb.org

This free, online 'how to' document is a step-by-step practical guide that details the characterization, design, construction and monitoring practices of constructed wetlands.

D Phytoforensics

Users Guide to the Collection and Analysis of Tree Cores to Assess the Distribution of Subsurface Volatile Organic Compounds

http://pubs.usgs.gov/sir/2008/5088

This free, online document is a technology transfer describing the process and techniques of using trees to track volatile subsurface organic contaminants. Case studies are included.

3 Plant lists

NDSU Phytoremediation Plant Database

http://www.ndsu.edu/pubweb/famulari_research/

This online database of potential phytoremediation plants species was compiled by Stevie Falmulari, a professor of Landscape Architecture, and her students at North Dakota State University from 2007 to 2011. Some of the plants included are from older studies and the plant species may no longer be viable, but it is the only known online database that includes photos of the plants with the species name and that is searchable by contaminant.

A Organics

PHYTOPET

PHYTOPET was a database developed by the University of Saskatchewan, Canada and its partners in the late 1990s as an interactive electronic database of plant species that degrade petroleum hydrocarbons in soil, sediment and water. This database is no longer available online but a copy may be obtained by contacting the University.

B Inorganics

Global Metallophyte Database (Plants that can tolerate high levels of metals)

http://www.metallophytes.com

Maintained by the International Serpentine Ecology Society and Centre for Mined Land Rehabilitation, this is a recently established database of metal-tolerant plants and will be enhanced and added to in coming years (Van der Ent et al., 2013).

PHYTOREM

PHYTOREM was a database developed by Environment Canada and its partners in 2003 as a worldwide interactive electronic database available on CD of more than 700 plants, lichens, algae, fungi and bryophytes with a demonstrated capacity to tolerate, accumulate or hyperaccumulate a range of 19 different metals. It is no longer available online, but may be requested by contacting Environment Canada.

Tropical nickel- and selenium-hyperaccumulating species

Roger D. Reeves. 'Tropical hyperaccumulators of metals and their potential for phytoextraction.' Plant and Soil 249 (2003), pp. 57–65.

This journal article is a review of many research studies and provides a list of many nickel- and selenium-hyperaccumulating species found in tropical environments.

Halophyte species with potential for metals accumulation

Eleni Manousaki and Nicolas Kalogerakis. 'Halophytes – an emerging trend in phytoremediation.' International Journal of Phytoremediation, 13 (2011), pp. 959–969.

This journal article is a review of several research studies and provides a list of some salt-accumulating plants that also have been considered for metals extraction in polluted soils.

C Air pollution

How to Grow Fresh Air: Plants that Purify Your Home or Office

Book by Dr. B. C. Wolverton (London/New York: Penguin, 1997)

This book lists the top 50 tropical plant species that can be grown indoors for air-quality improvement.

Mitigating New York City's Heat Island with Urban Forestry, Living Roofs, and Light Surfaces, Appendix A

Compiled by David J. Nowak of the USDA Forest Service (Syracuse, NY: USDA Forest Service, 2006)

This publication lists approximately 200 of the best urban tree species for the city of New York, ranked by functional attributes including: air pollution removal, air temperature reduction, tree shade, building energy conservation, carbon storage, pollen allergenicity and life span. The list considers the removal of air pollutants, but it also considers other important functions for urban trees.

305

Afterword

Dr. Lee Newman and Dr. Jason White

As in all aspects of life, there are tasks that can be completed very well by the individual, and tasks that are best accomplished by working as a member of a team. The discipline of phytotechnologies is very much of the second variety. The most effective and successful projects are those that bring together a suite of professionals, each with their own areas of expertise, to accomplish what none of them could do alone. For many years, it was difficult to get traditional remediation professionals to view phytoremediation as it truly is – a multidisciplinary team effort. Not only does phytoremediation work best as a multidisciplinary effort, but in fact that is the only way it works. To complete the field assessment, design the remedial strategy, implement the plan, and perform long-term maintenance and monitoring requires not only the accomplished engineer, but also the soil scientist, the agronomist, the forester, the analytical chemist and the plant physiologist to get it right. But as phytoremediation has moved beyond the laboratory and grown through the hype, more of these teams are working together to bring the knowledge and expertise to construct and implement remedial systems which are both biological and engineered.

However, even within this team of professionals, most if not all of these sites lacked an aesthetic and community engagement vision and perhaps a broader perspective beyond achieving a cleanup goal. Plants were selected for the job that they would do, thus monocultures were the norm. And while there certainly is a majestic beauty in rows of towering poplar trees cleaning the groundwater, or a field of mounding herbaceous plants accumulating heavy metals or nourishing the microbes that in turn are degrading organic contaminants, sites were designed solely and completely as engineered systems to achieve a regulatory goal. And for sites with large amounts of contamination, this is likely how the vision will remain.

However, not every site is a Superfund site, and not every site poses dramatic risks to the health of a person walking the site. In fact, the vast majority of sites are marginally to moderately contaminated. Some sites need final polishing steps; some need some long-term stabilization efforts. Some sites simply need an economical and resource-friendly solution. And perhaps most forgotten, many sites need to be brought back into the public sphere; to move from 'contaminated site' to 'useful site'.

And this is where the current team of expertise falls short. We as scientists and engineers are very good at designing installations to meet a regulatory target but not often are we very aesthetically creative or do we think long term about other relationships with larger human and natural systems. We see the beauty of the clean site, but not often beyond. This is where it becomes apparent that we need to include new team members, and the field of landscape architecture will provide the team members we need to embrace.

As a cautionary note, just as it was important for the engineers and the scientists to learn that we could not do this alone, it is equally important for the landscape architects to realize that they need to be a part of or build their own teams that bring together the required multidisciplinary expertise as well. And in a sentence, that is the goal of this text. By creating a single source containing the essence of phytotechnological approaches, including an historical perspective on both successes and failures, this text can serve as the primary tome for landscape architects, from the student to the practitioner, to guide effective, functional and aesthetic site design and implementation. This text serves to emphasize the point that no one person or group has the ability to do this alone, but that we need to consistently evaluate our teams and recognize when new knowledge is needed so as to achieve success.

In this, the landscape architects bring a whole new area of expertise to the table. By working with remediation experts, they can design sites that will not only meet the goals of regulatory cleanup or stabilization, but also bring the potential for the location to become a park, or a walking trail, or a nature preserve. They can work with the engineers and the scientists to design systems that enhance the remediation efficiency, promote plant productivity and biodiversity and, at the same time, open the lands back up for public use and appreciation. Lands taken from society can be returned.

Just as every site is not a candidate for phytotechnology, not every phytotechnology site is a candidate for a design scheme that would allow for public access and appreciation. But on the vast majority of sites where this approach is an option, it is time to open up the teams to accept and welcome new expertise; members that allow us not only to utilize a plant-based system for remediation, but to design that system in such a way as to bring the public back to these newly accessible lands.

Glossary

(* = Definition from ITRC, 2009)

absorption* The process of one substance actually penetrating into the structure of another substance. This is different from adsorption, in which one substance adheres to the surface of another substance.

adsorption* The physical process occurring when liquids, gases, or suspended matter adhere to the surfaces of, or in the pores of, an adsorbent material. Adsorption is a physical process which occurs without chemical reaction.

aerobic* An environment that has a partial pressure of oxygen similar to normal atmospheric conditions.

aliphatic Organic compounds joined together in straight chains, branched chains, or non-aromatic rings.

anaerobic* An environment without oxygen or air.

anion A negatively charged ion.

anoxic* An atmosphere greatly deficient in oxygen.

bacteria* A group of diverse and ubiquitous prokaryotic single-celled microorganisms.

bioaccumulation* Intracellular accumulation of environmental pollutants such as heavy metals by living organisms.

bioavailability The proportion of a contaminant that is available for uptake by a plant.

biodegradation* The breakdown of organic substances by microorganisms.

bioremediation* The process by which living organisms are used to degrade or transform hazardous organic contaminants.

brownfield An abandoned, idled, or underused industrial or commercial facility where expansion or redevelopment is complicated by a real or perceived environmental contamination.

capillary fringe* The porous material just above the water table which may hold water by capillarity (a property of surface tension that draws water upward) in the smaller soil void spaces.

cation A positively charged ion.

chelate* Chelating agents are used to remove ions from solutions and soil. The type of coordination compound in which a central metallic ion (CO^{2+}, Ni^{2+}, or Zn^{2+}) is attached by covalent bonds to two or more nonmetallic atoms in the same molecule, called 'ligands.'

The Comprehensive Environmental Response, Compensation, and Liability Act (CERCLA) See *Superfund*.

creosote* An antifungal wood preservative used frequently to treat telephone poles and railroad ties. Creosote consists of coal tar distillation products, including phenols and PAHs.

deed restriction A limitation on the use of a property that is recorded on the deed to the property. The limitations on use are legally enforceable against the owner of the property; but who may enforce the limitation depends on state law.

due diligence Evaluation of the environmental condition of a parcel of land, often as part of a real estate transaction. This is required in order for a purchaser to qualify for federal liability protection as an innocent purchaser. See also *environmental assessment*.

ecotype A genetically distinct geographic variety, population, or race within species.

environmental assessment A site evaluation or investigation conducted for purposes of determining the extent, if any, of contamination on a property. An assessment can be informal or formal, and can consist of several stages. For example, a Phase I assessment, or basic study of possible contamination at a site, is limited to collecting information about past and present site use and inspecting present conditions. A Phase II assessment can follow up a Phase I assessment with sampling and analysis of suspected contaminated areas of a site. A Phase III assessment can follow up a Phase II assessment either by gathering information on the exact extent of the contamination or by preparing plans and alternatives for site cleanup.

Environmental Protection Agency (EPA) Federal government regulatory agency in the United States responsible for enforcing laws pertaining to the natural environment and regulating the cleanup of contaminated sites.

enzyme* Protein that acts as a biological catalyst. These chemicals produced by living organisms bring about the digestion (breakdown) of organic molecules into smaller units that can be used by living cell tissues.

evapotranspiration* Water lost to the atmosphere from the ground surface, evaporation from the capillary fringe of the groundwater table, and the transpiration of groundwater by plants whose roots tap the capillary fringe of the groundwater table.

exsitu On site and using excavation to assemble a contaminated medium.

excluder A plant that will not extract a certain kind of contaminant but can live in soils elevated with that contaminant.

exudate* Soluble organic matter released from the roots of plants to enhance availability of nutrients or as a by-product of fine-root degradation.

greenfield Land that has not been previously used for site development or infrastructure.

greenhouse or lab study* Study conducted to evaluate the ability of green plants to grow in toxic soil or water environments. Greenhouse studies are normally conducted during treatability studies.

groundwater* Water found beneath the surface of the ground. Groundwater is primarily water which has seeped down from the surface by migrating through the interstitial spaces in soils and geologic formations.

halophytes Salt-tolerant plants that grow in saline environments such as salt water, soils with high salt, or come into contact with salt spray.

hot spots Specific areas where the level of contamination is very high.

hydrocarbon An organic compound consisting entirely of hydrogen and carbon.

hydrophobic* Repelling, tending not to combine with, or incapable of dissolving in water.

*in situ** In place, without excavation.

infrastructure The roads, utility lines, and other public amenities that support property use.

inorganic contaminants Inorganic pollutants are naturally occurring elements on the periodic table such as lead and arsenic. Human activities such as industry production and extraction mining create a release of inorganic pollutants into the environment, causing toxicity. These are elements, so they cannot be degraded and destroyed; instead they can sometimes be taken up and extracted by plants. If extraction is possible, the plants must be cut down and harvested to remove the pollutant from a site.

institutional controls Legal and administrative mechanisms designed to reduce exposure to contamination. Examples include: deed restrictions, easements, warning signs and notices, and zoning restrictions.

ion An atom or molecule in which the total number of electrons is not equal to the total number of protons, giving the atom a net positive or negative electrical charge.

Leaf Area Index (LAI) LAI is a measure of plant canopy thickness. A high LAI indicates that the plant generates significant leaf material between the ground and the top of the plant, so that a raindrop will hit a larger number of leaves or greater leaf area before it finally reaches the ground.

Licensed Site Professional (LSP) An engineer, environmental scientist, or geoscientist licensed by the State, who is qualified to assess contamination and conduct cleanups. An LSP certifies compliance with the MCP by issuing a final opinion at the completion of a cleanup.

log K_{ow}* The octanol-water partition coefficient, a dimensionless constant which provides a measure of how an organic compound will partition between an organic phase and water. A low log K_{ow} indicates that a chemical readily partitions into a water phase; a high log K_{ow} indicates that the chemical prefers to stay in the organic phase. It provides an indication of the quantity of the chemical that will be taken up by the plants.

metabolite The intermediates and products of chemical transformations generated within the cells of living organisms such as soil microbes and plants.

microorganism* Includes bacteria, algae, fungi, and viruses.

National Priorities List (NPL) The Environmental Protection Agency's list of the most serious uncontrolled or abandoned hazardous waste sites.

No-Further-Action (NFA) Letter A written statement by a state government that it has no present intention to take legal action or require additional cleanup by a party that satisfactorily cleans up a property under a state brownfield or voluntary cleanup program.

Nonresidential Use Standard A cleanup standard, usually expressed as a numerical ratio of parts of a specific contaminant to parts of the medium of concern (e.g., 5 parts of lead per million parts of soil) that describes the maximum concentration of the contaminant in the medium that will not present an unacceptable risk to the health of humans engaging in any activity other than residential or those other activities considered to be substantially similar to residential. The Nonresidential Use Standard is usually a less strict cleanup standard than the *Residential Use Standard*, and a site that meets the nonresidential standard is limited in its uses to nonresidential activities.

310

nutrients* Elements or compounds essential as raw materials for organism growth and development. Nitrogen, phosphorus, potassium, and numerous other mineral elements are essential plant nutrients.

organic contaminants Compounds that typically contain bonds of carbon, nitrogen, and oxygen, are man-made and foreign to living organisms. Many organic contaminants can be degraded with phytotechnologies, breaking them down into smaller, less toxic components. Organic contaminants may be degraded outside the plant in the root zone, taken into a plant, bound to the plant tissues, degraded to form non-toxic metabolites, or released to the atmosphere. If persistent, organic contaminants may not be able to be degraded by plants.

parts per billion* (ppb) A measure of proportion by weight which is equivalent to one unit weight of solute (dissolved substance) per billion unit weights of the solution. One liter of water weighs 1 billion micrograms, and 1 ppb is the equivalent of 1 microgram per liter (μg/L) when used for water analysis.

parts per million* (ppm) A measure of proportion by weight which is equivalent to one unit weight of solute (dissolved substance) per million unit weights of the solution. One liter of water weighs 1 million milligrams, and 1 ppm is equal to 1 milligram per liter (mg/L) when used for water analysis.

persistent Chemical which accumulates in the environment due to its stability.

petrochemicals Petroleum-derived chemicals.

phreatophyte A deep-rooted plant that obtains a significant portion of the water that it needs from groundwater. Phreatophytes are plants that often have their roots constantly in touch with moisture.

phytoaccumulation* The accumulation of pollutants on plant tissues.

phytobuffer An area planted before a contamination event occurs to preventatively attempt to treat pollution that may occur in the future.

phytodegradation* The process whereby plant-produced enzymes break down dissolved organic contaminants that are in the plant through the uptake of water.

phytoextraction* The uptake and accumulation of inorganic elements into the plant tissues.

phytoforensics Use of plant sampling to track subsurface contaminants.

phytoremediation* Use of plants to remediate contaminated soil, sediments, surface water, or groundwater.

phytosequestration* The ability of plants to hold and stabilize certain inorganic elements in the plant and the root zone.

phytostabilization* See *phytosequestration*.

phytotoxic* Harmful to plants.

phytovolatilization* The uptake and subsequent transpiration of volatile contaminants through the plant leaves.

polycyclic aromatic hydrocarbon (PAH) A hydrocarbon compound with multiple benzene rings. PAHs are typical components of asphalts, fuels, oils, and greases.

radionuclides Inorganic elements with an unstable nucleus, characterized by excess energy available.

recalcitrant Does not break down easily, persistent.

Residential Use Standard A cleanup standard, usually expressed as a numerical ratio of parts of a specific contaminant to parts of the medium of concern (e.g., 5 parts of lead per million parts of soil) that describes the maximum concentration of the contaminant in the medium that will not present an unacceptable risk to the health of humans residing on the site, or engaging in activities on the site that are considered to be substantially similar to residing on the site. The Residential Use Standard is usually the strictest cleanup standard, and a site that meets this standard can usually be used for any purpose.

The Resource Conservation and Recovery Act (RCRA) A federal statute that regulates the generation, transportation, storage, treatment, and disposal of hazardous waste. RCRA programs include the Corrective Action and Underground Storage Tank Programs.

restrictive covenant A specific type of deed restriction. For example, a restrictive covenant could prohibit commercial uses.

rhizodegradation* Biodegradation of organics by the soil organisms. Exuded plant products through phytosequestration can lead to enhanced biodegradation in the rhizosphere.

rhizofiltration* Trapping of contaminants by the roots of plants immersed in water and soils.

rhizosphere* Soil in the area surrounding plant roots that is influenced by the plant root. Typically a few millimeters or, at most, centimeters from the plant root. Important because this area is higher in nutrients and thus has a higher and more active microbial population.

risk assessment A study or evaluation that identifies and in many cases quantifies the potential harm posed to health and the environment by contamination on a property.

site assessment or site characterization Identification of contaminants and evaluation of the extent of contamination on a hazardous waste site.

sorbtion A physical and chemical process by which one substance becomes attached to another, including both absorbtion, adsorption and ion exchange.

Superfund The Comprehensive Environmental Response, Compensation, and Liability Act (CERCLA or 'Superfund'). A federal statute that governs the investigation and cleanup of sites contaminated with hazardous substances. The law establishes a trust fund that can be used by the government to clean up sites on the National Priorities List (NPL).

311

total petroleum hydrocarbon (TPH)   Term used for any mixture of hydrocarbons that are found in crude oil. There are several hundred of these compounds, but not all occur in any one sample. Crude oil is used to make petroleum products, which can range from gasoline to diesel to oil. Because there are so many different chemicals in crude oil and in other petroleum products, it is not practical to measure each one separately. However, it is useful to measure the total amount of TPH at a site.

toxic substances*   Chemical elements and compounds such as lead, benzene, dioxin, and others that have toxic (poisonous) properties when exposure by ingestion, inhalation, or absorption into the organism occurs. There is a large variation in the degree of toxicity among toxic substances and in the exposure levels that induce toxicity.

translocation*   Cellular transport through the plant vascular system (xylem) from roots to other plant tissues.

transpiration*   The plant-based process involving the uptake, transport, and eventual vaporization of water through the plant body.

treatment train   Remediation techniques executed in sequence to clean up a polluted site.

volatile organic compound* (VOC)   Synthetic organic chemical capable of becoming vapor at relatively low temperatures.

voluntary cleanups   Cleanups of identified contamination that are not court or agency ordered. Most states have voluntary cleanup programs that encourage voluntary cleanups and that may provide benefits if volunteers meet specified standards.

water table*   The level at the top of the zone of groundwater saturation.

zone of saturation*   The layer in the ground in which all available interstitial voids (cracks, crevices, and holes) are filled with water. The level at the top of this zone is the *water table*.

312

Bibliography

Abaga, N., Dousset, S., Munier-Lamy, C., and Billet, D. 2014. Effectiveness of Vetiver grass (Vetiveria zizanioides L. nash) for phytoremediation of endosulfan in two cotton soils from Burkina Faso. *International Journal of Phytoremediation* 16 (1), pp. 95–108, DOI, 10.1080/15226514.2012.759531.

Abhilash, P. C., and Singh, N. 2010. Effect of growing Sesamum indicum L. on enhanced dissipation of Lindane (1, 2, 3, 4, 5, 6-Hexachlorocyclohexane) from soil. *International Journal of Phytoremediation* 12, pp. 440–453.

Abhilash, P. C., Jamila, S., Singh, V., Singh, A., Singh, N., and Srivastava, S. C. 2008. Occurrence and distribution of hexachlorocyclohexane isomers in vegetation samples from a contaminated area. *Chemosphere* 72 (1), pp. 79–86.

Adamia, G., Ghoghoberidze, M., Graves, D., Khatisashvili, G., Kvesitadze, G., Lomidze, E., Ugrekhelidze, D., and Zaalishvili, G. 2006. Absorption, distribution and transformation of TNT in higher plants. *Ecotoxicology and Environmental Safety* 64, pp. 136–145.

Adesodun, J. K., Atayese, M. O., Agbaje, T., Osadiaye, B. A., Mafe, O., and Soretire, A. A. 2010. Phytoremediation potentials of sunflowers (Tithonia diversifolia and Helianthus annuus) for metals in soils contaminated with zinc and lead nitrates. *Water, Air, and Soil Pollution* 207 (1–4), pp. 195–201.

Adler, A., Karacic, A., and Weih, M. 2008. Biomass allocation and nutrient use in fast-growing woody and herbaceous perennials used for phytoremediation. *Plant and Soil* 305 (1–2), pp. 189–206.

Adler, T. 1996. Botanical cleanup crews, *Science News* 150, pp. 42–43.

AIBS (American Institute of Biological Sciences) 2013. Testimony in Support of FY 2014 Funding for the United States Geological Survey, United States Forest Service, and Environmental Protection Agency. Submitted to: Senate Committee on Appropriations, Subcommittee on Interior, Environment and Related, May 2013. Retrieved from http://www.actionbioscience.org.

Albright, R. D. 2013. *Cleanup of Chemical and Explosive Munitions: Locating, Identifying the Contaminants, and Planning for Environmental Cleanup of Land and Sea Military Ranges and Dumpsites*. Elsevier Science.

Albright III, V., and Coats, J. 2014. Disposition of atrazine metabolites following uptake and degradation of atrazine in Switchgrass. *International Journal of Phytoremediation* 16 (1), pp. 62–72, DOI, 10.1080/15226514.2012.759528.

Alexander, M. 1994, *Biodegradation and Bioremediation*. California: Academic Press, Inc.

Alexander, M. 2000. Aging, bioavailability, and overestimation of risk from environmental pollutants. *Environmental Science and Technology* 34, pp. 4259–4265.

Algreen, M., Trapp, S., and Rein, A. 2013. Phytoscreening and phytoextraction of heavy metals at Danish polluted sites using willow and poplar trees. *Environmental Science and Pollution Research*, epub. ahead of print, DOI, 10.1007/s11356-013-2085-z.

Allen, H. L., Brown, S. L., Chaney, R., Daniels, W. L., Henry, C. L., Neuman, D. R., Rubin, E., Ryan, J. A., and Toffey, W. 2007. The use of soil amendments for remediation, revitalization and reuse. US EPA 542-R-07-013.

Allen, R. G., Jensen, M. E., Wright, J. L., and Burman, R. D. 1989. Operational estimates of reference evapotranspiration. *Agronomy Journal* 81 (4), pp. 650–662.

Alvarenga, P., Varennes, A., and Cunha-Queda, A. 2014. The effect of compost treatments and a plant cover with Agrostis tenuis on the immobilization/mobilization of trace elements in a mine-contaminated soil. *International*

Journal of Phytoremediation 16 (2), pp. 138–154, DOI, 10.1080/15226514.2012.759533.

Anderson, L. L., Walsh, M., Roy, A., Bianchetti, C. M., and Merchan, G. 2010. The potential of Thelypteris palustris and Asparagus sprengeri in phytoremediation of arsenic contamination. *International Journal of Phytoremediation* 13 (2), pp. 177–184.

Anderson, T. A., and Walton, B. T. 1991. Fate of trichloroethylene in soil-plant systems. In *American Chemical Society, Division of Environmental Chemistry, Extended Abstracts*, pp. 197–200.

Anderson, T. A., and Walton, B. T. 1992. *Comparative Plant Uptake and Microbial Degradation of Trichloroethylene in the Rhizospheres of Five Plant Species: Implication for Bioremediation of Contaminated Surface Soils*. ORNL/TM-12017. Oak Ridge, TN: Oak Ridge National Laboratory, Environmental Sciences Division.

Anderson, T. A., and Walton, B. T. 1995. Comparative fate of 14C trichloroethylene in the root zone of plants from a former solvent disposal site. *Environmental Toxicology and Chemistry* 14 (12): pp. 2041–2047.

Anderson, T. A., Guthrie, E. A., and Walton, B. T. 1993. Bioremediation in the rhizosphere. *Environmental Science and Technology* 27, pp. 2630–2636.

Angelova, V., Ivanova, R., Delibaltova, V., and Ivanov, K. 2011. Use of sorghum crops for in situ phytoremediation of polluted soils. *Journal of Agricultural Science and Technology A* 1 (5), pp. 693–702.

Angle, J. S., and Linacre, N. A. 2005. Metal phytoextraction: a survey of potential risks. *International Journal of Phytoremediation* 7 (3), pp. 241–254.

Anjum, N. A., Pereira, M. E., Durate, A. C., Ahmad, I., Umar, S., and Khan, N. A. (Eds.). 2013. *Phytotechnologies: Remediation of Environmental Contaminants*. New York: CRC Press.

Antunes, M. S., Morey, K. J., Smith, J. J., Albrecht, K. D., Bowen, T. A. et al. 2011. Programmable ligand detection system in plants through a synthetic signal transduction pathway, *PLoS ONE* 6 (1), e16292. DOI, 10.1371/journal.pone.0016292.

Applied Natural Sciences, Inc. 1997. *Site Data Information*. Fairfield, OH: Applied Natural Sciences, Inc.

Aprill, W., and Sims, R. C. 1990. Evaluation of the use of prairie grasses for stimulating polycyclic aromatic hydrocarbon treatment in soils. *Chemosphere* 20, pp. 253–265.

Arnold, C. W., Parfitt, D. G., and Kaltreider, M. 2007. Phytovolatilization of oxygenated gasoline-impacted groundwater at an underground storage tank site via conifers. *International Journal of Phytoremediation* 9 (1–3), pp. 53–69.

Aronow, L., and Kerdel-Vegas, F. 1965. Seleno-cystathionine, a pharmacologically active factor in the seeds of *Lecythis ollaria*. *Nature* 205, pp. 1185–1186.

Arthur, W. J. III. 1982. Radionuclide concentrations in vegetation at a solid radioactive waste-disposal area in southeastern Idaho. *Journal of Environmental Quality*, 11 (3), pp. 394–399.

Astier, C., Gloaguen, V., and Faugeron, C. 2014. Phytoremediation of cadmium-contaminated soils by young Douglas Fir trees: effects of cadmium exposure on cell wall composition. *International Journal of Phytoremediation* 16, pp. 790–803.

ATC. 2013. Drycleaner Site Profiles Pinehurst Hotel Cleaners, Pinehurst, North Carolina website http://www.drycleancoalition.org/profiles/?id=2099 and NCDENR.org website http://portal.ncdenr.org/c/document_library/get_file?uuid=16e13771-0f66-427e-96d6-6bcf58a71384&groupId=38361. Both accessed 12 March 2013.

Atkinson, R. 1989. Kinetics and mechanisms of the gas-phase reactions of the hydroxyl radical with organic compounds. *Journal of Physical and Chemical Reference Data*, Monograph No. 1.

ATSDR. 2013. http://www.atsdr.cdc.gov/. Accessed 12 May 2013 for review of toxic substances including the following: arsenic, cadmium, selenium, molybdenum, fluorine, formaldehyde, nickel, lead, mercury, aluminum, chromium, zinc, copper, and cobalt.

Bačeva, K., Stafilov, T., and Matevski, V. 2013. Bioaccumulation of heavy metals by endemic Viola species from the soil in the vicinity of the As-Sb-Tl Mine "Allchar", Republic of Macedonia. *International Journal of Phytoremediation* 16, pp. 347–365.

Baker, A. J. M. 2000. Metal-accumulating plants. In I. Raskin and B. Ensley (Eds.) *Phytoremediation of Toxic Metals*. New York: J. Wiley, pp 193–229.

Baker, A. J. M. 2013. Keynote Milt Gordon presentation at the 10th Annual International Phytotechnologies Society Conference in Syracuse, NY, 2 October 2013.

Baker, A. J. M., and Brooks, R. R. 1989. Terrestrial higher plants which hyperaccumulate metal elements – a review of their distribution, ecology, and phytochemistry. *Biorecovery* 1, pp. 81–126.

Balouet, J. C., Burken, J. G., Karg, F., Vroblesky, D. A., Balouet, J. C., Smith, K. T., Grudd, H., Rindby, R., Beaujard, F., and Chalot, M. 2012. Dendrochemistry of multiple releases of chlorinated solvents at a former industrial site. *Environmental Science and Technology*, 46 (17), pp. 9541–9547. DOI: dx.doi.org/10.1021/

Banasova, V., and Horak, O. 2008. Heavy metal content in Thlaspi caerulescens J. et C. Presl growing on metalliferous and non-metalliferous soils in Central Slovakia. *International Journal of Environmental Pollution* 33, pp. 133–145.

314

Bani, A., Echevarria, G., Sulce, S., Morel, J. L., and Mullai, A. 2007. In-situ phytoextraction of Ni by a native population of Alyssum murale on an ultramafic site (Albania). *Plant and Soil* 293, pp. 79–89.

Banks, K., and Schwab, P. 1998. Phytoremediation in the field: Craney Island site. Presented at the 3rd Annual International Conference on Phytoremediation, Houston.

Bañuelos, G. S. 2000. Factors influencing field phytoremediation of selenium-laden soils. In N. Terry and G. S. Bañuelos (Eds.) *Phytoremediation of Contaminated Soil and Water.* Boca Raton, FL: CRC Press, pp. 41–59.

Bañuelos, G., and Meek, D. 1990. Accumulation of selenium in plants grown on selenium-treated soils. *Journal of Environmental Quality* 19 (4), 772–777.

Bañuelos, G. S., Ajwa, H. A., Wu, L., Guo, X., Akohoue, S., and Zambrzuski S. 1997. Selenium-induced growth reduction in Brassica land races considered for phytoremediation. *Ecotoxicology and Environmental Safety* 36, pp. 282–287.

Bañuelos, G. S., Ajwa, H. A., Wu, L., and Zambrzuski, S. 1998. Selenium accumulation by *Brassica napus* grown in Se-laden soil from different depths of Kesterson Reservoir. *Journal of Soil Contamination* 7, pp. 481–496.

Bañuelos, G., Lin, Z. Q., Arroyo, I., and Terry, N. 2005. Selenium volatilization in vegetated agricultural drainage sediment from the SanLuis Drain, Central California. *Chemosphere* 60 (9), pp. 1203–1213.

Bañuelos, G. S., Da Roche, J., and Robinson, J. 2010. Developing selenium enriched animal feed and biofuel from canola planted for managing Se-laden drainage waters in the Westside of central California. *International Journal of Phytoremediation* 12, pp. 243–253.

Barac, T., Weyens, N., Oeyen, L., Taghavi, S., van der Lelie, D., Dubin, D., Spliet, M., and Vangronsveld, J. 2009. Field note: hydraulic containment of a BTEX plume using poplar trees. *International Journal of Phytoremediation* 11 (5), pp. 416–424.

Basumatary, B., Saikia, R., Das, H. C., and Bordoloi, S. 2013. Field note: phytoremediation of petroleum sludge contaminated field using sedge species, Cyperus rotundus (Linn.) and Cyperus brevifolius (Rottb.) Hassk. *International Journal of Phytoremediation* 15 (9), pp. 877–888.

Batty, L. C., and Anslow, M. 2008. Effect of a polycyclic aromatic hydrocarbon on the phytoremediation of zinc by two plant species (Brassica juncea and Festuca arundinacea). *International Journal of Phytoremediation* 10 (3), pp. 236–251.

Bauddh, K., and Singh, R. P. 2011. Cadmium tolerance and its phytoremediation by two oil yielding plants Ricinus communis (L.) and Brassica juncea (L.) from the contaminated soil. *International Journal of Phytoremediation* 14, pp. 772–785.

Baumgartner, D. J. et al. 1996. *Plant Responses to Simulated and Actual Uranium Mill Tailings Contaminated Groundwater.* Report to UMTRA Project, US DOE, Environmental Research Laboratory, University of Arizona.

Beattie, G., and Seibel, J. 2007. Uptake and localization of gaseous phenol and P-cresol in plant leaves. *Chemosphere* 68, pp. 528–536.

Beckett, K. P., Freer-Smith, P. H., and Taylor, G. 1998. Urban woodlands: their role in reducing the effects of particulate pollution. *Environmental Pollution* 99 (1998), pp. 347–360.

Beckett, K. P., Freer-Smith, P. H., and Taylor, G. 2000. Effective tree species for local air quality management. *Journal of Arboriculture* 26 (1), pp. 12–19.

Bell, J. N. B., Minski, M. J., and Grogan, H. A. 1988. Plant uptake of radionuclides, *Soil Use Management* 4, pp. 76–84.

Berg, G. 2009. Plant-microbe interactions promoting plant growth and health: perspectives for controlled use of microorganisms in agriculture. *Appl. Microbiol. Biotechnol.* 84, 11–18.

Bes, C., and Mench, M. 2008. Remediation of copper-contaminated topsoils from a wood treatment facility using in situ stabilisation. *Environmental Pollution* 156 (3), pp. 1128–1138.

Bes, C. M., Jaunatre, R., and Mench, M. 2013. Seed bank of Cu-contaminated topsoils at a wood preservation site: impacts of copper and compost on seed germination. *Environmental Monitoring and Assessment* 185 (2), pp. 2039–2053.

Bes, C. M., Mench, M., Aulen, M., Gaste, H., and Taberly, J. 2010. Spatial variation of plant communities and shoot Cu concentrations of plant species at a timber treatment site. *Plant and Soil* 330 (1–2), pp. 267–280.

Best, E. P. H., Sprecher, S. L., Larson, S. L., Fredrickson, H. L., and Bader, D. F. 1999. Environmental behavior of explosives in groundwater from the Milan Army Ammunition Plant in aquatic and wetland plant treatments. Uptake and fate of TNT and RDX in plants. *Chemosphere* 39, pp. 2057–2072.

Bhadra, R., Wayment, D. G., Williams, R. K., Barman, S. N., Stone, M. B., Hughes, J. B., and Shanks, J. V. 2001. Studies on plant-mediated fate of the explosives RDX and HMX. *Chemosphere* 44, pp. 1259–1264.

Binet, P., Portal, J. M., and Lyval, C. 2000. Dissipation of 3–6 ring polycyclic aromatic hydrocarbons in the rhizosphere of ryegrass. *Soil Biology and Biochemistry* 32, pp. 2011–2017.

Black, H. 1995. Absorbing possibilities: phytoremediation. *Environmental Health Perspectives* 103, pp. 1106–1108.

Blanchfield, L. A., and Hoffman, L. G. 1984. *Environmental Surveillance for the INEL Radioactive Waste Management*

315

Complex and Other Areas. Annual Report 1983, EG&EG2312, INEL, August 1984.

Blaylock, M. 2008. *Arsenic Phytoextraction Phase 4 Field Verification Study Spring Valley FUDS, Operable Units 4 and 5, Washington, DC, 2007 Final Report*. Contract # W912DR-07-P-0205. Edenspace Systems Corporation. Chantilly, VA. Provided by Michael Blaylock, September 2013.

Blaylock, M. 2013. Edenspace, Inc. Personal Communication with Kate Kennen, September.

Blaylock, M. J., and Huang, J. W. 2000. Phytoextraction of metals. In I. Raskin and B. D. Ensley (Eds.) *Phytoremediation of Toxic Metals: Using Plants to Clean Up the Environment*, New York: John Wiley and Sons, Inc., pp. 53–70.

Blaylock, M. J., Salt, D. E., Dushenkov, S., Zakharova, O., Gussman, C., Kapulnik, Y., Ensley, B. D., and Raskin, I. 1997. Enhanced accumulation of Pb in Indian mustard by soil-applied chelating agents. *Environmental Science and Technology* 31, pp. 860–865.

Bluskov, S., Arocena, J. M., Omotoso, O. O., and Young, J. P. 2005 Uptake, distribution, and speciation of chromium in Brassica juncea. *International Journal of Phytoremediation* 7 (2), pp. 153–165.

Bogdevich, O., and Cadocinicov, O. 2010. Elimination of acute risks from obsolete pesticides in Moldova: phytoremediation experiment at a former pesticide storehouse. In *Application of Phytotechnologies for Cleanup of Industrial, Agricultural, and Wastewater Contamination*. NATO Science for Peace and Security Series C: Environmental Security 2010, pp. 61–85.

Boonsaner, M., Borrirukwisitsak, S., and Boonsaner, A. 2011. Phytoremediation of BTEX contaminated soil by Canna x generalis. *Ecotoxicology and Environmental Safety* 74 (6), pp. 1700–1707.

Boyd, R. S. 1998. Hyperaccumulation as a plant defensive strategy. In R. R. Brooks (Ed.) *Plants that Hyperaccumulate Heavy Metals*. Wallingford, UK: CAB International, pp. 181–201.

Boyd, R., Shaw, J., and Martens, S. 1994. Nickel hyperaccumulation defends Streptanthus polygaloides (Brassicaceae) against pathogens. *American Journal of Botany* 81 (3), pp. 294–300.

Brentner, L. B., Mukherji, S. T., Walsh, S. A., and Schnoor, J. L. 2010. Localization of hexahydro-1,3,5-trinitro-triazine (RDX) and 2,4,6-trinitrotoluene (TNT) in poplar and switchgrass plants using phosphor imager autoradiography. *Environmental Pollution* 158, pp. 470–475.

Briggs, G. G., Bromilow, R. H., and Evans, A. A. 1982. Relationship between lipophilicity and root uptake and translocation of non-ionised chemicals by barley. *Pesticide Science* 13, pp. 495–504.

Broadhurst, C. L., Chaney, R. L., Davis, A. P., Cox, A., Kumar, K., Reeves, R. D., and Green, C. E. 2013. Growth and cadmium phytoextraction by Swiss chard, maize, rice, Noccaea caerulescens and Alyssum murale in pH adjusted biosolids amended soils. *International Journal of Phytoremediation*, DOI, 10.1080/15226514.2013.828015.

Broadley, M. R., and Willey, N. J. 1997. Differences in root uptake of radiocaesium by 30 plant taxa. *Environmental Pollution*, 97 (1), p. 2.

Brooks, R.R., and Yang, X.H. 1984. Elemental levels and relationships in the endemic serpentine flora of the Great Dyke, Zimbabwe, and their significance as controlling factors for this flora. *Taxonomy* 33, pp. 392–399.

Brooks, R.R. et al. 1992. The Serpentine vegetation of Goiás State, Brazil. J. Proctor et al. (Eds.), In *The Vegetation of Ultramafic (Serpentine) Soils*, Intercept Ltd., Andover, UK, pp 67–81.

Brown, S., Sprenger, M., Maxemchuk, A., and Compton, H. 2005. An evaluation of ecosystem function following restoration with biosolids and lime addition to alluvial tailings deposits in Leadville, CO. *Journal of Environmental Quality* 34, pp. 139–148.

Brown, S., DeVolder, P., and Henry, C. 2007. Effect of amendment C:N ratio on plant diversity, cover and metal content for acidic Pb and Zn mine tailings in Leadville, CO. *Environmental Pollution* 149, pp. 165–172.

Brown, S., Svendson, A., and Henry, C. 2009. Restoration of high zinc and lead tailings with municipal biosolids and lime: field study. *Journal of Environmental Quality* 38, pp. 2189–2197.

Bunzl, K., and Kracke, W. 1984. Distribution of 210Pb, 210Po, stable lead, and fallout 137Cs in soil, plants, and moorland sheep on the heath. *Science of Total Environment* 39, pp. 143–59.

Burdette, L. J., Cook, L. L., and Dyer, R. S. 1988. Convulsant properties of cyclotrimethylenetrinitramine (RDX): spontaneous audiogenic, and amygdaloid kindled seizure activity. *Toxicology and Applied Pharmacology* 92 (3), pp. 436–444.

Burken, J. G. 2013. Presentation to Harvard University GSD 6335, Phyto Research Seminar: Remediation and Rebuilding Technologies in the Landscape. October 2013.

Burken, J. G. (Missouri S&T). 2014. Personal communication with Kate Kennen, April.

Burken, J. G., and Schnoor, J. L. 1997a. Uptake and fate of organic contaminants by hybrid poplar trees. In *Proceedings, 213th ACS National Meeting, American Chemical Society Environmental Division Symposia, San Francisco*, pp. 302–304 (Paper #106).

Burken, J. G., and Schnoor, J. L. 1997b. Uptake and metabolism of atrazine by poplar trees. *Environmental Science and Technology* 31, pp. 1399–1406.

Burken, J. G., and Schnoor, J. L. 1998. Predictive relationships for uptake of organic contaminants by hybrid poplar trees. *Environmental Science and Technology* 32, pp. 3379–3385.

Burken, J. G., Bailey, S. R., Shurtliff, M., and McDermott, J. 2009. Taproot Technology™: tree coring for fast, non-invasive plume delineations. *Remediation Journal* 19 (4), pp. 49–62.

Burken, J. G., Vroblesky, D. A., Balouet, J.-C. 2011. Phytoforensics, Dendrochemistry, and Phytoscreening: New Green Tools for Delineating Contaminants from Past and Present. *Environmental Science and Technology*. 2011, 45, 6218–6226.

Byers, H. G., 1935. Selenium occurrence in certain soils in the United States, with a discussion of related topics. *USDA Tech. Bull*. 482, pp. 1–47.

Byers, H. G. 1936. Selenium occurrence in certain soils in the United States, with a discussion of related topics. Second Report. *USDA Tech. Bull*. 530, pp. 1–78.

Byers, H. G. et al., 1938. Selenium occurrence in certain soils in the United States, with a discussion of related topics. Third Report. *USDA Tech. Bull*. 601, pp. 1–74.

Campbell, R., and Greaves, M. P. 1990. Anatomy and community structure of the rhizosphere. In *The Rhizosphere*. West Sussex, UK: Wiley & Sons.

Campbell, S., Arakaki, A., and Li, Q. 2009. Phytoremediation of heptachlor and heptachlor epoxide in soil by Cucurbitaceae. *International Journal of Phytoremediation* 11 (1), pp. 28–38.

Carbonell-Barrachina, A. A., Burlo-Carbonell, F., and Mataix-Beneyto, J. 1997. Arsenic uptake, distribution, and accumulation in bean plants: effect of arsenite and salinity on plant growth and yield. *Journal of Plant Nutrition* 20, pp. 1419–1430.

Carman, E. P., Crossman, T. L., and Gatliff, E. G. 1997. Phytoremediation of fuel oil-contaminated soil. *In Situ and On-Site Bioremediation* 4 (3), pp. 347–52. Columbus, OH: Battelle Press.

Carman, E. P., Crossman, T. L., and Gatliff, E. G. 1998. Trees stimulate remediation at fuel oil-contaminated site. *Soil and Groundwater Cleanup* Feb.–Mar., pp. 40–44.

Carman, E. P. et al. 2001. *Insitu Treatment Technology*, 2nd ed. New York: Lewis Publishers.

Cempel, M., and Nikel, G. 2006. Nickel: a review of its sources and environmental toxicology. *Polish Journal of Environmental Studies* 15 (3), pp. 375–382.

CH2MHill. 2011. International Phytotechnology Society Conference, Workshop 1, September 2011, Presentation by Mark Madison (CH2MHill), Curtis Stultz (POTW), Jason Smesrud (CH2MHill). Attended by Kate Kennen.

Chameides, W. L., Lindsay, R. W., Richardson, J., and Kiang, C. S. 1988. The role of biogenic hydrocarbons in urban photo- chemical smog: Atlanta as a case study. *Science* 241, pp. 1473.

Chaney, R. L. 2013. Personal correspondence via email and telephone with Kate Kennen, November.

Chaney, Rufus L., Angle, J. Scott, McIntosh, Marla S. , Reeves, Roger D. , Li, Yin-Ming, Brewer, Eric P. , Chen, Kuang-Yu , Roseberg, Richard J., Perner, Henrike, Synkowski, Eva Claire, Broadhurst, C. Leigh, Wang, S. and Baker, Alan J. M. 2005. Using hyperaccumulator plants to phytoextract soil Ni and Cd. In *Verlag der Zeitschrift für Naturforschung*, Tübingen 60c, pp. 190–198.

Chaney, R. L., Kukier, U., and Siebielec, G. 2003. Risk assessment for soil Ni, and remediation of soil-Ni phytotoxicity in situ or by phytoextraction. In *Proceedings Sudbury-2003 (Mining and the Environment III.) May 27–31, 2003. Laurentian University, Sudbury, Ontario, Canada*.

Chaney, R. L., Angle, J. S., Broadhurst, C. L., Peters, C. A., Tappero, R. V., and Sparks, D. L. 2007. Improved understanding of hyperaccumulation yields commercial phytoextraction and phytomining technologies. *Journal of Environmental Quality* 36, pp. 1429–1443.

Chaney, R. L., Broadhurst, C. L., and Centofanti, T. 2010. Phytoremediation of soil trace elements. In P. Hooda (Ed.) *Trace Elements in Soil*. Oxford: Wiley-Blackwell.

Chang, P., Gerhardt, K. E., Huang, X.-D., Yu, X.-M., Glick, B. R., Gerwing, P. D., and Greenberg, B. M. 2013. Plant growth-promoting bacteria facilitate the growth of barley and oats in salt-impacted soil: implications for phytoremediation of saline soils, *International Journal of Phytoremediation* 16 (7–12), pp. 1133–1147.

Chang, Y., Kwon, Y., Kim, S., Lee, I., and Bae, B. 2003. Ehanced degradation of 2,4,6-trinitrotoluene (TNT) in a soil column planted with Indian mallow (*Abutillon avicennae*). *Journal of Bioscience and Bioengineering* 97, pp. 99–103.

Chappell, J. 1997. *Phytoremediation of TCE using Populus*. Status report prepared for the US Environmental Protection Agency Technology Innovation Office under the National Network of Environmental Management Studies.

Chard, J. K., Orchard, B. J., Pajak, C. J., Doucette, W. J., and Bugbee, B. 1998. Design of a plant growth chamber for studies on the uptake of volatile organic compounds. *Proceedings, Conference on Hazardous Waste Research*, Snow Bird, Utah, p. 75 (Abstract P42).

317

Chardot, V., Massoura, S. T., Echevarria, G., Reeves, R. D., and Morel, J. 2005. Phytoextraction potential of the nickel hyperaccumulators Leptoplax emarginata and Bornmuellera tymphaea. *International Journal of Phytoremediation* 7 (4), pp. 323–335.

Chen, B., Jakobsen, I., Roos, P., and Zhu, Y.-G. 2005. Effects of the mycorrhizal fungus Glomus intraradices on uranium uptake and accumulation by Medicago truncatula L. from uranium-contaminated soil. *Plant and Soil* 275 (1), pp. 349–359.

Chen, L., Long, X., Zhang, Z., Zheng, X., Rengel, Z., and Liu, Z. 2011. Cadmium accumulation and translocation in two Jerusalem artichoke (Helianthus tuberosus L.) cultivars. *Pedosphere* 21 (5), pp. 573–580.

Chen, Z., Setagawa, M., Kang, Y., Sakurai, K., Aikawa, Y., and Iwasaki, K. 2009. Zinc and cadmium uptake from a metalliferous soil by a mixed culture of *Anthyrium yokoscense* and *Arabis flagellosa*. *Soil Science Plant Nutrition* 55, pp. 315–324.

Cherian, S., and Oliveira, M. M. 2005. Critical review transgenic plants in phytoremediation: recent advances and new possibilities. *Environmental Science and Technology* 39 (24), pp. 9377–9390.

Chintakovid, W., Visoottiviseth, P., Khokiattiwong, S., and Lauengsuchonkul, S. 2008. Potential of the hybrid marigolds for arsenic phytoremediation and income generation of remediators in Ron Phibun District, Thailand. *Chemosphere* 70 (8), pp. 1532–1537.

Ciabotti, J. 2004. *Understanding Environmental Contaminants: Lessons Learned and Guidance to Keep Your Rail-Trail Project on Track*. Washington, DC: Rails to Trails Conservancy.

Ciurli, A., Lenzi, L., Alpi, A., and Pardossi, A. 2014. Arsenic uptake and translocation by plants in pot and field experiments. *International Journal of Phytoremediation* 16 (7–8), Special Issue: The 9th International Phytotechnology Society Conference – Hasselt, Belgium 2012, pp. 804–823, DOI, 10.1080/15226514.2013.856850.

Clu-In. 2014. Website Trenton Magic Marker Case Study. http://clu-in.org/products/phyto/search/phyto_details.cfm?ProjectID=69. Accessed 1 March 2014.

Clulow, F. V., Lim, T. P., Dave, N. K., and Avadhanula, R. 1992. Radium-226 levels and concentration ratios between water, vegetation, and tissues of Ruffed grouse (*Bonasa umbellus*) from a watershed with uranium tailings near Elliot Lake, Canada. *Environmental Pollution* 77, pp. 39–50.

Coleman, J., Hench, K., Sexstone, A., Bissonnette, G., and Skousen, J. 2001. Treatment of domestic wastewater by three plant species in constructed wetlands. *Water, Air, and Soil Pollution* 128 (3–4), pp. 283–295.

Conger, R. M. 2003. Black Willow (*Salix Nigra*) Use in phytoremediation techniques to remove the herbicide Bentazon from shallow groundwater. Dissertation, Louisiana State University, Baton Rouge.

Conger, R. M., and Portier, R. J. 2006. Before-after control-impact paired modeling of groundwater bentazon treatment at a phytoremediation site. *Remediation Journal* 17 (1), pp. 81–96.

Cook, R., and Hesterberg, D. 2013. Comparison of trees and grasses for rhizoremediation of petroleum hydrocarbons. *International Journal of Phytoremediation* 15 (9), pp. 844–860.

Cook, R. L., Landmeyer, J. E., Atkinson, B., Messier, J., and Nichols, E. G. 2010. Field note: successful establishment of a phytoremediation system at a petroleum hydrocarbon contaminated shallow aquifer: trends, trials, and tribulations. *International Journal of Phytoremediation* 12 (7), pp. 716–732.

Cortes-Jimenez, E., Mugica-Alvarez, V., Gonzalez-Chavez, M., Carrillo-Gonzalez, R., Gordillo, M., and Mier, M. 2013. Natural revegetation of alkaline tailing heaps at Taxco, Guerrero, Mexico. *International Journal of Phytoremediation* 15 (2), pp. 127–141.

Coughtery, P. J., Kirton, J. A., and Mitchell, N. G. 1989. Transfer of radioactive cesium from soil to vegetation and comparison with potassium in upland grasslands. *Environmental Pollution* 62, pp. 281–315.

Couto, M. N. P. F. S., Basto, M. C. P., and Vasconcelos, M. T. S. D. 2012. Suitability of Scirpus maritimus for petroleum hydrocarbons remediation in a refinery environment. *Environmental Science and Pollution Research* 19 (1), pp. 86–95.

Craw, D., Rufaut, C., Haffert, L., and Paterson, L. 2007. Plant colonization and arsenic uptake on high arsenic mine wastes, New Zealand. *Water, Air, and Soil Pollution* 179 (1–4), pp. 351–364.

Cunningham, S. D., and Berti, W. R. 1993. Remediation of contaminated soils with green plants: an overview. *In Vitro Cellular and Developmental Biology – Plant* 29 (4), pp. 207–212.

Cunningham, S. D., and Lee, C. R. 1995. Phytoremediation: plant based remediation of contaminated soils and sediments. In H. D. Skipper and R. F. Turco (Eds.) *Bioremediation: Science and Applications*. Madison: Soil Science Society of America, pp. 145–156.

Cutright, T., Gunda, N., and Kurt, F. 2010. Simultaneous hyperaccumulation of multiple heavy metals by Helianthus annuus grown in a contaminated sandy-loam soil. *International Journal of Phytoremediation* 12, pp. 562–573.

Dahlman, R. C., Auerbach, S. I., and Dunaway, P. B. 1969. *Environmental Contamination by Radioactive Materials*.

318

Vienna: International Atomic Energy Agency and World Health Organization.

Dahmani-Muller, H., Van Oort, F., Gelie, B., and Balabane, M. 2000. Strategies of heavy metal uptake by three plant species growing near a metal smelter. *Environmental Pollution* 109 (2), pp. 231–238.

Danh, L. T., Truong, P., Mammucari, R., Tran, T., and Foster, N. 2009. Vetiver grass, Vetiveria zizanioides: a choice plant for phytoremediation of heavy metals and organic wastes. *International Journal of Phytoremediation* 11 (8), pp. 664–691.

Danh, L. T., Truong, P., Mammucari, R., and Foster, N. 2014. A critical review of the arsenic uptake mechanisms and phytoremediation potential of Pteris vittata. *International Journal of Phytoremediation* 16 (5), pp. 429–453.

Darlington, A. 2013. Presentation to Harvard University GSD 6335, Phyto Practicum Research Seminar. October.

Darlington, A. 2014. Nedlaw Living Walls, Inc. Personal communication with Kate Kennen, February.

Das, D., Datta, R., Markis, K. C., and Sakar, D. 2010. Vetiver grass is capable of removing TNT from soil in the presence of urea. *Environmental Pollution* 158, pp. 1980–1983.

Davenport, C. 2013. EPA funding reductions have kneecapped environmental enforcement. *National Journal*, NJ Daily. 3 March 2013. Retrieved from: http://www.nationaljournal.com/daily/epa-funding-reductions-have-kneecapped-environmental-enforcement-20130303.

Davis, L. C., Muralidharan, N., Visser, V. P., Chaffin, C., Fateley, W. G., Erickson, L. E., and Hammaker, R. M. 1994. Alfalfa plants and associated microorganisms promote biodegradation rather than volatilization of organic substances from groundwater. In T. A. Anderson and J. R. Coats (Eds.) *Bioremediation Through Rhizosphere Technology*, ACS Symposium Series 563. Washington, D C: American Chemical Society.

Davis, L. C., Erickson, L. E., Narayanan, N., and Zhang, Q. 2003. Modeling and design of phytoremediation. In S. C. McCutcheon and J. L. Schnoor (Eds.) *Phytoremediation: Transformation and Control of Contaminants*. New York: Wiley, pp. 663–694.

Del Tredici, P. 2010. *Wild Urban Plants of the Northeast: A Field Guide*. Ithaca, NY: Comstock Publishing Associates.

Delorme, E. A. 2000. Phytoremediation of phosphorus-enriched soils. *International Journal of Phytoremediation* 2 (2), pp. 173–181.

Department of the Airforce. 2010 September. Final Technical Report on Phytostabilization at Travis Air Force Base, California. Prepared by Parsons, 1700 Broadway, Suite 900, Denver, CO 80290.

Detroit Future City. 2012. Detroit Strategic Framework Plan prepared by The Detroit Works Project Long Term Planning Steering Committee, Director of Projects, Dan Kinkead.

Dickinson, N. M., Baker, A. J., Doronila, A., Laidlaw, S., and Reeves, R. 2009. Phytoremediation of inorganics: realism and synergies. *International Journal of Phytoremediation* 11, pp. 97–114.

Ding, C., Zhang, T., Wang, X., Zhou, F., Yang, Y., and Yin, Y. 2013. Effects of soil type and genotype on cadmium accumulation by rootstalk crops: implications for phytomanagement. *International Journal of Phytoremediation* 16, pp. 1018–1030.

Dodge, C. J., and Francis, A. J. 1997. Biotransformation of binary and ternary citric acid complexes of iron and uranium. *Environmental Science and Technology* 31, pp. 3062–3067.

Dominguez, M., Madrid, F., Maranon, T., and Murillo, J. 2009. Cadmium availability in soil and retention in oak roots: potential for phytostabilization. *Chemosphere* 76 (4), pp. 480–486.

Dominguez-Rosado, E., Pichtel, J., and Coughlin, M. 2004. Phytoremediation of soil contaminated with used motor oil: I. Enhanced microbial activities from laboratory and growth chamber studies. *Environmental Engineering Science* 21 (2), pp. 157–168.

Dosnon-Olette, R., Couderchet, M., Oturan, M., Oturan, N., and Eullaffroy, P. 2011. Potential use of Lemna minor for the phytoremediation of isoproturon and glyphosate. *International Journal of Phytoremediation* 13 (6), pp. 601–612.

Doucette, W. 2014. Personal communication with Dr. William Doucette, Utah Water Research Laboratory, Utah State University, 8200 Old Main Hill, Logan, Utah 84322-8200, United States.

Doucette, W., Klein, H., Chard, J., Dupont, R., Plaehn, W., and Bugbee, B. 2013. Volatilization of trichloroethylene from trees and soil: measurement and scaling approaches. *Environmental Science and Technology* 47, pp. 5813–5820.

Duringer, J. M., Craig, A. M., Smith, D. J., and Chaney, R. L. 2010. Uptake and transformation of soil [C-14]-trinitrotoluene by cool-season grasses. *Environmental Science and Technology* 44 (16), pp. 6325–6330.

Dushenkov, S., Kapulnik, Y., Blaylock, M., Sorochinsky, B., Raskin, I., and Ensley, B. 1997a. Phytoremediation: a novel approach to an old problem. In D. L. Wise (Ed.) *Global Environmental Biotechnology*, Amsterdam: Elsevier Science B.V., pp. 563–572.

Dushenkov, S., Vasudev, D., Kapulnik, Y., Gleba, D., Fleisher, D., Ting, K. C., and Ensley, B. 1997b. Removal of uranium

319

from water using terrestrial plants. *Environmental Science and Technology* 31, pp. 3468–3474.

Dutton, M. V., and Humphreys, P. N. 2005. Assessing the potential of short rotation coppice (Src) for cleanup of radionuclide contaminated sites. *International Journal of Phytoremediation* 7 (4), pp. 279–293.

Dzierżanowski, K., and Gawroński, S. 2011. Use of trees for reducing particulate matter pollution in air. *Natural Sciences* 1(2), pp. 69–73.

Dzierżanowski, K., Popek, R., Gawroński, H., Saebo, A., and Gawroński, S. 2011. Deposition of particulate matter of different size fractions on leaf surfaces and in waxes of urban forest species. *International Journal of Phytoremediation* 13 (10), pp. 1037–1046.

Ebbs, S., Hatfield, S., Nagarajan, V., and Blaylock, M. 2009. A comparison of the dietary arsenic exposures from ingestion of contaminated soil and hyperaccumulating Pteris ferns used in a residential phytoremediation project. *International Journal of Phytoremediation* 12, pp. 121–132.

Efe, S. I., and Okpali, A. E. 2012. Management of petroleum impacted soil with phytoremediation and soil amendments in Ekpan Delta State, Nigeria. *Journal of Environmental Protection* 3, pp. 386–393.

Eisner, S., and CH2MHill. 2011. International Phytotechnology Society Conference, Workshop 1, Site Visit by Kate Kennen, September 2011, Information provided by Stephanie Eisner (City of Salem), Mark Madison (CH2MHill), and Jason Smesrud (CH2MHill).

El-Gendy, A. S., Svingos, S., Brice, D., Garretson, J. H., and Schnoor, J. 2009. Assessments of the efficacy of a long-term application of a phytoremediation system using hybrid poplar trees at former oil tank farm sites. *Water Environment Research* 81 (5), pp. 486–498.

Entry, J. A., and Emmingham, W. H. 1995. Sequestration of 137Cs and 90Sr from soil by seedlings of *Eucalyptus tereticornis. Canadian Journal of Forest Research* 25, pp. 1044–1047.

Entry, J. A., and Watrud, L. S. 1998. Potential remediation of Cs-137 and Sr-90 contaminated soil by accumulation in Alamo switchgrass. *Water, Air and Soil Pollution* 104, pp. 339–352.

Entry, J. A., Rygiewicz, P. T., and Emmingham, W. H. 1993. Accumulation of cesium-137 and strontium-90 in Ponderosa pine and Monterey pine seedlings. *Journal of Environmental Quality* 22, pp. 742–746.

Entry, J. A., Vance, N. C., Hamilton, M. A., Zabowski, D., Watrud, L. S., and Adriano, D. C. 1996. Phytoremediation of soil contaminated with low concentrations of radionuclides. *Water, Air, and Soil Pollution* 88, pp. 167–176.

Environment Canada. 2001. Priority Substances List Assessment Report: Road Salts. Report. Environment Canada, Health Canada.

Erickson, L. E. et al. 1999. Simple plant-based design strategies for volatile organic pollutants. *Environmental Progress* 18 (4), pp. 231–242.

Euliss, K., Ho, C., Schwab, A. P., Rock, S., and Banks, A. K. 2008. Greenhouse and field assessment of phytoremediation for petroleum contaminants in a riparian zone. *Bioresource Technology* 99 (6), pp. 1961–1971.

Evangelou, M. W. H., Ebel, M., and Schaeffer, A. 2007. Chelate assisted phytoextraction of heavy metals from soil. Effect, mechanism, toxicity, and fate of chelating agents. *Chemosphere* 68, p. 989.

Evangelou, M. W., Robinson, B. H., Gunthardt-Goerg, M. S., and Schulin, R. 2013. Metal uptake and allocation in trees grown on contaminated land: implications for biomass production. *International Journal of Phytoremediation* 15 (1), pp. 77–90.

Faison, B. D. 2004. Biological treatment of metallic pollutants. In A. Singh and O. P. Ward (Eds.) *Applied Bioremediation and Phytoremediation*. New York: Springer, pp. 81–114.

Ferro, A. M., Sims, R. C., and Bugbee, B. 1994. Hycrest Crested wheatgrass accelerates the degradation of pentachlorophenol in soils. *Journal of Environmental Quality* 23, pp. 272–279.

Ferro, A. M., Kennedy, J., and Knight, D. 1997. Greenhouse-scale evaluation of phytoremediation for soils contaminated with wood preservatives. *In Situ and On-Site Bioremediation* 4 (3), pp. 309–314. Columbus, OH: Battelle Press.

Ferro, A. M., Rock, S. A., Kennedy, J., and Herrick, J. J. 1999. Phytoremediation of soils contaminated with wood preservatives: greenhouse and field evaluations. *International Journal of Phytoremediation* 1 (3), pp. 289–306.

Ferro, A., Kennedy, J. and LaRue, J. 2013. Phytoremediation of 1,4-dioxane-containing recovered groundwater. *International Journal of Phytoremediation* 15:10, pp. 911–923.

Ferro, A., Kennedy, J., Kjelgren, R., Rieder, J., and Perrin, S. 1999. Toxicity assessment of volatile organic compounds in poplar trees. *International Journal of Phytoremediation* 1 (1), pp. 9–17.

Ferro, A. M., Adham, T., Berra, B., and Tsao, D. 2013. Performance of deep-rooted phreatophytic trees at a site containing total petroleum hydrocarbons. *International Journal of Phytoremediation* 15 (3), pp. 232–244.

Ferro, A. M., Chard, B., Gefell, M., Thompson, B., and Kjelgren, R. 2000. Phytoremediation of organic solvents in groundwater: Pilot study at a Superfund site. In G. B. Wickramanayake, A. R. Gavaskar, B. C. Alleman, and V. S. Maga (Eds.), *Bioremediation and phytoremediation of chlorinated and recalcitrant compounds*, vol. C2–4, Columbus: Battelle Press, pp. 461–466.

320

Ferro, A. M., Kennedy, J., and LaRue, J. C. 2013. Phytoremediation of 1,4- Dioxane-containing recovered groundwater. *International Journal of Phytoremediation* 15 (10), pp. 911–923.

Ficko, S. A., Rutter, A., and Zeeb, B. A. 2010. Potential for phytoextraction of PCBs from contaminated soils using weeds. *Science of the Total Environment* 408 (16), pp. 3469–3476.

Ficko, S., Rutter, A., and Zeeb, B. 2011. Phytoextraction and uptake patterns of weathered polychlorinated biphenyl-contaminated soils using three perennial weed species. *Journal of Environmental Quality* 40 (6), pp. 1870–1877.

Field, C. B., Campbell, J. E., and Lobell, D. B. 2008. Biomass energy: the scale of the potential resource. *Trends in Ecology & Evolution* 23, pp. 65–72.

Fiorenza, S. (BP) and Thomas, F. (Phytofarms) 2004. Phytoscapes plant testing program. Unpublished draft provided on 11 November 2013 by Dr. David Tsao, BP Corporation North America, Inc. 150 W. Warrenville Rd. Naperville, IL 60563 USA.

Flathman, P. E., and Lanza, G. R. 1998. Phytoremediation: current views on an emerging green technology. *Journal of Soil Contamination* 7, pp. 415–432.

Fletcher, J. S., and Hegde, R. S. 1995. Release of phenols by perennial plant roots and their potential importance in bioremediation. *Chemosphere* 31, pp. 3009–3016.

Forman, R. T., and Alexander, L. E. 1998. Roads and their major ecological effects. *Annual Review of Ecology and Systematics* 29, pp 207–231.

Forman, R. T. et al. 2003. *Road Ecology: Science and Solutions*. Washington, DC: Island Press.

Francesconi, K., Visoottiviseth, P., Sridokchan, W., and Goessler, W. 2002. Arsenic species in an arsenic hyperaccumulating fern, *Pityrogramma calomelanos*: a potential phytoremediator of arsenic-contaminated soils. *Journal of the Science of the Total Environment* 284, pp. 27–35.

Freeman, J., Zhang, L., Marcus, M., Fakra, S., McGrath, S., and Pilon-Smits, E. 2006. Spatial imaging, speciation, and quantification of selenium in the hyperaccumulator plants Astragalus bisulcatus and Stanleya pinnata. *Plant Physiology* 142 (1), pp. 124–134.

Freeman, J. L. 2014. Correspondence with Kate Kennen, 13 January, including copy of PowerPoint.

Freeman, J. L., and Banuelos, G. S. 2011. Selection of salt and boron tolerant selenium hyperaccumulator Stanleya pinnata genotypes and characterization of Se phytoremediation from agricultural drainage, *Environmental Science and Technology* 45, pp. 9703–9710.

French, C. E., Rosser, S. J., Davies, G. J., Nicklin, S., and Bruce, N. C. 1999. Biodegradation of explosives by transgenic plants expressing pentaerythritol tetranitrate reductase. *Nature Biotechnology* 17, pp. 491–494.

Friesl, W., Friedl, J., Platzer, K., Horak, O., and Gerzabek, M. 2006. Remediation of contaminated agricultural soils near a former Pb/Zn smelter in Austria: batch, pot and field experiments. *Environmental Pollution* 144 (1), pp. 40–50.

Fu, D., Teng, Y., Shen, Y., Sun, M., Tu, C., Luo, Y., Li, Z., and Christie, P. 2012. Dissipation of polycyclic aromatic hydrocarbons and microbial activity in a field soil planted with perennial ryegrass. *Frontiers of Environmental Science & Engineering* 6 (3), pp. 330–335.

Gao, J., Garrison, A. W., Hoehamer, C., Mazur, C., and Wolfe, N. L. 1998. Bioremediation of organophosphate pesticides using axenic plant tissue cultures and tissue extracts. Poster abstract at 3rd Annual International Conference on Phytoremediation, Houston.

Gardea-Torresdey, J. L., Sias, S., Tiemann, K. J., Hernandez, A., Rodriguez, O., and Arenas, J. 1998. Evaluation of northern Chihuahuan Desert plants for phytoextraction of heavy metals from contaminated soils. *Proceedings, Conference on Hazardous Waste Research*, Snow Bird, Utah, pp. 26–27 (Abstract 39).

Gaston, L. A., Eilers, T. L., Kovar, J. L., Cooper, D., and Robinson, D. L. 2003. Greenhouse and field studies on hay harvest to remediate high phosphorus soil. *Communications in Soil Science and Plant Analysis* 34 (15–16), pp. 2085–2097.

Gatliff, E. G. 1994. Vegetative remediation process offers advantages over traditional pump-and-treat technologies. *Remediation* 4 (3), pp. 343–52.

Gatliff, E. G. 2012. Personal communication with Kate Kennen regarding Tree Well System, April.

Gawronski, S. 2010. Presentation at International Phytotechnologies Conference, Parma, Italy, 2010.

Gawronski, S., Greger, M., and Gawronska, H. 2011. Plant taxonomy and metal phytoremediation. In I. Sherameti and A. Varma (Eds.) *Detoxification of Heavy Metals, Soil Biology* 30, DOI, 10.1007/978-3-642-21408-0_5, Berlin, Heidelberg: Springer-Verlag.

GCSAA. 2013. Golf Course Superintendents Associate of America, http://www.gcsaa.org/. Accessed 15 January 2014.

Gerhardt, K., Huang, X.-D., Glick, B. R., and Greenberg, B. M. 2009. Phytoremediation and rhizoremediation of organic soil contaminants. *Plant Science* 176, pp. 20–30.

Ghaderian, S.M., Mohtadi, A., Rahiminejad, M.R., and Baker, A.J.M. 2007. Nickel and other metal uptake and accumulation by species of *Alyssum* (Brassicaceae) from the ultramafics of Iran. *Environmental Pollution* 145, pp. 293–298.

321

Ghnaya, T., Nouairi, I., Slama, I., Messedi, D., Grignon, C., Adbelly, C., and Ghorbel, M. H. 2005. Cadmium effects on growth and mineral nutrition of two halophytes: Sesuvium portulacastrum and Mesembryanthemum crystallinum. *Journal of Plant Physiology* 162, pp. 1133–1140.

Gilbert, E. S., and Crowley, D. E. 1997. Plant compounds that induce polychlorinated biphenyl biodegradation by Arthrobacter sp. strain B1B. *Applied Environmental Microbiology* 63 (5), pp. 1933–1938.

Gisbert, C., Almela, C., Vélez, D., López-Moya, J. R., De Haro, A., Serrano, R., Montoro, R., and Navarro-Aviñó, J. 2008. Identification of As accumulation plant species growing on highly contaminated soils. *International Journal of Phytoremediation* 10, pp. 185–196.

Glass, D. J. 1999. *U.S. and International Markets for Phytoremediation, 1999–2000.* Needham, MA: D. Glass Associates.

Godish, T., and Guindon, C. 1989. An assessment of botanical air purification as a formaldehyde mitigation measure under dynamic laboratory chamber conditions. *Environmental Pollution* 61, pp. 13–20.

Godsy, E. M., Warren, E., and Paganelli, V. V. 2003. The role of microbial reductive dechlorination of TCE at the phytoremediation site at the Naval Air Station, Fort Worth, Texas. *International Journal of Phytoremediation* 5 (1), pp. 73–87.

Gomes, P., Valente, T., Pamplona, J., Sequeira Braga, M. A., Pissarra, J., Grande Gil, J. A., and De La Torre, M. L. 2013. Metal uptake by native plants and revegetation potential of mining sulfide-rich waste-dumps. *International Journal of Phytoremediation* 16, pp. 1087–1103.

Gorbachevskaya, O., Kappis, C., Schreiter, H., and Endlicher, W. 2010. Das grüne Gleis – vegetationstechniche, ökologische und ökonomische Aspekte der Gleisbettbegrünung. *Berliner Geographische Arbeiten* 116.

Gordon, M. P., Choe, N., Duffy, J., Ekuan, G., Heilman, P., Muiznieks, I., Newman, L., Ruszaj, M., Shurtleff, B., Strand, S., and Wilmoth, J. 1997. Phytoremediation of trichloroethylene with hybrid poplars. In E. L. Kruger, T. A. Anderson and J. R. Coats (Eds.) *Phytoremediation of Soil and Water Contaminants*, American Chemical Society Symposium Series 664. Washington, DC: American Chemical Society, pp. 177–185.

Gouthu, S., Arie, T., Ambe, S., and Yamaguchi, I. 1997. Screening of plant species for comparative uptake abilities of radioactive Co, Rb, Sr and Cs from soil. *Journal of Radioanalytical and Nuclear Chemistry* 222, pp. 247–251.

Greger, M., and Landberg, T. 1999. Use of willow in phytoextraction. *International Journal of Phytoremediation* 1 (2), pp 115–123.

Gunther, T., Dornberger, U., and Jones, D. 1996. Effects of ryegrass on biodegradation of hydrocarbons in soil. *Chemosphere* 33, pp. 203–215.

Guo, P., Wang, T., Liu, Y., Xia, Y., Wang, G., Shen, Z., and Chen, Y. 2013. Phytostabilization potential of evening primrose (Oenothera glazioviana) for copper-contaminated sites. *Environmental Science and Pollution Research* 21 (1), pp. 1–10.

Gupta, D. K., Srivastava, A., and Singh, V. P. 2008. EDTA enhances lead uptake and facilitates phytoremediation by Vetiver grass. *Journal of Environmental Biology* 29 (6), pp. 903–906.

Guthrie Nichols, E. 2013. Personal communication with Dr. Elizabeth Guthrie Nichols, Department of Forestry and Environmental Resources, North Carolina State University, Raleigh, NC.

Guthrie Nichols, E., Cook, R. L, Landmeyer, J. E., Atkinson, B., Malone, D. R., Shaw, G., and Woods, L. 2014. Phytoremediation of a petroleum-hydrocarbon contaminated shallow aquifer in Elizabeth City, North Carolina, USA. *Remediation Journal* 24 (2), pp. 29–46.

Hagler, G. S. W., Thoma, E. D., and Baldauf, R. W. 2010. High-resolution monitoring of carbon monoxide and ultrafine particle concentrations in a near-road environment. *Journal of the Air and Waste Management Association* 60 (3), pp. 328–336.

Haith, D. A., and Rossi, F. S. 2003. Risk assessment of pesticide runoff from turf. *Journal of Environment Quality* 3 (2), pp. 447–455.

Hall, J., Soole, K., and Bentham, R. 2011. Hydrocarbon phytoremediation in the family Fabaceae – a review. *International Journal of Phytoremediation* 13 (4), pp. 317–333.

Hannink, N., Rosser, S. J., French, C. E., Basran, A., Murray, J. A. H., Nicklin, S., and Bruce, N. C. 2001. Phytodetoxification of TNT by transgenic plants expressing a bacterial nitroreductase. *Nature Biotechnology* 19, pp. 1168–1172.

Hanson, R., Lindblom, S. D., Loeffler, M. L., and Pilon-Smits, E. A. H. 2004. Selenium protects plants from phloem-feeding aphids due to both deterrence and toxicity. *New Phytologist* 162, pp. 655–662.

Harris, M. 2008. After the burial. In *Grave Matters, a Journey through the Modern Funeral Industry to a Natural Way of Burial.* New York: Scribner, chapter 2.

Harper, G. 2013. Green roof runoff characterization: nutrient loading and erosion control on a newly planted green roof. Presentation at 10th Annual International Phytotechnologies Society Conference, Syracuse, NY, 3 October 2013. Work completed at Missouri S&T with Lea Ahrens, Missouri S&T, Joel Burken, Missouri S&T, Eric Showalter, Missouri S&T.

Harvey, G. 1998. How to evaluate the efficacy and cost at the fields scale. Presented at the 3rd Annual International Conference on Phytoremediation, Houston.

Hayhurst, S. C., Doucette, W. J., Orchard, B. J., Pajak, C. J., Bugbee, B., and Koerner, G. 1998. Phytoremediation of trichloroethylene: a field evaluation. In *Proceedings, Conference on Hazardous Waste Research,* Snow Bird, Utah, p. 74 (Abstract P40).

Henderson, K. L. D., Belden, J. B., Zhao, S., and Coats, J. R. 2006. Phytoremediation of pesticide wastes in soil. *Zeitschrift für Naturforschung Section C – a Journal of Biosciences* 61 (3–4), pp. 213–221.

Hettiarachchi, G. 2011. Soil contaminants in urban environments: their bioavailability and transfer. Presentation to Harvard GSD 9108 Phytotechnologies Research Seminar by Dr. Hettiarachchi, Kansas State University, Department of Agronomy, April 2011.

Hewamanna, R., Samarakoon, C. M., and Karunaratne, P. A. V. N. 1988. Concentration and chemical distribution of radium in plants from monazite-bearing soils. *Environmental and Experimental Botany* 28, pp. 137–43.

Hill, J. 2014. University of Toronto, Canada. Personal communication with Kate Kennen, February.

Hinchman, R. R., Negri, M. C., and Gatliff, E. G. 1997. *Phytoremediation: Using Green Plants to Clean Up Contaminated Soil, Groundwater, and Wastewater.* Submitted to the US Department of Energy, Assistant Secretary for Energy Efficient and Renewable Energy under Contract W-31-109-Eng-38.

Hoagland, R. E., Zablotowicz, R. M., and Locke, M. A. 1994. Propanil metabolism by rhizosphere microflora. In T. A. Anderson and J. R. Coats (Eds.) *Bioremediation through Rhizosphere Technology*, ACS Symposium Series 563. Washington, DC: American Chemical Society.

Hogan, C. M. 2010. Heavy metal. In E. Monosson and C. Cleveland (Eds.) *Encyclopedia of Earth.* Washington, DC: National Council for Science and the Environment.

Hong, M. S., Farmayan, W. F., Dortch, I. J., Chiang, C. Y., McMillan, S. K., and Schnoor, J. L. 2001. Phytoremediation of MTBE from a groundwater plume. *Environmental Science and Technology* 35, pp. 1231–1239.

Hsu, T. S., and Bartha, R. 1979. Accelerated mineralization of two organophosphate insecticides in the rhizosphere. *Applied Environmental Microbiology* 37, pp. 36–41.

Hu, Y., Nan, Z., Jin, C., Wang, N., and Luo, H. 2013. Phytoextraction potential of Poplar (Populus alba L. var. pyramidalis Bunge) from calcareous agricultural soils contaminated by cadmium. *International Journal of Phytoremediation* 16, pp. 482–495.

Huang X.-D., El-Alawi, Y., Penrose, D. M., Glick, B. R., Greenberg, B. M. 2004. A multi-process phytoremediation system for removal of polycyclic aromatic hydrocarbons from contaminated soils. *Environ. Pollut.* 130, 465–476.

Hue, N. V. 2013. Arsenic chemistry and remediation in Hawaiian soils. *International Journal of Phytoremediation* 15, pp. 105–116.

Hughes, J. B., Shanks, J., Vanderford, M., Lauritzen, J., and Bhadra, R. 1997. Transformation of TNT by aquatic plants and plant tissue cultures. *Environmental Science and Technology* 31, pp. 266–271.

Hülster, A., Muller, J. F., and Marschner, H. 1994. Soil-plant transfer of poly-chlorinated-p-dioxins and dibenzofurans to vegetables of the cucumber family (Cucurbitaceae). *Environmental Science and Techonology* 28, pp. 1110–1115.

Hultgren, J., Pizzul, L., Pilar Castillo, M., and Granhall, U. 2009. Degradation of PAH in a creosote-contaminated soil. A comparison between the effects of willows (Salix viminalis), wheat straw and a nonionic surfactant. *International Journal of Phytoremediation* 12 (1), pp. 54–66.

Hutchinson, S. L., Schwab, A. P., and Banks, M. K. 2003. Biodegration of petroleum hydrocarbons in the rhizosphere. In S. C. McCutcheon and J. L. Schnoor (Eds.) *Phytoremediation: Transformation and Control of Contaminants.* Hoboken: John Wiley, pp. 355–386.

Isleyen, M., Sevim, P., Hawthorne, J., Berger, W., and White, J. C. 2013. Inheritance profile of weathered Chlordane and P,P -DDTs accumulation by Cucurbita pepo hybrids. *International Journal of Phytoremediation* 15, pp. 861–876.

Israr, M., Jewell, A., Kumar, D., and Sahi, S.V. 2011. Interactive effects of lead, copper, nickel, and zinc on growth, metal uptake, and antioxidative metabolism of Sesbania drummondii. *Journal of Hazardous Materials* 186, pp. 1520–1526.

ITRC (Interstate Technology & Regulatory Council). 2003. WTLND-1 *Technical and Regulatory Guidance Document for Constructed Treatment Wetlands*. Washington, DC: Interstate Technology & Regulatory Council, http://www.itrcweb.org.

ITRC (Interstate Technology & Regulatory Council). 2005. WTLND-2 *Technical and Regulatory Characterization, Design, Construction, and Monitoring of Mitigation Wetlands*. Washington, DC: Interstate Technology & Regulatory Council, http://www.itrcweb.org.

ITRC (Interstate Technology & Regulatory Council). 2009. PHYTO-3 *Phytotechnology Technical and Regulatory Guidance and Decision Trees, Revised*. Washington, DC: Interstate Technology & Regulatory Council, Phytotechnologies Team, http://www.itrcweb.org.

Jaffré, T., and Schmid, M. 1974. Accumulation du Nickel par une Rubiacée de Nouvelle Calédonie, Psychotria douarrei

323

(G. Beauvisage) Däniker. *Compt. Rend. Acad. Sci* (Paris) Sér. 278, pp. 1727–1730.

Jeffers, P. M., and Liddy, C. D. 2003. Treatment of atmospheric halogenated hydrocarbons by plants and fungi. In S. C. McCutcheon and J. L. Schnoor (Eds.) *Phytoremediation: Transformation and Control of Contaminates*. New York: Wiley, pp. 409–427.

Ji, P., Song, Y., Sun, T., Liu, Y., Cao, X., Xu, D., Yang, X., and McRae, T. 2011. In-situ cadmium phytoremediation using Solanum nigrum L.: the bio-accumulation characteristics trail. *International Journal of Phytoremediation* 13 (10), pp. 1014–1023.

Johansson, L., Xydas, C., Messios, N., Stoltz, E., and Greger, M. 2005. Growth and Cu accumulation by plants grown on Cu containing mine tailings in Cyprus. *Applied Geochemistry* 20 (1), pp. 101–107.

Jones, S. A., Lee, R. W., and Kuniansky, E. L. 1999, Phytoremediation of trichloroethene (TCE) using cottonwood trees. In Leeson, A., and B. C. Alleman (Eds.) *Phytoremediation and Innovative Strategies for Specialized Remedial Applications, The Fifth International In Situ and On-Site Bioremediation Symposium, San Diego, California, April 19–22*, Columbus, OH: Battelle Press, v. 6, pp. 101–108.

Just, C. L., and Schnoor J. L. 2004. Phytophotolysis of hexahydro-1,3,5-trinitro-1,3,5-triazine (RDX) in leaves of Reed canary grass. *Environmental Science and Technology* 38 (1), pp. 290–295.

Kachout, S. S., Mansoura, A. B., Mechergui, R., Leclerc, J. C., Rejeb, M. N., and Ouerghi, Z. 2012. Accumulation of Cu, Pb, Ni and Zn in the halophyte plant Atriplex grown on polluted soil. *Journal of the Science of Food and Agriculture* 92 (2), pp. 336–342.

Kadlec, R. H., and Knight, R. L. 1996. *Treatment Wetlands*. Boca Raton, FL: CRC Press, Lewis Publishers.

Kadlec, R. H., and Knight, R. L. 1998. *Creating and Using Wetlands for Wastewater and Stormwater Treatment and Water Quality Improvement, Part I. Treatment Wetlands*. Madison, WI: University of Wisconsin–Madison Department of Engineering.

Kadlec, R. H., and Wallace, S. D. 2009. *Treatment Wetlands*, 2nd ed. Boca Raton, FL: CRC Press.

Kaimi, J. E., Mukaidani, T., and Tamak, M. 2007. Screening of twelve plant species for phytoremediation of petroleum hydrocarbon contaminated soil. *Plant Production Science* 10 (2), pp. 211–218, DOI, 10.1626/pps.10.211.

Karimi, N., Ghaderian, S. M., Raab, A., Feldmann, J., and Meharg, A. A. 2009. An arsenic-accumulating, hypertolerant brassica, Isatis capadocica. *New Phytologist* 184 (1), pp. 41–47.

Karthikeyan, R., Kulakow, P. A., Leven, B. A., and Erickson, L. E. 2012. Remediation of vehicle wash sediments contaminated with hydrocarbons: a field demonstration. *Environmental Progress & Sustainable Energy* 31 (1), pp. 139–146.

Keeling, S. M., Stewart, R. B., Anderson, C. W., and Robinson, B. H. 2003. Nickel and cobalt phytoextraction by the hyperaccumulator Berkheya coddii: implications for polymetallic phytomining and phytoremediation. *International Journal of Phytoremediation* 5, pp. 235–244.

Keiffer, C. H., and Ungar, I. A. 1996. *Bioremediation of Brine Contaminated Soils*. Final Report, PERF Project #91-18.

Kelepertsis, A. E. et al. 1990. The use of the genus *Alyssum* as a reliable geobotanical-biogeochemical indicator in geological mapping of altruabasic rocks in Greece. *Praktika tis Akademias Athinon* 65, pp. 170–176.

Kelsey, J. W., Colino, A., Koberle, M., and White, J. C. 2006. Growth conditions impact 2,2-bis (p-chlorophenyl)-1,1-dichloroethylene (p,p -DDE) accumulation by Cucurbita pepo. *International Journal of Phytoremediation* 8 (3), pp. 261–271.

Kersten, W. J., Brooks, R. R., Reeves, R. D., and Jaffre, T. 1979. Nickel uptake by New Caledonian species of *Phyllanthus. Taxon* 28 (5–6), pp. 529–534.

Kertulis-Tartar, G., Ma, L., Tu, C., and Chirenje, T. 2006. Phytoremediation of an arsenic-contaminated site using Pteris vittata L.: a two-year study. *International Journal of Phytoremediation* 8 (4), pp. 311–322.

Kiker, J. H., Larson, S., Moses, D. D., and Sellers, R. 2001. Use of engineered wetlands to phytoremediate explosives contaminated surface water at the Iowa Army Ammunition Plant, Middletown, Iowa. *Proceedings of the 2001 International Containment and Remediation Technology Conference and Exhibition*. http:// www.containment.fsu.edu/cd/content/pdf/416.pdf.

Kirkwood, N. 2001. *Manufactured Sites: Rethinking the Post-Industrial Landscape*. London, New York: Spon Press.

Kirkwood, N. 2002. Here come the hyper-accumulators, cleaning toxic sites from the roots up. *Harvard Design Magazine*, Fall–Winter.

Kirkwood, N., Hollander, J., and Gold, J. 2010. *Principles of Brownfield Regeneration, Cleanup, Design and Reuse of Derelict Land*. Washington, DC: Island Press.

Klein, H. A. 2011. Measuring the removal of trichloroethylene from phytoremediation sites at Travis and Fairchild Air Force bases. M.Sc. thesis, University of Utah.

Knight, S. H., and Beath, O. A., 1937. The occurrence of selenium and seleniferous vegetation in Wyoming. *Wyoming Agric. Exper. Stn. Bull.* 221.

Knott, S. G. et al. 1958. Selenium poisoning in horses in North Queensland. *Queensland Dept. Agric., Div. Animal Ind., Bull.* 41, pp. 1–16.

Kocon, A., and Matyka, M. 2012. Phytoextractive potential of Miscanthus giganteus and Sida hermaphrodita growing under moderate pollution of soil with Zn and Pb. *Journal of Food, Agriculture and Environment* 10 (2), pp. 1253–1256.

Kolbas, A., Mench, M., Herzig, R., Nehnevajova, E., and Bes, C. 2011. Copper phytoextraction in tandem with oilseed production using commercial cultivars and mutant lines of sunflower. *International Journal of Phytoremediation* 13 (suppl), pp. 55–76.

Kolbas A., Mench M., Marchand L., Herzig R., and Nehnevajova, E. 2014. Phenotypic seedling responses of a metal-tolerant mutant line of sunflower growing on a Cu-contaminated soil series. *Plant and Soil* 376, pp. 377–397, DOI, 10.1007/s11104-013-1974-8.

Komisar, S. J., and Park, J. 1997. Phytoremediation of diesel-contaminated soil using Alfalfa. *In Situ and On-Site Bioremediation* 4 (3), pp. 331–335. Columbus, OH: Battelle Press.

Kostick, D. S. 2010. *2008 Minerals Yearbook: Salt*. US Geological Survey.

Kothe, E., and Varma, A. (Eds.). 2012. *Bio-Geo Interactions in Metal-Contaminated Soils*. Springer ebook, Chapter 1, DOI, 10.1007/978-3-642-23327-2_1.

Kratochvil, R., Coale, F., Momen, B., Harrison, M., Pearce, J., and Schlosnagle, S. 2006. Cropping systems for phytoremediation of phosphorus-enriched soils. *International Journal of Phytoremediation* 8 (2), pp. 117–130.

Kruckeberg, A., Peterson, P., and Samiullah, Y. 1993. Hyperaccumulation of nickel by Arenaria rubella (Caryophyllaceae) from Washington State. *Madrono* 42, pp. 458–469

Kruger, T., Anderson, A., and Coats, J. R. (Eds.). 1997. *Phytoremediation of Soil and Water Contaminants*. ACS Symposium Series No. 664. Washington, DC: American Chemical Society.

Kühl, K. 2010. *The Field Guide to Phytoremediation*. New York: youarethecity.

Kumar, P. B. A. N., Dushenkov, V., Motto, H., and Raskin, I. 1995. Phytoextraction: the use of plants to remove heavy metals from soils. *Environmental Science and Technology* 29, pp. 1232–1238.

Kupper, H., Lombi, E., Zhao, F. J., Wieshammer, G., and McGrath, S. P. 2001 Cellular compartmentation of nickel in the hyperaccumulators Alyssum lesbiacum, Alyssum Bertolonii and Thlaspi goesingense. *Journal of Experimental Botany* 52, pp. 2291–2300.

Kuzovkina, Y. A., and Volk, T. A. 2009. The characterization of willow (Salix L.) varieties for use in ecological engineering applications: co-ordination of structure, function and autecology. *Ecological Engineering* 35 (8), pp. 1178–1189.

Lai, H.-Y., Chen, S.-W., and Chen, Z.-S. 2008. Pot experiment to study the uptake of Cd and Pb by three Indian Mustards (Brassica juncea) grown in artificially contaminated soils. *International Journal of Phytoremediation* 10, pp. 91–105.

Landmeyer, J. E. 2001. Monitoring the effect of poplar trees on petroleum-hydrocarbon and chlorinated-solvent contaminated ground water. *International Journal of Phytoremediation* 3 (1), pp. 61–85.

Landmeyer, J. E. 2001. *Introduction to Phytoremediation of Contaminated Groundwater: Historical Foundation, Hydrologic Control, and Contaminant Remediation*. Springer ebook.

Landmeyer, J. E. 2012. *Phytoremediation of Contaminated Groundwater*. New York: Springer.

Lanphear, B. P., Matte, T. D., Rogers, J., Clickner, R. P., Dietz, B., Bornschein, R. L., Succop, P., Mahaffey, K. R., Dixon, S., Galke, W., Rabinowitz, M., Farfel, M., Rohde, C., Schwartz, J., Ashley, P., and Jacobs, D. E. 1998. The contribution of lead-contaminated house dust and residential soil to children's blood lead levels: a pooled analysis of 12 epidemiologic studies. *Environmental Research* 79, pp. 51–68.

Lasat, M. M. 2000. The use of plants for the removal of toxic metals from contaminated soil. Online publication, http://nepis.epa.gov/Adobe/PDF/9100FZE1.PDF.

Lasat, M. M., and Kochian, L. V. 2000. Physiology of Zn Hyperaccumulation in Thlaspi caerulescens. In N. Terry and G. S. Bañuelos (Eds.) *Phytoremediation of Contaminated Soil and Water*. Boca Raton, FL: CRC Press.

Lasat, M. M. et al. 1997. Potential phytoextraction of Cs from contaminated soil. *Plant and Soil*, 195, pp. 99–106.

Lasat, M. M., Pence, N. S., and Kochian, L. V. 2001. Zinc phytoextraction in Thlaspi caerulescens. *International Journal of Phytoremediation* 3 (1), pp. 129–144.

Lee, I., Baek, K., Kim, H., Kim, S., Kim, J., Kwon, Y., Chang, Y., and Bae, B. 2007. Phytoremediation of soil co-contaminated with heavy metals and TNT using four plant species. *Journal of Environmental Science and Health Part A* 42 (13), pp. 2039–2045.

Lee, K. Y., and Doty, S. L. 2012. Phytoremediation of chlorpyrifos by populus and salix. *International Journal of Phytoremediation* 14 (1), pp. 48–61.

Lee, L. S., Suresh, P., Rao, C. et al. 1992. Equilibrium partitioning of polycyclic aromatic hydrocarbons from coal tar into water. *Environmental Science and Technology* 26, pp. 2110–2115.

Lee, S.-H., Lee, W.-S., Lee, C.-H., and Kim, J.-G. 2008. Degradation of phenanthrene and pyrene in rhizosphere of grasses and legumes. *Journal of Hazardous Materials* 153, pp. 892–898.

Leewis, M.-C., Reynolds, C. M., and Leigh, M. B. 2013. Long-term effects of nutrient addition and phytoremediation

325

on diesel and crude oil contaminated soils in subarctic Alaska. *Cold Regions Science and Technology*, DOI, 10.1016/j.coldregions.2013.08.011.

Lefevre, I., Marchal, G., Meerts, P., Correal, E., and Lutts, S. 2009. Chloride salinity reduces cadmium accumulation by the Mediterranean halophyte species Atriplex halimus L. *Environmental and Experimental Botany* 65, pp. 142–152.

Leigh, M. B. 2014. Personal correspondence via email with Kate Kennen, February.

Leigh, M. B., Prouzova, P. et al. 2006. Polychlorinated biphenyl (PCB)-degrading bacteria associated with trees in a PCB-contaminated site. *Applied and Environmental Microbiology* 72 (4), pp. 2331–2342.

Lewis, J., Qvarfort, U., and Sjostrom, J. 2013. Betula pendula: a promising candidate for phytoremediation of TCE in northern climates. *International Journal of Phytoremediation*, DOI, 140528074112008.

Li, J. T., Liao, B., Lan, C.Y., Qiu, J. W., Shu, W. S. 2007. Nickel and cadmium in carambolas marketed in Guangzhou and Hong Kong, China: implications for human health. *Science Total Environment* 388, pp. 405–412.

Li, N., Li, Z., Fu, Q., Zhuang, P., Guo, B., and Li, H. 2013. Agricultural technologies for enhancing the phytoremediation of cadmium-contaminated soil by Amaranthus hypochondriacus L. *Water, Air, & Soil Pollution* 224 (9), pp. 1–8.

Li, T., Di, Z., Islam, E., Jiang, H., and Yang, X. 2011. Rhizosphere characteristics of zinc hyperaccumulator Sedum alfredii involved in zinc accumulation. *Journal of Hazardous Materials* 185 (2), pp. 818–823.

Licht, I. 2012. Presentation to Harvard University GSD 9108, Phyto Research Seminar: Remediation and Rebuilding Technologies in the Landscape. 20 February.

Licht, L., and Isebr, S. J. 2005. Linking phytoremediated pollutant removal to biomass economic opportunities. *Biomass and Bioenergy* 28 (2), pp. 203–218.

Limmer, M. A., Balouet, J.-C., Karg, F., V., D. A., Burken, J. G. 2011. Phytoscreening for Chlorinated Solvents Using Rapid in Vitro SPME Sampling: Application to Urban Plume in Verl, Germany. *Environmental Science and Technology*. 2011, 45, 8276–8282.

Limmer, M. A., and Burken, J. G. 2014. Plant translocation of organic compounds: molecular and physicochemical predictors. *Environmental Science and Technology Letters* 1 (2), pp. 156–161.

Limmer, M.A., Martin, G., Watson, C.J., Martinez, C., Burken, and J.G. 2014. Phytoscreening: A comparison of *in planta* portable GC-MS and *in vitro* analyses. *Groundwater Monitoring and Remediation* 34, no. 1, pp. 49–56.

Limmer, M.A., Shetty, M., Markus, S.A., Kroeker, R., Parker, B. and Burken, J.G. 2013. Directional phytoscreening:

contaminant gradients in trees for plume delineation. *Environmental Science and Technology*, 47 (16), pp. 9069–9076.

Lin, C.-C., Lai, H.-Y., and Chen, Z.-S. 2010. Bioavailability assessment and accumulation by five garden flower species grown in artificially cadmium-contaminated soils. *International Journal of Phytoremediation* 12, pp. 454–467.

Linn, W. et al. 2004. State Coalition for Remediation of Drycleaners–SCRD Conducting Contamination Assessment Work at Drycleaning Sites, revised October 2010, p. 26.

Liu, L., Wu, L., Li, N., Luo, Y., Li, S., Li, Z., Han, C., Jiang, Y., and Christie, P. 2011. Rhizosphere concentrations of zinc and cadmium in a metal contaminated soil after repeated phytoextraction by Sedum plumbizincicola. *International Journal of Phytoremediation* 13 (8), pp. 750–764.

Liu, W., Luo, Y., Teng, Y., and Li, Z. 2010. Phytoremediation of oilfield sludge after prepared bed bioremediation treatment. *International Journal of Phytoremediation* 12 (3), pp. 268–278.

Llewellyn, D., and Dixon, M. A. 2011. Can plants really improve indoor air quality? In B. Grodzinski, W. A. King, and R. Yada (Eds.) *Comprehensive Biotechnology*, M. M. Young (Ed.), *Agricultural and Related Biotechnologies*, 2nd ed., vol. 4, Oxford: Elsevier, pp. 331–338.

Loehr, R. C., and Webster, M. T. 1996. Behavior of fresh vs aged chemicals in soil. *Journal of Soil Contamination* 5, pp. 361–384.

Lu, L., Tian, S., Yang, X., Peng, H., and Li, T. 2013. Improved cadmium uptake and accumulation in the hyperaccumulator Sedum alfredii: the impact of citric acid and tartaric acid. *Journal of Zhejiang University SCIENCE B* 14 (2), pp. 106–114.

Lu, Y., Li, X., He, M., and Zeng, F. 2013. Behavior of native species Arrhenatherum elatius (Poaceae) and Sonchus transcaspicus (Asteraceae) exposed to a heavy metal-polluted field: plant metal concentration, phytotoxicity, and detoxification responses. *International Journal of Phytoremediation* 15, pp. 924–937.

Lugtenberg, B. and Kamilova, F. 2009. Plant-Growth-Promoting Rhizobacteria. *Annual Review of Microbiology* 63, 541–556.

Lugtenberg, B. J. and Dekkers, L. C. 1999. What makes Pseudomonas bacteria rhizosphere competent? *Environ. Microbiol.* 1, 9–13.

Lunney, A. I., Zeeb, B. A., and Reimer, K. J. 2004. Uptake of weathered DDT in vascular plants: potential for phytoremediation. *Environmental Science and Techonology* 38, pp. 6147–6154.

Ma, J., Chu, C., Li, J., and Song, B. 2009. Heavy metal pollution in soils on railroad side of Zhengzhou-Putian section of Longxi-Haizhou Railroad, China. *Pedosphere* 19 (1), pp. 121–128.

Ma, L. Q., Komar, K. M., Tu, C., Zhang, W. H., Cai, Y., and Kennelley, E. D. 2001. A fern that hyperaccumulates arsenic – a hardy, versatile, fast-growing plant helps to remove arsenic from contaminated soils. *Nature* 409, pp. 579–579.

Ma, T. T., Teng, Y., Luo, Y. M., and Christie, P. 2013. Legume-grass intercropping phytoremediation of phthalic acid esters in soil near an electronic waste recycling site: a field study. *International Journal of Phytoremediation* 15 (2), pp. 154–167.

Ma, X. M., and Burken, J. G. 2003. TCE diffusion to the atmosphere in phytoremediation applications. *Environmental Science and Technology* 37 (11), pp. 2534–2539.

Ma, X., Richter, A. R., Albers, S., and Burken, J. G. 2004. Phytoremediation of MTBE with hybrid poplar trees. *International Journal of Phytoremediation* 6 (2), pp. 157–167.

Macci, C., Doni, S., Peruzzi, E., Bardella, S., Filippis, G., Ceccanti, B., and Masciandaro, G. 2012. A real-scale soil phytoremediation. *Biodegradation* 24 (4), pp. 521–538.

Macek, Tomas et al. 2004. Phytoremediation of metals and inorganic pollutants. In A. Singh and O. P. Ward (Eds.) *Applied Bioremediation and Phytoremediation*. New York: Springer, pp. 134–158.

Macklon, A. E. S., and Sim, A. 1990. Cortical cell fluxes of cobalt in roots and transport to the shoots of Ryegrass seedlings. *Physiologia Plantarum* 80, pp. 409–416.

Madejón, P., Ciadamidaro, L., Marañón, T., and Murillo, J. M. 2012. Long-term biomonitoring of soil contamination using poplar trees: accumulation of trace elements in leaves and fruits. *International Journal of Phytoremediation* 15, pp. 602–614.

Mahdieh, M., Yazdani, M., and Mahdieh, S. 2013. The high potential of Pelargonium roseum plant for phytoremediation of heavy metals. *Environmental Monitoring and Assessment*, pp. 1–5.

Mahmud, R., Inoue, N., Kasajima, S.-Y., and Shaheen, R. 2008. Assessment of potential indigenous plant species for the phytoremediation of arsenic-contaminated areas of Bangladesh. *International Journal of Phytoremediation* 10, pp. 119–132.

Maila, M. P., Randima, P., and Cloete, T. E. 2005. Multispecies and monoculture rhizoremediation of polycyclic aromatic hydrocarbons (PAHs) from the soil. *International Journal of Phytoremediation* 7 (2), pp. 87–98.

Mandal, A., Purakayastha, T., Patra, A., and Sanyal, S. 2012. Phytoremediation of arsenic contaminated soil by Pteris vittata L. II. Effect on arsenic uptake and rice yield. *International Journal of Phytoremediation* 14 (6), pp. 621–628.

Manousaki, E., and Kalogerakis, N. 2011. Halophytes – an emerging trend in phytoremediation. *International Journal of Phytoremediation* 13, pp. 959–969.

Manousaki, E., Galanaki, K., Papadimitriou, L., and Kalogerakis, N. 2014. Metal phytoremediation by the halophyte Limoniastrum monopetalum (L.) Boiss: two contrasting ecotypes. *International Journal of Phytoremediation* 16 (7–8), pp. 755–769.

Marcacci, S. and Schwitzguébel, J.-P. 2007. Using plant phylogeny to predict detoxification of triazine herbicides. In *Phytoremediation: Methods and Reviews*. Totowa, N.J. : Humana Press.

Marchand, L., Mench, M., March and, C., Le Coustumer, P., Kolbas, A., and Maalouf, J. 2011. Phytotoxicity testing of lysimeter leachates from aided phytostabilized Cu-contaminated soils using duckweed (Lemna minor). *Science of the Total Environment* 410 pp. 146–153, DOI, 10.1016/j.scitotenv.2011.09.049.

Marecik, R., Bialas, W., Cyplik, P., Lawniczak, L., and Chrzanowski, L. 2012. Phytoremediation potential of three wetland plant species toward atrazine in environmentally relevant concentrations. *Polish Journal of Environmental Studies* 21 (3), pp. 697–702.

Margolis, L., and Robinson, A. 2007. Toxic filtration via fungi. In *Living Systems: Innovative Materials and Technologies for Landscape Architecture*. Boston: Birkhauser, pp. 166–167.

Markis, K. C., Shakya, K. M., Datta, R., Sarkar, D., and Pachanoor, D. 2007a. High uptake of 2,4,6-trinitrotoluene by Vetiver grass – potential for phytoremediation? *Environmental Pollution* 146, pp. 1–4.

Markis, K. C., Shakya, K. M., Datta, R., and Pachanoor, D. 2007b. Chemically catalyzed uptake of 2,4,6-trinitrotoluene by *Vetiveria zizanoides*. *Environmental Pollution* 148, pp. 101–106.

Martin, H. W., Young, T. R., Kaplan, D. I., Simon, L., and Adriano, D. C. 1996. Evaluation of three herbaceous index plant species for bioavailability of soil cadmium, chromium, nickel, and vanadium. *Plant and Soil* 182, pp. 199–207.

Massoura, S., Echevarria, G., Leclerc-Cessac, E., and Morel, J. 2005. Response of excluder, indicator, and hyperaccumulator plants to nickel availability in soils. *Soil Research* 42 (8), pp. 933–938.

Mattina, M. I., Iannucci-Berger, W., Dykas, L., and Pardus, J. 1999. Impact of long-term weathering, mobility, and land use on chlordane residues in soil. *Environmental Science and Technology* 33, pp. 2425–2431.

Mattina, M. J. I., Eitzer, B. D., Iannucci-Berger, W., Lee, W. Y., and White, J. C. 2004. Plant uptake and translocation of highly weathered, soil-bound technical chlordane residues: data from field and rhizotron studies. *Environmental Toxicology and Chemistry*, 23, pp. 2756–2762.

327

McCray, C.W.R. and Hurwood, I.S. 1963. Selenosis in northwestern Queensland associated with a marine Cretaceous formation. *Queensland J. Agric. Sci.* 20, pp. 475–498.

McCutcheon, S. C., and Schnoor, J. L. (Eds.). 2003. *Phytoremediation: Transformation and Control of Contaminants*. Hoboken, NJ: Wiley-Interscience, Inc.

McGrath, S. P., Dunham, S. J., and Correll, R. L. 2000. Potential for phytoextraction of zinc and cadmium from soils using hyperaccumulator plants. In N. Terry and G. S. Bañuelos, (Eds.) *Phytoremediation of Contaminated Soil and Water*. Boca Raton, FL: CRC Press.

McIntyre, T. C. 2003. Databases and protocol for plant and microorganism selection: hydrocarbons and metals. In S. C. McCutcheon and J. L. Schnoor (Eds.) *Phytoremediation: Transformation and Control of Contaminants*. Hoboken, NJ: John Wiley, pp. 887–904.

Meeinkuirt, W., Pokethitiyook, P., Kruatrachue, M., Tanhan, P., and Chaiyarat, R. 2012. Phytostabilization of a Pb-contaminated mine tailing by various tree species in pot and field trial experiments. *International Journal of Phytoremediation* 14 (9), pp. 925–938.

Meera, M., and Agamuthu, P. 2012. Phytoextraction of As and Fe using Hibiscus cannabinus l. from soil polluted with landfill leachate. *International Journal of Phytoremediation* 14 (2), pp. 186–199.

Meers, E., Van Slycken, S., Adriaensen, K., Ruttens, A., Vangronsveld, J., Du Laing, G., Witters, N., Thewys, T., and Tack, F. 2010. The use of bio-energy crops (Zea mays) for 'phytoattenuation' of heavy metals on moderately contaminated soils: A field experiment. *Chemosphere* 78 (1), pp. 35–41.

Meng, L., Qiao, M., and Arp, H. P. H. 2011. Phytoremediation efficiency of a PAH-contaminated industrial soil using ryegrass, white clover, and celery as mono- and mixed cultures. *Journal of Soils and Sediments* 11 (3), pp. 482–490.

Metcalf, R. L. 2002. Insect control. In *Ullmann's Encyclopedia of Industrial Chemistry*. Weinheim: Wiley-VCH, DOI, 10.1002/14356007.a14_263.

Miller, R., Khan, Z., and Doty, S. 2011. Comparison of trichloroethylene toxicity, removal and degradation by varieties of Populus and Salix for improved phytoremediation applications. *Journal of Bioremediation and Biodegradation* 7 p. 2.

Mingorance, M., Leidi, E., Valdès, B., and Oliva, S. 2012. Evaluation of lead toxicity in Erica andevalensis as an alternative species for revegetation of contaminated soils. *International Journal of Phytoremediation* 14 (2), pp. 174–185.

Ministry of Environment. 2006. *Environmental Best Management Practices for Urban and Rural Land Development in British Columbia: Air Quality BMPs and Supporting Information*. British Columbia Ministry of Environment.

Mirka, M. A., Clulow, F. V., Dave, N. K., and Lim, T. P. 1996. Radium-226 in Cattails, *Typha latifolia*, and bone of muskrat, *Ondatra zibethica* (L.), from a watershed with uranium tailings near the city of Elliot Lake, Canada. *Environmental Pollution* 91, pp. 41–51.

Mohanty, M., and Patra, H. K. 2012. Phytoremediation potential of Paragrass – an in situ approach for chromium contaminated soil. *International Journal of Phytoremediation* 14 (8), pp. 796–805.

Monaci, F., Leidi, E., Mingorance, M., Valdes, B., Oliva, S., and Bargagli, R. 2011. Selective uptake of major and trace elements in Erica andevalensis, an endemic species to extreme habitats in the Iberian Pyrite Belt. *Journal of Environmental Sciences* 23 (3), pp. 444–452.

Moore, M. T., and Kroeger, R. 2010. Effect of three insecticides and two herbicides on rice (Oryza sativa) seedling germination and growth. *Archives of Environmental Contamination and Toxicology* 59 (4), pp. 574–581.

Moral, R., Navarro-Pedreno, J., Gomez, I., and Mataix, J. 1995. Effects of chromium on the nutrient element content and morphology of tomato. *Journal of Plant Nutrition* 18, pp. 815–822.

Morey, K., Antunes, J., Albrecht, M. S., Bowen, K. D., Troupe, T. A., Havens, J. F., Medford, K. L., and June, I. Developing a synthetic signal transduction system in plants. 2011. In C. Voigt (Ed.), *Methods in Enzymology* 497, *Synthetic Biology, Part A*. Burlington, VA: Academic Press, pp. 581–602.

Morikawa, H., and Ozgur, C. E. 2003. Basic processes in phytoremediation and some applications to air pollution control. *Chemosphere* 52, pp. 1554–1558.

Morikawa, H., Takahashi, M., and Kawamura, Y. 2003. Metabolism and genetics of atmospheric nitrogen dioxide control using pollutant-philic plants. In S. C. McCutcheon and J. L. Schnoor (Eds.) *Phytoremediation: Transformation and Control of Contaminates*. New York: Wiley, pp. 465–486.

Morrey, D. R., et al. 1989. Studies on serpentine flora: Preliminary analyses of soils and vegetation associated with serpentine rock formations in the south-eastern Transvaal. *S. Afr. J. Bot.* 55, pp. 171–177.

Moxon, A. L. et al. 1939. Selenium in rocks, soils and plants. *S. Dakota Agric. Exper. Stn. Rev. Tech. Bull.* 2, pp. 1–94.

Moyers, B. 2007. Rachel Carson and DDT. 21 September. http://www.pbs.org/moyers/journal/09212007/profile2.html. Accessed 11 November 2013.

Mueller, J. G., Cerniglia, C. E., and Pritchard, P. H. 1996. Bioremediation of environments contaminated by polycyclic aromatic hydrocarbons. In R. L. Crawford and D. L. Crawford (Eds.) *Bioremediation: Principles and*

328

Applications. Cambridge, UK: Cambridge University Press, pp. 1215–1294.

Mukherjee, I., and Kumar, A. 2012. Phytoextraction of endosulfan a remediation technique. *Bulletin of Environmental Contamination and Toxicology* 88 (2), pp. 250–254.

Murakami, M., Ae, N., and Isikawa, S. 2007. Phytoextraction of cadmium by rice, soybean, and maize. *Environmental Pollution* 145, pp. 96–103.

Muralidharan, N., Davis, L. C., and Erickson, L. E. 1993. Monitoring the fate of toluene and phenol in the rhizosphere. In R. Harrison (Ed.) *Proceedings, 23rd Annual Biochemical Engineering Symposium*, University of Oklahoma, Norman.

National Research Council. 2003. R. Luthy (chair), R. Allen-King, S. Brown, D. Dzombak, S. Fendorf, J. Geisy, J. Hughes, S. Luoma, L. Malone, C. Menzie, S. Roberts, M. Ruby, T. Schultz, and B. Smets. *Bioavailabililty of Contaminants in Soils and Sediments*. Washington, DC: National Academy of Sciences.

Nedunuri, K., Lowell, C., Meade, W., Vonderheide, A., and Shann, J. 2009. Management practices and phytoremediation by native grasses. *International Journal of Phytoremediation* 12 (2), pp. 200–214.

Negri, M. C., and Hinchman, R. R. 2000. The use of plants for the treatment of radionuclides. In I. Raskin and B. D. Ensley (Eds.) *Phytoremediation of Toxic Metals*. New York: John Wiley & Sons, Inc., pp. 107–132.

Negri, M. C., Hinchman, R. R., and Johnson, D. O. 1998. An overview of Argonne National Laboratory's phytoremediation program. Presented at the Petroleum Environmental Research Forum's Spring General Meeting, Argonne National Laboratory, Argonne, IL.

Negri, M. C. et al. 2003. Root development and rooting at depths. In S. C. McCutcheon and J. L. Schnoor (Eds.) *Phytoremediation: Transformation and Control of Contaminants*. Hoboken, NJ: John Wiley & Sons, Inc., pp. 233–262.

Nehnevajova, E., Herzig, R., Federer, G., Erismann, K. H., and Schwitzguébel, J. P. 2005. Screening of sunflower cultivars for metal phytoextraction in a contaminated field prior to mutagenesis. *International Journal of Phytoremediation* 7 (4), pp. 337–349.

Nehnevajova, E., Herzig, R., Federer, G., Erismann, K. H., and Schwitzguébel, J. P. 2007. Chemical mutagenesis – a promising technique to increase metal concentration and extraction in sunflowers. *International Journal of Phytoremediation* 9 (2), pp. 149–165.

Nelson, S. 1996. Summary of the Workshop on Phytoremediation of Organic Contaminants. Fort Worth, TX.

Nepovim, A., Hebner, A., Soudek, P., Gerth, A., Thomas, H., Smreck, S., and Vanek, T. 2005. Degradation of 2,4,6-trinitrotoluene by selected helophytes. *Chemosphere* 60, pp. 1454–1461.

Newman, L. A., Bod, C., Cortellucci, R., Domroes, D., Duffy, J., Ekuan, G., Fogel, D., Heilman, P., Muiznieks, I., Newman, T., Ruszaj, M., Strand, S. E., and Gordon, M. P. 1997a. Results from a pilot-scale demonstration: phytoremediation of trichloroethylene and carbon tetrachloride. Abstract for the 12th Annual Conference on Contaminated Soils, Amherst, MA.

Newman, L. A., Strand, S. E., Choe, N., Duffy, J., Ekuan, G., Ruszaj, M., Shurtleff, B. B., Wilmoth, J., Heilman, P., and Gordon, M. P. 1997b. Uptake and biotransformation of trichloroethylene by hybrid poplars. *Environmental Science and Technology* 31, pp. 1062–1067.

Newman, L. A., Gordon, M. P., Heilman, P., Cannon, D. L., Lory, E., Miller, K., Osgood, J., and Strand, S. E. 1999a. Phytoremediation of MTBE at a California naval site. *Soil and Groundwater Cleanup* Feb.–Mar., pp. 42–45.

Newman, L. A., Wang, X., Muiznieks, I. A., Ekuan, G., Ruszaj, M., Cortellucci, R., Domroes, D., Karscig, G., Newman, T., Crampton, R. S., Hashmonay, R. A., Yost, M. G., Heilman, P. E., Duffy, J., Gordon, M. P., and Strand, S. E. 1999b. Remediation of trichloroethylene in an artificial aquifer with trees: a controlled field study. *Environmental Science and Technology* 33 (13), pp. 2257–2265.

NHDES (New Hampshire Department of Environmental Services). 2006. Environmental fact sheet – ethylene glycol and propylene glycol: health information summary, http://des.nh.gov/. Accessed 5 March 2014.

Niazi, N., Singh, B., Van Zwieten, L., and Kachenko, A. 2011. Phytoremediation potential of Pityrogramma calomelanos var. austroamericana and Pteris vittata L. grown at a highly variable arsenic contaminated site. *International Journal of Phytoremediation* 13 (9), pp. 912–932.

Nowak, D. J. 1994. Air pollution removal by Chicago's urban forest. General technical report NE-186. In: McPherson, E.G. (Ed.), Chicago's Urban Forest Ecosystem: Results of the Chicago Urban Forest Climate Project. United States Department of Agriculture, Forest Service, Northeastern Forest Experimental Station, Randnor, PA, pp. 63–81.

Nowak, D. J. 2002. The effects of urban trees on air quality. Available at: http://www.nrs.fs.fed.us/units/urban/local-resources/downloads/Tree_Air_Qual.pdf. Accessed 11 December 2013.

Nowak, D. J. 2006. Appendix A: tree species selection list for New York City. In *Mitigating New York City's Heat Island with Urban Forestry, Living Roofs, and Light Surfaces: New York City Regional Heat Island Initiative Final Report*. Syracuse, NY: USDA Forest Service.

329

Nowak, D. J., Crane, D. E., and Stevens, J. C. 2006. Air pollution removal by urban trees and shrubs in the United States. *Urban Forestry & Urban Greening* 4, pp. 115–123.

Nowak, D. J., Hirabayashi, S., Bodine, A. and Greenfield, E. 2014. Tree and forest effects on air quality and human health in the United States. *Environ. Pollut.* 193, 119–129.

Office of Underground Storage Tanks (OUST) 2011. What is the History of the Federal Underground Storage Tank Program? April 11. http://www.epa.gov/oust/faqs/genesis1.htm. Accessed April–May 2011.

Olette, R., Couderchet, M., Biagianti, S., and Eullaffroy, P. 2008. Toxicity and removal of pesticides by selected aquatic plants. *Chemosphere* 70 (8), pp. 1414–1421.

Olson, P. E., and Fletcher, J. S. 2000. Ecological recovery of vegetation at a former industrial sludge basin and its implications to phytoremediation. *Environmental Science and Pollution Research* 7, pp. 1–10.

Olson, P. E. , Reardon, K. F. , and Pilon-Smits, E. A. H. 2003. Ecology of rhizosphere bioremediation. In S. C. McCutcheon and J.L. Schnoor (Eds.). *Phytoremediation: Transformation and Control of Contaminants*. Hoboken, NJ: Wiley-Interscience, pp. 317–353.

Olsen, R. A. 1994. The transfer of radiocaesium from soil to plants and fungi in seminatural ecosystems. *Studies in Environmental Science* 62, pp. 265–286.

OSHA. 2013. Health and safety topics: lead, https://www.osha.gov/SLTC/lead/. Accessed 23 November 2013.

Otabbong, E. 1990. Chemistry of Cr in some Swedish soils. *Plant and Soil* 123, pp. 89–93.

Ouyang, Y. 2005. Phytoextraction: simulating uptake and translocation of arsenic in a soil–plant system. *International Journal of Phytoremediation* 7 (1), pp. 3–17.

Padmavathiamma, P. K., and Li, L. Y. 2009. Phytoremediation of metal-contaminated soil in temperate humid regions of British Columbia, Canada. *International Journal of Phytoremediation* 11, pp. 575–590.

Parrish, Z. D., Banks, M. K., and Schwab, A. P. 2004. Effectiveness of phytoremediation as a secondary treatment for polycyclic aromatic hydrocarbons (PAHs) in composted soil. *International Journal of Phytoremediation* 6 (2), pp. 119–137.

Parsons, 2010 (September). Technical report phytostabilization at Travis Air Force Base, California prepared for: Air Force Center for Engineering and the Environment Restoration Branch, Technology Transfer Office (TDV) Brooks City-Base, Texas and Travis Air Force Base California Contract Number FA8903-08-C-8016.

Paterson, K. G., and Schnoor, J. L. 1992. Fate of alachlor and atrazine in riparian zone field site. *Water Environment Research* 64, pp. 274–283.

Perez-Esteban, J., Escolastico, C., Moliner, A., Masaguer, A., and Ruiz-Fernandez, J. 2013. Phytostabilization of metals in mine soils using Brassica juncea in combination with organic amendments. *Plant and Soil* 377 (1–2), pp. 97–109.

Perrino, E. V., Brunetti, G., and Farrag, K. 2013. Plant communities in multi-metal contaminated soils: a case study in the national park of Alta Murgia (Apulia Region – Southern Italy). *International Journal of Phytoremediation* 16, pp. 871–888.

Perrino, E. et al. 2014. Plant communities in multi-metal contaminated soils: a case study in the National Park of Alta Murgia (Apulia Region-Southern Italy). *International Journal of Phytoremediation* 16 (9), pp. 871–888, DOI, 10.1080/15226514.2013.798626.

Phaenark, C., Pokethitiyook, P., Kruatrachue, M., and Ngernsansaruary, C. 2009. Cadmium and zinc accumulation in plants from the Padaeng zinc mine area. *International Journal of Phytoremediation* 11, pp. 479–495.

Phytokinetics, Inc. 1998. *Using Plants to Clean Up Environmental Contaminants*. Logan, UT: Phytokinetics, Inc.

Phytotech, Inc. 1997. *Phytoremediation Technical Summary*. Monmouth, NJ: Phytotech, Inc.

Pieper, D. H., and Reineke, W. 2000. Engineering bacteria for bioremediation. *Current Opinion in Biotechnology* 11, pp. 379–388.

Pierzynski, G., Schnoor, J. L., Youngman, A., Licht, L., and Erickson, L. 2002. Poplar trees for phytostabilization of abandoned zinc-lead smelter. *Practice Periodical of Hazardous, Toxic, and Radioactive Waste Management* 6 (3), pp. 177–183.

Pignattelli, S., Colzi, I., Buccianti, A., Cecchi, L., Arnetoli, M., Monnanni, R., Gabbrielli, R., and Gonnelli, C. 2012. Exploring element accumulation patterns of a metal excluder plant naturally colonizing a highly contaminated soil. *Journal of Hazardous Materials* 227 pp. 362–369.

Pilon-Smits, E. 2005. Phytoremediation. *Annual Review of Plant Biology* 56 (1), p. 15.

Pilon-Smits, E. A. H., and Freeman, J. L. 2006. Environmental cleanup using plants: biotechnological advances and ecological considerations. *Frontiers in Ecology and the Environment* 4, pp. 203–210.

Pinder, J. E. III, McLeod, K. W., Alberts, J. J., Adriano, D. C., and Corey, J. C. 1984. Uptake of 244Cm, 238Pu, and other radionuclides by trees inhabiting a contaminated floodplain. *Health Physics* 47, pp. 375–384.

Popek, R., Gawrońska, H., Wrochna, M., Gawronski, S., and Saebo, A. 2013. Particulate matter on foliage of 13 woody species: deposition on surfaces and phytostabilisation in waxes – a 3-year study. *International Journal of Phytoremediation* 15 (3), pp. 245–256.

330

Porta, M., and Zumeta, E. 2002. Implementing the Stockholm Treaty on Persistent Organic Pollutants. *Occupational and Environmental Medicine* 10 (59), pp. 651–652.

Potter, S. T. 1998. Computation of the hydraulic performance of a phyto-cover using the HELP model and the water balance method. Presented at the 3rd Annual International Conference on Phytoremediation, Houston.

Pradhan, S. P., Conrad, J. R., Paterek, J. R., and Srinistava, V. J. 1998. Potential of phytoremediation for treatment of PAHs in soil at MGP sites. *Journal of Soil Contamination* 7, pp. 467–480.

Prasad, M. N. V. 2005. Nickelophilous plants and their significance in phytotechnologies. *Brazilian Journal of Plant Physiology* 17 (1), pp. 113–128.

Pulford, I., Riddell-Black, D., and Stewart, C. 2002. Heavy metal uptake by willow clones from sewage sludge-treated soil: the potential for phytoremediation. *International Journal of Phytoremediation* 4 (1), pp. 59–72.

Purakayastha, T. J., Viswanath, T., Bhadraray, S., Chhonkar, P. K., Adhikari, P. P., and Suribabu, K. 2008. Phytoextraction of zinc, copper, nickel and lead from a contaminated soil by different species of Brassica. *International Journal of Phytoremediation* 10, pp. 61–72.

Qadir, M., Steffen, D., Yan, F., and Schubert, S. 2003. Sodium removal from a calcareous saline-sodic soil through leaching and plant uptake during phytoremediation. *Land Degradation & Development* 14, pp. 301.

Qiu, R., Fang, X., Tang, Y., Du, S., Zeng, X., and Brewer, E. 2006. Zinc hyperaccumulation and uptake by Potentilla griffithii Hook. *International Journal of Phytoremediation*, 8 (4), pp. 299–310.

Qiu, X., Shah, S. I., Kendall, E. W., Sorenson, D. L., Sims, R. C., and Engelke M. C. 1994. Grass enhanced bioremediation for clay soils contaminated with polynuclear aromatic hydrocarbons. In T. A. Anderson and J. R. Coats, (Eds.) *Bioremediation through Rhizosphere Technology*. ACS Symposium Series No. 563. Washington, DC: American Chemical Society, pp. 142–157.

Qiu, X., Leland, T. W., Shah, S. I., Sorensen, D. L., and Kendall E. W. 1997. Field study: grass remediation for clay soil contaminated with polycyclic aromatic hydrocarbons. In E. L. Kruger, T. A. Anderson, and J. R. Coats (Eds.) *Phytoremediation of Soil and Water Contaminants*. Washington, DC: American Chemical Society, pp. 189–199.

Radwan, S. S., Dashti, N., and El-Nemr, I. M. 2005. Enhancing the growth of Vicia faba plants by microbial inoculation to improve their phytoremediation potential for oily desert areas. *International Journal of Phytoremediation* 7 (1), pp. 19–32.

Ramaswami, A., Carr, P., and Burkhardt, M. 2001. Plant-uptake of uranium: hydroponic and soil system studies. *International Journal of Phytoremediation* 3 (2), pp. 189–201.

Rascio, N. 1977. Metal accumulation by some plants growing on zinc-mine deposits. *Oikos* 29, pp. 250–253.

Raskin, I., and Ensley, B. D. (Eds.). 2000. *Phytoremediation of Toxic Metals: Using Plants to Clean Up the Environment.* New York: Wiley.

Reddy, B. R., and Sethunathan, N. 1983. Mineralization of Parathion in the rice rhizosphere. *Applied and Environmental Microbiology* 45, pp. 826–829.

Reeves, R. D. 2006. Hyperaccumulation of trace elements by plants. In J.-L. Morel et al. (Eds.) *Phytoremediation of Metal-Contaminated soils*. Netherlands: Springer.

Reeves, R.D., and Brooks, R.R. 1983. European species of Thlaspi L. (Cruciferae) as indicators of Nickel and Zinc. *Journal of Geochemical Explorations* 18, pp. 275–283.

Reeves, R.D., Baker, A.J.M, Borhidi, A., Berazaín, R. 1996. Nickel-accumulating plants from the ancient serpentine soils of Cuba. *New Phytologist* 133, pp. 217–224.

Reiche, N., Lorenz, W., and Borsdorf, H. 2010. Development and application of dynamic air chambers for measurement of volatilization fluxes of benzene and MTBE from constructed wetlands planted with common reed. *Chemosphere* 79 (2), pp. 162–168.

Reilley, K., Banks, M. K., and Schwab, A. P. 1993. Dissipation of polycyclic aromatic hydrocarbons in the rhizosphere. *Journal of Environmental Quality* 25, pp. 212–219.

Reynolds, C. M. 2012 (March). Presentation to Harvard University GSD 9108, Phyto Research Seminar: Remediation and Rebuilding Technologies in the Landscape.

Reynolds, C. M., and Koenen, B. A. 1997. Rhizosphere-enhanced bioremediation. *Military Engineering* 586, pp. 32–33.

Reynolds, C. M., Koenen, B. A., Carnahan, J. B., Walworth, J. L., and Bhunia, P. 1997. Rhizosphere and nutrient effects on remediating subartic soils. In B. C. Alleman, and A. Leeson (Eds.) *In Situ and On-Site Bioremediation: Volume 1, Cold Region Applications*. Columbus, OH: Battelle Press.

Reynolds, C. M., Koenen, B. A., Perry, L. B., and Pidgeon, C. S. 1997b. Initial field results for rhizosphere treatment of contaminated soils in cold regions. In. H. K. Zubeck, C. R. Woolard, D. M. White, and T. S. Vinson (Eds.) *International Association of Cold Regions Development Studies*. Anchorage: American Society of Civil Engineers, pp. 143–146.

331

Reynolds, C. M., Pidgeon, C. S., Perry, L. B., Gentry, T. J., and Wolf, D. C. 1998. Rhizosphere-enhanced benefits for remediating recalcitrant petroleum compounds. Poster abstract #51 at the 14th Annual Conference on Contaminated Soils, Amherst, MA.

Reynolds, C. M., Perry, L. B., Pidgeon, C. S, Koenen, B. A., Pelton, D. K., and Foley, K. L. 1999. Plant-based treatment of organic-contaminated soils in cold climates. In *Edmonton 99, Proceedings of Assessment and Remediation of Contaminated Sites in Arctic and Cold Climates Workshop*, Edmonton, Alberta, Canada, 3–4 May, pp. 166–172.

Ribeiro, H., Almeida, C., Mucha, A., and Bordalo, A. 2013. Influence of different salt marsh plants on hydrocarbon degrading microorganisms abundance throughout a phenological cycle. *International Journal of Phytoremediation* 15 (8), pp. 715–728.

Rice, P. J., Anderson, T. A., and Coats, J. R. 1996a. The use of vegetation to enhance biodegradation and reduce off-site movement of aircraft deicers. Abstract 054 at the 212th American Chemical Society National Meeting, Orlando, FL.

Rice, P. J., Anderson, T. A., and Coats, J. R. 1996b. Phytoremediation of herbicide-contaminated water with aquatic plants. Presented at the 212th American Chemical Society National Meeting, Orlando, FL.

Roberts, B.A. 1992. The ecology of serpentine areas, Newfoundland, Canada. B.A. Roberts and J. Proctor (Eds.), In *The Ecology of Areas with Serpentinized Rocks-a World View*. Kluwer Academic Publishers, Dordrecht, pp. 75–113.

Robinson, B. H., Brooks, R. R., Howes, A. W., Kirkman, J. H., and Gregg, P. E. H. 1997a. The potential of the high-biomass nickel hyperaccumulator Berkheya coddii for phytoremediation and phytomining. *Journal of Geochemical Exploration* 60, pp. 115–126.

Robinson, B. H., Chiarucci, A., Brooks, R. R., Petit, D. et al. 1997b. The nickel hyperaccumulator plant Alyssum betrolonii as a potential agent for phytoremediation and phytomining of nickel. *Journal of Geochemical Exploration* 59, pp. 75–96.

Robinson, T. W. 1958. Phreatophytes. US Geological Survey Water Supply Paper 1423, available at http://pubs.er.usgs.gov/usgspubs/wsp/wsp1423.

Robson, D. B., Knight, J. D., Farrell, R. E., and Germida, J. J. 2003. Ability of cold-tolerant plants to grow in hydrocarbon-contaminated soil, *International Journal of Phytoremediation* 5 (2), pp. 105–123.

Rock, S. 2000. Lecture delivered at the Phytoremediation, State of the Science Conference, Boston, MA, May.

Rock, S. 2010. EPA phytotechnologies fact sheets, Office of Superfund Remediation and Technology Innovation. http://www.epa.gov/tio/download/remed/phytotechnologies-factsheet.pdf.

Rock, S. (US EPA). 2014. Personal communication with Kate Kennen, April 2014.

Rog, C. 2013. Presentation to Harvard University GSD 6335, Phyto Practicum Research Seminar. 17 October.

Romeh, A. 2009. Phytoremediation of water and soil contaminated with Imidacloprid pesticide by Plantago major, L. *International Journal of Phytoremediation* 12 (2), pp. 188–199.

Rosario, K., Iverson, S. L., Henderson, D. A., Chartrand, S., McKeon, C., Glenn, E. P., and Maier, R. M. 2007. Bacterial community changes during plant establishment at the San Pedro River mine tailings site. *Journal of Environmental Quality* 36 (5), pp. 1249–1259.

Rosen, C. J., and Horgan, B. P. 2013. Preventing pollution problems from lawn and garden fertilizers. *University of Minnesota Extension*. N.p., n.d. Web. 10 December. http://www1.extension.umn.edu/garden/yard-garden/lawns/preventingpollutionproblems/.

Rosenfeld, A. H., Akbari, H., Romm, J. J. and Pomerantz, M., 1998. Cool communities: strategies for heat island mitigation and smog reduction. *Energy and Buildings* 28, 51–62.

Rosenfeld, I., and Beath, O.A. 1964. *Selenium: Geobotany, Biochemistry, Toxicity, and Nutrition*, Academic Press, New York.

Rosser, S. I., French, C. E., and Bruce, N. C. 2001. Special symposium: Engineering plants for the phytodetoxification of explosives. *In Vitro Cellular & Developmental Biology* 37, pp. 330–333.

Rotkittikhun, P., Kruatrachue, M., Chaiyarat, R., Ngernsansaruay, C., Pokethitiyook, P., Paijitprapaporn, A., and Baker, A. J. M. 2006. Uptake and accumulation of lead by plants from the Bo Ngam lead mine area in Thailand. *Environmental Pollution* 144 (2), pp. 681–688, DOI, 10.1016/j.envpol.2005.12.039.

Rouhi, A. M. 1997. Plants to the rescue. *Chemical and Engineering News* 75 (2), pp. 21–23.

Roux Associates, Inc. 2014. Communication with Amanda Ludlow, Roux Associates, Inc, 209 Shafter Street, Islandia, New York 11749.

Roy, S., Labelle, S., Mehta, P., Mihoc, A., Fortin, N., Masson, C., Leblanc, R., Chateauneuf, G., Sura, C., Gallipeau, C. et al. 2005. Phytoremediation of heavy metal and PAH-contaminated brownfield sites. *Plant and Soil* 272 (1–2), pp. 277–290.

Rozema, J., and Flowers T. 2008. Crops for a salinized world. *Science* 322, pp. 1478–1480.

Rune, O., and Westerbergh, A. 1991. Phytogeographic aspects of the serpentine flora of Scandinavia. In A. J. M. Baker, J. Proctor, and R. D. Reeves (Eds.) *The Vegetation of Ultramafic (Serpentine) Soils. Proceedings of the First International Conference on Serpentine Ecology. University of California, Davis, 19–22 June 1991.* Andover, UK: Intercept.

Ruttens, A., Boulet, J., Weyens, N., Smeets, K., Adriaensen, K., Meers, E., Van Slycken, S., Tack, F., Meiresonne, L., Thewys, T. et al. 2011. Short rotation coppice culture of willows and poplars as energy crops on metal contaminated agricultural soils. *International Journal of Phytoremediation* 13 (suppl), pp. 194–207.

Rylott, E. L., Budarina, M. V., Barker, A., Lorenz, A., Strand, S. E., and Bruce, N. C. 2011. Engineering plants for the phytoremediation of RDX in the presence of the co-contaminating explosive TNT. *New Phytologist* 192 (2), pp. 405–413.

Rylott, E. 2012. Presentation on explosives at 9th International Phytotechnologies Society Conference, 12 September 2012, Hasselt University, Belgium.

Rylott, E. R., Jackson, G., Sabbadin, F., Seth-Smith, H. M. B., Edwards, J., Chong, C. S., Strand, S. E., Grogan, G., and Bruce, N. C. 2010. The explosive-degrading cytochrome P450 XplA: biochemistry, structural features and prospects for bioremediation. *Biochimica et Biophysica Acta 1814* (2011), pp. 230–236.

Saboora, A. et al. 2006. Salinity (NaCl) tolerance of wheat genotypes at germination and early seedling growth. *Journal of Biological Science* 9 (11), pp. 2009–2021.

Saebo, A., Popek, R., Nawrot, B., Hanslin, H., Gawronska, H., and Gawronski, S. 2012. Plant species differences in particulate matter accumulation on leaf surfaces. *Science of the Total Environment* 427, pp. 347–354.

Salem. 2013. City of Salem Website. http://www.cityofsalem.net/DEPARTMENTS/PUBLICWORKS/WASTEWATERTREATMENT/Pages/default.aspx. Accessed 15 December 2013.

Salido, A., Hasty, K., Lim, J., and Butcher, D. 2003. Phytoremediation of arsenic and lead in contaminated soil using Chinese brake ferns (Pteris vittata) and Indian mustard (Brassica juncea). *International Journal of Phytoremediation* 5 (2), pp. 89–103.

Salt, C. A., Mayes, R. W., and Elston, D. A. 1992. Effects of season, grazing intensity, and diet composition on the radiocaesium intake by sheep on reseeded hill pasture. *Journal of Applied Ecology* 29, pp. 378–387.

Salt, D. E., Blaylock, M., Kumar, P. B. A. N., Dushenkov, V., Ensley, B. D., Chet, I., and Raskin, I. 1995. Phytoremediation: a novel strategy for the removal of toxic metals from the environment using plants. *Biotechnology* 13, pp. 468–474.

Sand Creek. 2013. Personal communication with Christopher Rog, Bart Sexton, and Mark Dawson, Sand Creek Consultants, 108 E. Davenport I Rhinelander, WI 54501.

Sandermann, H. 1994. Higher plant metabolism of xenobiotics: The 'green liver' concept. *Pharmacogenetics* 4, pp. 225–241.

Sangster Research Laboratories LOG KOW Databank. [online] Available at: http://logkow.cisti.nrc.ca/logkow/index.jsp. Accessed 3 February 2014.

Saraswat, S., and Rai, J. P. N. 2009 Phytoextraction potential of six plant species grown in multi-metal contaminated soil. *Chemistry and Ecology* 25, pp. 1–11.

Sarma, H. 2011. Metal hyperaccumulation in plants: a review focusing on phytoremediation technology. *Journal of Environmental Science and Technology* 4 (2), pp. 118–138.

Sass, J. B., and Colangelo, A. 2006. European Union bans atrazine, while the United States negotiates continued use. *International Journal of Occupational and Environmental Health* 12 (3), pp. 260–267.

Sattler R. (Law Office of Posternak, Blankstein & Lund, Boston, MA). 2010. Lecture delivered 21 September at Harvard University, Cambridge, MA.

Schnoor, J. L. 1997. *Phytoremediation.* Ground-Water Remediation Technologies Analysis Center Technology Evaluation Report TE-98-01.

Schnoor, J. L. 2007. EPA's research budget. *Environmental Science and Technology* 41(7), pp. 2071–2072.

Schwab, A. P., and Banks, M. K. 1994. Biologically mediated dissipation of polyaromatic hydrocarbons in the root zone. In T. A. Anderson and J. R. Coats (Eds.) *Bioremediation through Rhizosphere Technology.* ACS Symposium Series 563. Washington, DC: American Chemical Society.

Schwartz, C., Sirguey, C., Peronny, S., Reeves, R. D., Bourgaud, F., and Morel, J. L. 2006. Testing of outstanding individuals of Thlaspi caerulescens for cadmium phytoextraction. *International Journal of Phytoremediation* 8 (4), pp. 339–357.

Scott, K. I., McPherson, E. G. and Simpson, J. R., 1998. Air pollutant uptake by Sacramento's urban forest. *Journal of Arboriculture* 24, 224–234.

Selamat, S. N., Abdullah, S. R. S., and Idris, M. 2013. Phytoremediation of lead (Pb) and arsenic (As) by Melastoma malabathricum L. from contaminated soil in separate exposure. *International Journal of Phytoremediation* 16, pp. 694–703.

Shahsavari, E., Adetutu, E. M., Anderson, P. A., and Ball, A. S. 2013. Tolerance of selected plant species to petrogenic hydrocarbons and effect of plant rhizosphere on the microbial removal of hydrocarbons in contaminated soil. *Water, Air and Soil Pollution* 224 (4), p. 1495.

333

Sharma, N. C., Starnes, D. L., and Sahi, S. V. 2007. Phytoextraction of excess soil phosphorus. *Environmental Pollution* 146 (1), pp. 120–127.

Shay, S. D., and Braun C. L. 2004. Demonstration-site development and phytoremediation processes associated with trichloroethene (TCE) in ground water, Naval Air Station-Joint Reserve Base Carswell Field, Fort Worth, Texas. In *U.S. Geological Survey Fact Sheet 2004-3087*, http://pubs.usgs.gov/fs/2004/3087/pdf/FS_2004-3087.pdf. Accessed 14 September 2009.

Sheehan, E., Burken, J.G., Limmer, M.A., Mayer, P., and Gosewinkel, U. 2012. Time weighted SPME analysis for in-planta phytomonitoring analysis. *Environmental Science and Technology* 46(6), pp. 3319–3325. DOI: 10.1021/es2041898.

Shetty, M., Limmer, M.A., Waltermire, K.W., Morrison, G.C., and Burken, J.G. 2014. *In planta* passive sampling devices for assessing subsurface chlorinated solvents. *Chemosphere*, Vol. 104, pp. 149–154.

Shi, G., and Cai, Q. 2009. Cadmium tolerance and accumulation in eight potential energy crops. *Biotechnology Advances* 27 (5), pp. 555–561.

Shirdam, R., and Tabrizi, A. M. 2010. Total petroleum hydrocarbon (TPHs) dissipation through rhizoremediation by plant species. *Polish Journal of Environmental Studies* 19 (1), pp. 115–122.

Shuttleworth, K. L., and Cerniglia, C. E. 1995. Environmental aspects of PAH biodegradation. *Applied Biochemistry and Biotechnology* 54, pp. 291–302.

Silveira, M. L., Vendramini, J. M. B., Sui, X., Sollenberger, L., and O'Connor, G. A. 2013. Screening perennial warm-season bioenergy crops as an alternative for phytoremediation of excess soil P. *Bioenergy Research* 6 (2), pp. 469–475.

Simmons, R. W., Chaney, R. L., Angle, J. S., Kruatrachue, M., Klinphoklap, S., Reeves, R. D., and Bellamy, P. 2014. Towards practical cadmium phytoextraction with Noccaea caerulescens, *International Journal of Phytoremediation*, DOI, 10.1080/15226514.2013.876961.

Singh, S., Eapen, S., Thorat, V., Kaushik, C. P., Raj, K., and D'Souza, S. F. 2008. Phytoremediation of (137)cesium and (90)strontium from solutions and low-level nuclear waste by Vetiveria zizanoides. *Ecotoxicology and Environmental Safety* 69, pp. 306–311.

Smesrud, J. 2012. Communication with CH2MHill. Phone conversation with Jason Smesrud, CH2MHill on 7 August 2012 and PowerPoint provided by Jason Smesrud, CH2MHill on 7 August 2011 via email and CH2M Hill 2011 Project Cutsheet.

Smith, A. E., and Bridges, D. C. 1996. Movement of certain herbicides following application to simulated golf course greens and fairways. *Crop Science* 36 (6), p. 1439.

Smith, K. E., Schwab, A. R. et al. 2007. Phytoremediation of polychlorinated biphenyl (PCB)-contaminated sediment: a greenhouse feasibility study. *Journal of Environmental Quality* 36 (1), pp. 239–244.

Smith, K. E., Putnam, R. A., Phaneuf, C., Lanza, G. R., Dhankher, O. P., and Clark, J. M. 2008. Selection of plants for optimization of vegetative filter strips treating runoff from turfgrass. *Journal of Environmental Quality* 37 (5), pp. 1855–1861.

Smith, K. E., Schwab, A. P., and Banks, M. K. 2008. Dissipation of PAHs in saturated, dredged sediments: a field trial. *Chemosphere* 72 (10), pp. 1614–1619, DOI, 10.1016/j.chemosphere.2008.03.020.

Smith, M. J., Flowers, T. H., Duncan, H. J., and Alder, J. 2006. Effects of polycyclic aromatic hydrocarbons on germination and subsequent growth of grasses and legumes in freshly contaminated soil and soil with aged PAHs residues. *Environmental Pollution* 141, pp. 519–525.

Soreanu, G., Dixon, M., and Darlington, A. 2013. Botanical biofiltration of indoor gaseous pollutants: a mini-review. *Chemical Engineering Journal* 229, pp. 585–594.

Soudek, P., Tykva, R., and Vanek, T. 2004. Laboratory analyses of Cs-137 uptake by sunflower, reed and poplar. *Chemosphere* 55, pp. 1081–1087.

Soudek, P., Tykva, R., Vankova, R., and Vanek, T. 2006a. Accumulation of radioiodine from aqueous solution by hydroponically cultivated sunflower (Helianthus annuus L.). *Environmental and Experimental Botany* 57, pp. 220–225.

Soudek, P., Valenova, S., Vavrikova, Z., and Vanek, T. 2006b. Cs-137 and Sr-90 uptake by sunflower cultivated under hydroponic conditions. *Journal of Environmental Radioactivity* 88, pp. 236–250.

Soudek, P., Petrova S., Vankova, R., Song, J., and Vanek, T. 2014. Accumulation of heavy metals using Sorghum sp. *Chemosphere* 104, pp. 15–24.

Speir, T. W., August, J. A., and Feltham, C. W. 1992. Assessment of the feasibility of using CCA (copper, chromium and arsenic)-treated and boric acid-treated sawdust as soil amendments, I. Plant growth and element uptake. *Plant and Soil* 142, pp. 235–48.

Spence, P. L., Osmund, D. L., Childres, W., Heitman, J., and Robarge, W. P. 2012. Effects of lawn maintenance on nutrient losses via overland flow during natural rainfall events. *Journal of the American Water Resources Association* 48 (5), pp. 909.

334

Spriggs, T., Banks, M. K., and Schwab, P. 2005. Phytoremediation of polycyclic aromatic hydrocarbons in manufactured gas plant-impacted soil. *Journal of Environmental Quality* 34 (5), pp. 1755–1762.

Stanhope, A., Berry, C. J., and Brigmon, R. L. 2008. Field note: phytoremediation of chlorinated ethenes in seepline sediments: tree selection. *International Journal of Phytoremediation* 10, pp. 529–546.

Stomp, A. M., Han, K. H., Wilbert, S., Gordon, M. P., and Cunningham, S. D. 1994. Genetic strategies for enhancing phytoremediation. *Annals of the New York Academy of Sciences* 721, pp. 481–491.

Stoops, R. 2014. Correspondence via email and telephone with Kate Kennen, February.

Strand, S. E., Doty, S. L., and Bruce, N. 2009. Engineering transgenic plants for the sustained containment and in situ treatment of energetic materials. *Strategic Research and Development Program, Project ER*. Final Report.

Stritsis, C., Steingrobe, B., and Claassen, N. 2013. Cadmium dynamics in the rhizosphere and Cd uptake of different plant species evaluated by a mechanistic model. *International Journal of Phytoremediation* 16, pp. 1104–1118.

Stroud J. L., Paton G. I., Semple K. T. 2007. Microbe-aliphatic hydrocarbon interactions in soil: implications for biodegradation and bioremediation. *Journal Applied Microbiology* 102, 1239–1253

Strycharz, S., and Newman, L. 2009a. Use of native plants for remediation of trichloroethylene: II. Coniferous trees. *International Journal of Phytoremediation* 11 (2), pp. 171–186.

Strycharz, S., and Newman, L. 2009b. Use of native plants for remediation of trichloroethylene: I. Deciduous trees. *International Journal of Phytoremediation* 11 (2), pp. 150–170.

Stultz, C., and CH2MHill. 2011. International Phytotechnology Society Conference, Workshop 1, Site visit by Kate Kennen, September 2011. Information provided by Curtis Stultz (POTW), Mark Madison (CH2MHill), and Jason Smesrud (CH2MHill).

Subramanian, M., and Shanks J. V. 2003. Role of plants in the transformation of explosives. In S. C. McCutcheon and J. L. Schnoor (Eds.) *Phytoremediation: Transformation and Control of Contaminants*. New York: Wiley, chapter 12.

Sun, M., Fu, D., Teng, Y., Shen, Y., Luo, Y., Li, Z., and Christie, P. 2011. In situ phytoremediation of PAH-contaminated soil by intercropping alfalfa (Medicago sativa L.) with tall fescue (Festuca arundinacea Schreb.) and associated soil microbial activity. *Journal of Soils and Sediments* 11 (6), pp. 980–989.

Syc, M., Pohorely, M., Kamenikova, P., Habart, J., Svoboda, K., and Puncochar, M. 2012. Willow trees from heavy metals phytoextraction as energy crops. *Biomass and Bioenergy* 37 pp. 106–113.

Szabolcs, I. 1994. Soils and salinization. In M. Pessarakli (Ed.) *Handbook of Plant and Crop Stress*. New York: Marcel Dekker, pp. 3–11.

Takahashi, M., Higaki, A., Nohno, M., Kamada, M., Okamura, Y., Matsui, K., Kitani, S., and Morikawa, H. 2005. Differential assimilation of nitrogen dioxide by 70 taxa of roadside trees at an urban pollution level. *Chemosphere* 61 (5), pp. 633–639.

Tang, S., and Willey, N. J. 2003. Uptake of 134 Cs by four species from Asteraceae and two varieties from the Chenopodiaceae grown in two types of Chinese soil. *Plant and Soil* 250 (1), pp. 75–81.

Techer, D., Martinez-Chois, C., Laval-Gilly, P., Henry, S., Bennasroune, A., D'Innocenzo, M., and Falla, J. 2012. Assessment of Miscanthus x giganteus for rhizoremediation of long term PAH contaminated soils. *Applied Soil Ecology* 62, pp. 42–49.

Teixeira, S., Vieira, M. N., Marques, J. E., and Pereira, R. 2013. Bioremediation of an iron-rich mine effluent by Lemna minor. *International Journal of Phytoremediation* 16, pp. 1228–1240.

Terry, N., and Banuelos, G. (Eds.). 2000. *Phytoremediation of Contaminated Soil and Water*. New York: Lewis Publishers.

Terry, N., Zayed, A. M., de Souza, M. P., and Tarun, A. S. 2000. Selenium in higher plants. *Annual Review of Plant Physiology and Plant Molecular Biology* 51, pp. 401–432.

Thewys, T., Witters, N., Van Slycken, S., Ruttens, A., Meers, E., Tack, F., and Vangronsveld, J. 2010. Economic viability of phytoremediation of a cadmium contaminated agricultural area using energy maize. Part I: Effect on the farmer's income. *International Journal of Phytoremediation* 12 (7), pp. 650–662.

Thewys, T., Witters, N., Meers, E., and Vangronsveld, J. 2010a. Economic viability of phytoremediation of a cadmium contaminated agricultural area using energy maize. Part II: Economics of anaerobic digestion of metal contaminated maize in Belgium. *International Journal of Phytoremediation* 12 (7), pp. 663–679.

Thomas, J., Cable, E., Dabkowski, R., Gargala, S., McCall, D., Pangrazzi, G., Pierson, A., Ripper, M., Russell, D., and Rugh, C. 2013. Native Michigan plants stimulate soil microbial species changes and PAH remediation at a legacy steel mill. *International Journal of Phytoremediation* 15 (1), pp. 5–23.

Thompson, P. 1997. Phytoremediation of munitions (RDX, TNT) waste at the Iowa Army Ammunition Plant with hybrid poplar trees. Ph.D. thesis, University of Iowa, Iowa City, IA.

335

Thompson, P. L., Moses, D., and Howe, K. M. 2003. Phytorestoration at the Iowa Army Ammunition Plant. In S. C. McCutcheon and J. L. Schnoor (Eds.) *Phytoremediation: Transformation and Control of Contaminants*. New York: John Wiley and Sons, Inc.

Thompson, P. L., Ramer, L. A., and Schnoor, J. L. 1999. 1,3,5-trinitro-1,3,5-triazaine (RDX) translocation in hybrid poplar trees. *Environmental Toxicology and Chemistry* 18 (2), pp. 279–284.

Thompson, P., Ramer, L., and Schnoor, J. 1999. Uptake and transformation of TNT by hybrid poplar trees. *Environmental Science and Technology* 32 (7), pp. 975–980.

Tiemann, K. J., Gardea-Torresdey, J. L., Gamez, G., and Dokken, K. 1998. Interference studies for multi-metal binding by *Medicago sativa* (Alfalfa). In *Proceedings, Conference on Hazardous Waste Research*, Snow Bird, Utah, p. 42 (Abstract 67).

Tischer, S. and Hübner, T. 2002. Model trials for phytoremediation of hydrocarbon-contaminated sites by the use of different plant species. *International Journal of Phytoremediation* 4:3, pp. 187–203.

Todd, J. 2013 (August). John Todd Ecological Design, Presentation to Cape Cod Commission, Falmouth, MA, http://www.toddecological.com.

Toland, T. 2013. The use of native plants on an intensive green roof: initial results. Presentation at 10th annual International Phytotechnologies Society Conference, Syracuse, NY, 3 October 2013, Work completed at SUNY ESF with collaborators: Donald Leopold, SUNY ESF; Doug Daley, SUNY ESF; Darren Damone, Andropogon Associates.

Tome, F. V., Rodriguez, P. B., and Lozano, J. C. 2008. Elimination of natural uranium and Ra-226 from contaminated waters by rhizofiltration using Helianthus annuus L. *Science of the Total Environment* 393, pp. 351–357.

Trapp, S., and McFarlane, C. (Eds.). 1995. *Plant Contamination: Modeling and Simulation of Organic Processes*. Boca Raton, FL: Lewis.

Tsao, D. T. 1997. *Development of Phytoremediation Technology Developments*. Technology Assessment and Development (HEM) Status Report 0497.

Tsao, D. 2003. *Phytoremediation*. New York: Springer.

Tsao, K., and Tsao, D. 2003 (March 28) *Analysis of Phytoscapes Species for BP Retail Sites*. BP Group Environmental Management Company. Report published as Capstone Project paper to Benedictine University, Lile, IL by Kim Tsao. Property of Atlantic Richfield Company, permission to use given by Dr. David Tsao on 11 November 2013.

Tsao, D. 2014. Personal communication with Kate Kennen, November.

Turgut, C. 2005. Uptake and modeling of pesticides by roots and shoots of parrotfeather (Myriophyllum aquaticum). *Environmental Science and Pollution Research*, 12 (6), pp. 342–346.

Ulam, A. 2012. 'Phyto your life: phytoremediation provides a sustainable approach to building landscapes on brownfields. *Landscape Architecture Magazine*, March, pp. 52–58.

Ulriksen, C., Ginocchio, R., Mench, M., and Neaman, A. 2012. Lime and compost promote plant re-colonization of metal-polluted, acidic soils. *International Journal of Phytoremediation* 14 (8), pp. 820–833.

US EPA. 1988. Closed Landfills – Federal Register, 30 August 1988, 53 (168).

US EPA. 1991. *R.E.D. Facts: Potassium Bromide*. http://www.epa.gov/oppsrrd1/REDs/factsheets/0342fact.pdf. Accessed 12 November 2013.

US EPA. 1996. *Be a Grower, Not a Mower*. Burlington, VT: US Environmental Protection Agency.

US EPA. 2002. *Cost and Performance Report, Phytoremediation at the Magic Marker and Fort Dix Site*. February 2002. Office of Solid Waste and Emergency Response Technology Innovation Office. http://costperformance.org/pdf/MagicMarker-Phyto.pdf. Accessed 13 March 2014.

US EPA. 2005a. *Use of Field-Scale Phytotechnology for Chlorinated Solvents, Metals, Explosives and Propellants, and Pesticides*. EPA 542-R-05-002, 2005, http://clu-in.org/download/remed/542-r-05-002.pdf. Accessed 14 September 2009.

US EPA. 2005b. *Evaluation of Phytoremediation for Management of Chlorinated Solvents and Groundwater*. EPA 542-R-05-001. Remediation Technologies Development Forum Phytoremediation of Organics Action Team, Chlorinated Solvents Workgroup.

US EPA. 2006. *Lindane Voluntary Cancellation and RED Addendum Fact Sheet* http://www.epa.gov/oppsrrd1/REDs/factsheets/lindane_fs_addendum.htm. Accessed 11 November 2013.

US EPA. 2010. EPA's Petroleum Brownfields Action Plan: two years later. Office of Underground Storage Tanks and Office of Brownfields and Land Revitalization. http://www.epa.gov/oust/pubs/petrobfactionplan2year.pdf.

US EPA. 2000. J-Field phytoremediation project field events and activities through July 31, 2000. Aberdeen Proving Ground, Edgewood, Maryland. August 31, 2000.

US EPA. 2013a. Basic information about nitrate in drinking water. http://water.epa.gov/drink/contaminants/basicinformation/nitrate.cfm. Accessed 11 November 2013.

336

US EPA. 2013b. Research and development: trees and air pollution. http://www.epa.gov/ORD/sciencenews/scinews_trees-and-air-pollution.htm. Accessed 27 November 2013.

US EPA. 2013c. Contaminated sites clean-up information: persistent organic pollutants (POPs). http://www.cluin.org/contaminantfocus/default.focus/sec/Persistent_Organic_Pollutants_(POPs)/cat/Overview/. Accessed 3 December 2013.

US EPA. 2014a. Technical factsheet on: XYLENES. http://www.epa.gov/safewater/pdfs/factsheets/voc/tech/xylenes.pdf. Accessed 3 February 2014.

US EPA. 2014b. Technical factsheet on: ETHYLBENZENE. http://www.epa.gov/ogwdw/pdfs/factsheets/voc/tech/ethylben.pdf. Accessed 3 February 2014.

US EPA. 2014c. Indoor air quality. http://www.epa.gov/region1/communities/indoorair.html. Accessed 7 March 2014.

US EPA. 2014d. Active landfills. http://www.epa.gov/lmop/projects-candidates/index.html. Accessed 27 April 2014.

US EPA. 2014e. Trees and air pollution. http://www.epa.gov/ord/sciencenews/scinews_trees-and-air-pollution.htm, Accessed April 2014.

US EPA. 2014f. Contaminant human health effects, http://www.epa.gov, search by contaminant, March 2014.

US EPA. 2014g. Trenton Magic Marker Site. http://www.epa.gov/region2/superfund/brownfields/mmark.htm. Accessed 1 March 2014.

US PIRG Education Fund. 2004. More highways, more pollution: road-building and air pollution in America's cities. http://research.policyarchive.org/5542.pdf. Accessed 27 November 2013.

Van Aken, B., Yoon, J. M., Just, C. L., and Schnoor, J. L. 2004. Metabolism and mineralization of hexahydro 1,3,5-trinitro-1,3,5-triazine inside poplar tissues (Populus deltoides x nigra DN-34). *Environmental Science and Technology* 38, pp. 4572–4579.

Van der Ent, A., Baker, A., Reeves, R., Pollard, A., and Schat, H. 2013. Hyperaccumulators of metal and metalloid trace elements: facts and fiction. *Plant and Soil* 362 (1–2), pp. 319–334.

Van Dillewijn, P., Couselo, J. L., Corredoira, E., Delgado, A., Wittich, R., Ballester, A., and Ramos, J. L. 2008. Bioremediation of 2,4,6-trinitrotoluene by bacterial nitroreductase expressing transgenic aspen. *Environmental Science and Technology*, 42, pp. 7405–7410.

Van Slycken, S., Witters, N., Meiresonne, L., Meers, E., Ruttens, A., Van Peteghem, P., Weyens, N., Tack, F. M., and Vangronsveld, J. 2013. Field evaluation of willow under short rotation coppice for phytomanagement of metal-polluted agricultural soils. *International Journal of Phytoremediation* 15 (7), pp. 677–689.

Vandenhove, H., Goor, F., Timofeyev, S., Grebenkov, A., and Thiry, Y. 2004. Short rotation coppice as alternative land use for Chernobyl-contaminated areas of Belarus. *International Journal of Phytoremediation* 6 (2), pp. 139–156, DOI, 10.1080/16226510490454812.

Vanek, T., Nepovim, A., Podlipna, R., Hebner, A., Vavrikova, Z., Gerth, A., Thomas, H., and Smrcek, S. 2006. Phytoremediation of explosives in toxic wastes. *Soil and Water Pollution Monitoring, Protection and Remediation* 69, pp. 455–465.

Vardoulakis, S., Fisher, B. E. A., Pericleous, K., and Gonzalez-Flesca, N. 2003. Modelling air quality in street canyons: a review. *Atmospheric Environment* 37, pp. 155–182.

Vasiliadou, S., and Dordas, C. 2009. Increased concentration of soil cadmium effects on plant growth, dry matter accumulation, Cd, and Zn uptake of different tobacco cultivars (Nicotiana tabacum L.). *International Journal of Phytoremediation* 11, pp. 115–130.

Vasudev, D., Ledder, T., Dushenkov, S., Epstein, A., Kumar, N., Kapulnik, Y., Ensley, B., Huddleston, G., Cornish, J., Raskin, I., Sorochinsky, B., Ruchko, M., Prokhnevsky, A., Mikheev, A., and Grodzinsky, D. 1996. Removal of radionuclide contamination from water by metal-accumulating terrestrial plants. Presented at the In Situ Soil and Sediment Remediation Conference, New Orleans.

Vervaeke, P., Luyssaert, S., Mertens, J., Meers, E., Tack, F. M., and Lust, N. 2003. Phytoremediation prospects of willow stands on contaminated sediment: a field trial. *Environmental Pollution* 126 (2), pp. 275–282.

Videa-Peralta, J.R., and Ramon, J. 2002. Feasibility of using living alfalfa plants in the phytoextraction of cadmium(II), chromium(VI), copper(II), nickel(II), and zinc(II): Agar and soil studies. (Ph.D. thesis). The University of Texas, El Paso, AAT 3049704.

Viessman, W., Lewis, G. L., and Knapp, J. W. 1989. *Introduction to Hydrology*, 3rd ed. New York: Harper & Row.

Vila, M., Lorber-Pascal, S., and Laurent, F. 2007a. Fate of RDX and TNT in agronomic plants. *Environmental Pollution* 148, pp. 148–154.

Vila, M., Mehier, S., Lorber-Pascal, S., and Laurent, F. 2007b. Phytotoxicity to and uptake of RDX by rice. *Environmental Pollution* 145, pp. 813–817.

Vila, M., Lorber-Pascal, S., and Laurent, F. 2008. Phytotoxicity to and uptake of TNT by rice. *Environmental Geochemistry and Health* 30 (2), pp. 199–203.

Volk, T. (SUNY ESF) 2014. State University of New York, College of Environmental Science and Forestry. Photographs provided and personal communication with Kate Kennen, April.

337

Volkering, F., Breure, A. M., and Rulkens, W. H. 1998. Microbiological aspects of surfactant use for biological soil remediation. *Biodegradation* 8, pp. 401–417.

Von Caemmerer, S., and Baker, N. 2007. The biology of transpiration: from guard cells to globe. *Plant Physiology* 143, p. 3.

Wang, A. S., Angle, J. S., Chaney, R. L., Delorme, T. L., and Reeves, R. D. 2006. Soil pH effects on uptake of Cd and Zn by Thlaspi caerulescens. *Plant and Soil* 281, pp. 325–337.

Wang, C. H., Lyon, D. Y., Hughes, J. B., and Bennett, G. N. 2003. Role of hydroxylamine intermediates in the phytotransformation of 2,4,6-trinitrotoluene by *Myriophyllum aquaticum*. *Environmental Science and Technology* 37, pp. 3595–3600.

Wang, H. B., Ye, Z. H., Shu, W. S., Li, W. C., Wong, M H., and Lan, C. Y. 2006. Arsenic uptake and accumulation in fern species growing at arsenic-contaminated sites of Southern China: field surveys. *International Journal of Phytoremediation* 8 (1), pp. 1–11.

Wang, K., Huang, H., Zhu, Z., Li, T., He, Z., Yang, X., and Alva, A. 2013. Phytoextraction of metals and rhizoremediation of PAHs in co-contaminated soil by co-planting of Sedum alfredii with Ryegrass (Lolium perenne) or Castor (Ricinus communis). *International Journal of Phytoremediation* 15 (3), pp. 283–298.

Wang, Q., Zhang, W., Li, C., and Xiao, B. 2012. Phytoremediation of atrazine by three emergent hydrophytes in a hydroponic system. *Water Science and Technology* 66 (6), pp. 1282–1288.

Wang, X,. Newman, L. A., Gordon, M. P., and Strand, S. E. 1999. Biodegradation by poplar trees: results from cell culture and field experiments. In A. Leeson, and B. C. Alleman (Eds.) *Phytoremediation and Innovative Strategies for Specialized Remedial Applications, The Fifth International In Situ and On-Site Bioremediation Symposium, San Diego, California, April 19–22.* Columbus, OH: Battelle Press, v. 6, p. 133–138.

Wang, X., White, J. C., Gent, M. P., Iannucci-Berger, W., Eitzer, B. D., and Mattina, M. I. 2004. Phytoextraction of weathered p, p'-DDE by zucchini (Cucurbita pepo) and cucumber (Cucumis sativus) under different cultivation conditions. *International Journal of Phytoremediation* 6 (4), pp. 363–385.

Wargo, J., Alderman, N., and Wargo, L. 2003. *Risks from Lawn-Care Pesticides: Including Inadequate Packaging and Labeling.* North Haven, CT: Environmental & Human Health, Inc.

Warsaw, A., Fernandez, R. T., Kort, D. R., Cregg, B. M., Rowe, B., and Vandervoort, C. 2012. Remediation of metalaxyl, trifluralin, and nitrate from nursery runoff using container-grown woody ornamentals and phytoremediation areas. *Ecological Engineering* 47, pp. 254–263. http://

dx.doi.org.ezpprod1.hul.harvard.edu/10.1016/j.ecoleng.2012.06.036.

Wattiau, P. 2002. Microbial aspects in bioremediation of soils polluted by polyaromatic hydrocarbons. *Focus on Biotechnology* 3A, pp. 2–22.

Wei, S., and Zhou, Q. 2006. Phytoremediation of cadmium-contaminated soils by Rorippa globosa using two-phase planting. *Environmental Science and Pollution Research* 13 (3), pp. 151–155.

Wei, S., and Zhou, Q. 2008. Screen of Chinese weed species for cadmium tolerance and accumulation characteristics. *International Journal of Phytoremediation* 10, pp. 584–597.

Wei, S., Zhou, Q., Wang, X., Cao, W., Ren, L., and Song, Y. 2004. Potential of weed species applied to remediation of soils contaminated with heavy metals. *Journal of Environmental Sciences* 16 (5), pp. 868–873.

Wei, S., Clark, G., Doronila, A. I., Jin, J., and Monsant, A. C. 2012. Cd hyperaccumulative characteristics of Australia ecotype Solanum nigrum L. and its implication in screening hyperaccumulators. *International Journal of Phytoremediation* 15, pp. 199–205.

Weston Solutions. 2014. Website for DiamlerChrysler Forge Site http://www.westonsolutions.com/projects/technology/rt/phytoremediation.htm. Accessed 16 March 2014.

Weyens, N., Taghavi, S., Barac, T., van der Lelie, D., Boulet, J., Artois, T., Carleer, R., and Vangronsveld, J. 2009. Bacteria associated with oak and ash on a TCE-contaminated site: characterization of isolates with potential to avoid evapotranspiration of TCE. *Environmental Science and Pollution Research* 16, pp. 830–843.

Weyens, N., van der Lelie, D., Artois, T., Smeets, K., Taghavi, S., Newman, L., Carleer, R., and Vangronsveld, J. 2009. Bioaugmentation with engineered endophytic bacteria improves contaminant fate in phytoremediation. *Environmental Science and Technology* 43, pp. 9413–9418.

White, J. 2000. Phytoremediation of weathered p, p'-DDE residues in soil. *International Journal of Phytoremediation* 2 (2), pp. 133–144.

White, J. C. 2010. Phytoremediation and persistent organic pollutants. Presentation to Phyto Seminar, Harvard University, Graduate School of Design.

White, J. C., and Newman, L. A. 2011. Phytoremediation of soils contaminated with organic pollutants. In B. Xing, N. Senesi and P. M. Huang (Eds.) *Biophysico-Chemical Processes of Anthropogenic Organic Compounds in Environmental Systems.* Hoboken, NJ: Wiley.

White, P. J., Bowen, H. C., Marshall, B., and Broadley, M. R. 2007. Extraordinarily high leaf selenium to sulfur ratios define 'Se-accumulator' plants. *Annals of Botany* 100 (1), pp. 111–118.

338

Whitfield Aslund, M. L., Zeeb, B. A., Rutter, A., and Reimer, K. J. 2007. In situ phytoextraction of polychlorinated biphenyl (PCB) contaminated soil. *Science of the Total Environment* 374 (1), pp. 1–12.

Whitfield Aslund, M. L., Rutter, A., Reimer, K. J., and Zeeb, B. A. 2008. The effects of repeated planting, planting density, and specific transfer pathways on PCB uptake by Cucurbita pepo grown in field conditions. *Science of the Total Environment* 405 (1), pp. 14–25.

Widdowson, M., Shearer, S., Andersen, R., and Novak, J. 2005. Remediation of polycyclic aromatic hydrocarbon compounds in groundwater using poplar trees. *Environmental Science and Technology* 39 (6), pp. 1598–1605.

Wild, H. 1970. The vegetation of Nickel-bearing soils. *Kirkia* 7, pp. 271–275.

Wilkomirski, B., Sudnik-Wojcikowska, B., Galera, H., Wierzbicka, M., and Malawska, M. 2011. Railway transportation as a serious source of organic and inorganic pollution. *Water, Air and Soil Pollution* 218 (1–4), pp. 333–345.

Willey, N., Hall, S., and Mudigantia, A. 2001. Assessing the potential of phytoremediation at a site in the UK contaminated with 137Cs. *International Journal of Phytoremediation* 3 (3), pp. 321–333.

Wilste, C. C., Rooney, W. L., Chen, Z., Schwab, A. P., and Banks, M. K. 1998. Greenhouse evaluation of agronomic and crude oil phytoremediation potential among alfalfa genotypes. *Journal of Environmental Quality* 27, pp. 169–173.

Witters, N., Mendelsohn, R., Van Slycken, S., Weyens, N., Schreurs, E., Meers, E., Tack, F., Carleer, R., and Vangronsveld, J. 2012a. Phytoremediation, a sustainable remediation technology? Conclusions from a case study. I: Energy production and carbon dioxide abatement. *Biomass and Bioenergy* 39 pp. 454–469.

Witters, N., Mendelsohn, R., Van Passel, S., Van Slycken, S., Weyens, N., Schreurs, E., Meers, E., Tack, F., Vanheusden, B., and Vangronsveld, J. 2012b. Phytoremediation, a sustainable remediation technology? II: Economic assessment of CO_2 abatement through the use of phytoremediation crops for renewable energy production. *Biomass and Bioenergy* 39 pp. 470–477.

Wojtera-Kwiczor, J., Zukowska, W., Graj, W., Malecka, A., Piechalak, A., Ciszewska, L., Chrzanowski, L., Lisiecki, P., Komorowicz, I., Baralkiewicz, D. et al. 2013. Rhizoremediation of diesel-contaminated soil with two rapeseed varieties and petroleum degraders reveals different responses of the plant defense mechanisms. *International Journal of Phytoremediation* 16 (7–8) Special Issue: The 9th International Phytotechnnology Society

Conference – Hasselt, Belgium 2012, pp. 770–789, DOI, 10.1080/15226514.2013.856848.

Wolverton, B. C., Johnson, A., and Bounds, K. 1989. *Interior Landscape Plants for Indoor Air Pollution Abatement*, Final Report NASA (NASA-TM-101760), National Aeronautics and Space Administration.

Wood, B., Chaney, R., and Crawford, M. 2006. Correcting micronutrient deficiency using metal hyperaccumulators: *Alyssum* biomass as a natural product for nickel deficiency correction. In *HortScience* 41 (5), pp. 1231–1234.

Woodburn. 2013. Woodburn, Oregon Wastewater Treatment Facility, http://www.woodburn-or.gov/?q=waste_water. Accessed 15 December 2013.

Woodward, R. 1996. Summary of the Workshop on Phytoremediation of Organic Contaminants. Fort Worth, TX.

World Health Organization (WHO). 2002. *The World Health Report 2002: Reducing Risks, Promoting Healthy Life*. Geneva: WHO.

Wu, C., Liao, B., Wang, S.-L., Zhang, J., and Li, J.-T. 2010. Pb and Zn accumulation in a Cd-hyperaccumulator (Viola baoshanensis). *International Journal of Phytoremediation* 12, pp. 574–585.

Wu, Q., Wang, S., Thangavel, P., Li, Q., Zheng, H., Bai, J., and Qiu, R. 2011. Phytostabilization potential of Jatropha curcas L. in polymetallic acid mine tailings. *International Journal of Phytoremediation* 13 (8), pp. 788–804.

Xiaomei, L., Qitang, W., and Banks, M. K. 2005. Effects of simultaneous establishment of Sedum alfredii and Zea mays on heavy metal accumulation in plants. *International Journal of Phytoremediation* 7 (1), pp. 43–53.

Xing, Y., Peng, H., Gao, L., Luo, A., and Yang, X. 2013. A compound containing substituted indole ligand from a hyperaccumulator Sedum alfredii Hance under Zn exposure. *International Journal of Phytoremediation* 15 (10), pp. 952–964.

Xu, L., Zhou, S., Wu, L., Li, N., Cui, L., Luo, Y., and Christie, P. 2009. Cd and Zn tolerance and accumulation by Sedum jinianum in east China. *International Journal of Phytoremediation* 11 (3), pp. 283–295.

Yanai, J., Zhao, F. J., McGrath, S. P., and Kosaki, N. 2006. Effect of soil characteristics on Cd uptake by the hyperaccumulator Thlaspi caerulescens. *Environmental Pollution* 139, pp. 67–175.

Yancey, N. A., McLean, J. E., Grossl, P., Sims, R. C., and Scouten, W. H. 1998. Enhancing cadmium uptake in tobacco using soil amendments. In *Proceedings, Conference on Hazardous Waste Research*, Snow Bird, Utah, pp. 25–26 (Abstract 38).

Yang, J., Yu, Q., and Gong, P. 2008. Quantifying air pollution removal by green roofs in Chicago. *Atmospheric Environment* 42 (31), pp. 7266–7273.

339

Yang, X., Long, X., Ye, H., He, Z., Calvert, D., and Stoffella, P. 2004. Cadmium tolerance and hyperaccumulation in a new Zn-hyperaccumulating plant species (Sedum alfredii Hance). *Plant and Soil* 259 (1–2), pp. 181–189.

Yateem, A. 2013. Rhizoremediation of oil-contaminated sites: a perspective on the Gulf War environmental catastrophe on the State of Kuwait. *Environmental Science and Pollution Research* 20 (1), pp. 100–107.

Yoon, J., Cao, X., Zhou, Q., and Ma, L. 2006. Accumulation of Pb, Cu, and Zn in native plants growing on a contaminated Florida site. *Science of the Total Environment* 368 (2), pp. 456–464.

Yoon, J. M., Oh, B.-T., Just, C. L., and Schnoor, J. L. 2002. Uptake and leaching of octahydro-1,3,5,7-tetranitro-1,3,5,7-tetrazocine by hybrid poplar trees. *Environmental Science and Technology* 36 (21), pp. 4649–4655.

Yu, X., and Gu, J. 2006. Uptake, metabolism, and toxicity of methyl tert-butyl ether (MTBE) in weeping willows. *Journal of Hazardous Materials* 137 (3), pp. 1417–1423.

Zalesny Jr., R. S., and Bauer, E. O. 2007. Evaluation of Populus and Salix continuously irrigated with landfill leachate I. Genotype-specific elemental phytoremediation. *International Journal of Phytoremediation* 9 (4), pp. 281–306.

Zand, A. D., Nabibidendi, G., Mehrdadi, N., Shirdam, R., and Tabrizi, A. M. 2010. Total petroleum hydrocarbon (TPHs) dissipation through rhizoremediation by plant species. *Polish Journal of Environmental Studies* 19 (1), pp. 115–122.

Zayed, A., Pilon-Smits, E., de Souza, M., Lin, Z., and Terry, N. 2000. Remediation of selenium-polluted soils and waters by phytovolatilization. In N. Terry and G. S. Bañuelos, (Eds.) *Phytoremediation of Contaminated Soil and Water*. Boca Raton, FL: CRC Press.

Zeeb, B. A., Amphlett, J. S., Rutter, A., and Reimer, K. J. 2006. Potential for phytoremediation of polychlorinated biphenyl-(PCB)-contaminated soil. *International Journal of Phytoremediation* 8 (3), pp. 199–221.

Zhang, X., Liu, J., Huang, H., Chen, J., Zhu, Y., and Wang, D. 2007. Chromium accumulation by the hyperaccumulator plant Leersia hexandra Swartz. *Chemosphere* 67 (6), pp. 1138–1143.

Zhang, Z., Sugawara, K., Hatayama, M., Huang, Y., and Inoue, C. 2014. Screening of As-accumulating plants using a foliar application and a native accumulation of As. *International Journal of Phytoremediation* 16 (3), pp. 257–266, DOI, 10.1080/15226514.2013.773277.

Zhao, F., Dunham, S., and McGrath, S. 2002. Arsenic hyperaccumulation by different fern species. *New Phytologist* 156 (1), pp. 27–31.

Zhao, F., Jiang, R., Dunham, S., and McGrath, S. 2006. Cadmium uptake, translocation and tolerance in the hyperaccumulator Arabidopsis halleri. *New Phytologist* 172 (4), pp. 646–654.

Zhua, Y., Hinds, W. C., Shen, S., Kim, S., and Sioutas, C. 2002. Study of ultrafine particles near a major highway with heavy-duty diesel traffic. *Atmospheric Environment* 36, pp. 4323–4335.

Zhuang, P., Yang, Q., Wang, H., and Shu, W. 2007. Phytoextraction of heavy metals by eight plant species in the field. *Water, Air and Soil Pollution* 184 (1–4), pp. 235–242.

Zia, M. H., Eton, E., Codling, B., Kirk, G., Scheckel, C., and Chaney, R. L. 2011. In vitro and in vivo approaches for the measurement of oral bioavailability of lead (Pb) in contaminated soils: a review. *Environmental Pollution* 159 (2011) 2320–2327.

Index

345

T - #0637 - 071024 - C378 - 276/216/22 - PB - 9780415814157 - Gloss Lamination